T0320337

Experiment Design for Civil Engineering

Experiment Design for Civil Engineering provides guidance to students and practicing civil engineers on how to design a civil engineering experiment that will produce useful and unassailable results. It includes a long list of complete experiment designs that students can perform in the laboratory at most universities and that many consulting engineers can do in corporate laboratories. These experiments also provide a way to evaluate a new design against an existing experiment to determine what information is most appropriate in each section and how to format the data for the most effective outcome. Interpretation of output data is discussed, along with uncertainty, as well as optimal presentation of the data to others.

The content of the first 8 chapters is similar in format to authors' recent title, *Experiment Design for Environmental Engineering: Methods and Examples* (CRC Press, 2022) and has been revised for civil engineers. This textbook:

- Fills in the gap in ABET requirements to teach experiment design.
- Provides a standardized approach to experiment design that can work for any experiment.
- Includes completed experiment designs suitable for college laboratory and professional applications.
- Shows how to organize experimental data as it is collected to optimize usefulness.
- Provides templates for design of the experiment and for presenting the resulting data to technical and nontechnical audiences or clients.

Experiment Design for Civil Engineering

Methods and Examples

Francis J. Hopcroft and Abigail J. Charest

CRC Press is an imprint of the
Taylor & Francis Group, an **informa** business

Designed cover image: ©Unsplash

First edition published 2023
by CRC Press
6000 Broken Sound Parkway NW, Suite 300, Boca Raton, FL 33487–2742

and by CRC Press
4 Park Square, Milton Park, Abingdon, Oxon, OX14 4RN

CRC Press is an imprint of Taylor & Francis Group, LLC

ISBN: 978-1-032-38200-5 (hbk)
ISBN: 978-1-032-38770-3 (pbk)
ISBN: 978-1-003-34668-5 (ebk)

DOI: 10.1201/9781003346685

Typeset in Times
by Apex CoVantage, LLC

Contents

Figures

Tables

Preface

Designing experiments for civil engineering projects differs from designing experiments for social science studies and other disciplines where experimental design is more commonly encountered in the context of social research projects. With civil engineering projects, the interconnectedness of nature and the interaction of environmental factors with the outcome data can significantly affect the decisions made on the bases of those outcomes. This book describes a method for preparing a civil engineering experiment to demonstrate specific design phenomena or to document expected construction outcomes. It includes the methodology, the background, and potential interferences. The final five chapters provide successful experiments that can be implemented to demonstrate specific civil engineering topics. These experiments are specifically geared to a university laboratory but may also be applicable to commercial or municipal applications or field experimentation.

Acknowledgments

The work and contributions to the success of this book by the Wentworth Institute of Technology students who tried out many of them in the laboratory over the years are greatly appreciated and gratefully acknowledged. In addition, Mr. Peter Pengo, a Wentworth Civil Engineering student, provided excellent CAD drawing expertise to the project, producing all the CAD drawings in this book. That effort is also very much appreciated and gratefully acknowledged.

Author Biographies

Prof. Francis J. Hopcroft recently retired from teaching civil and environmental engineering after 23 years in the classroom and about 40 years of consulting in the field. He is the author of five environmental engineering and hazardous waste management books, the coauthor of 23 such books, and a contributor to a dozen or more professional manuals of practice. A graduate of Northeastern University and the University of Maine at Portland-Gorham, he has been registered as a Professional Engineer in all six New England states and as a Licensed Site Professional in the Commonwealth of Massachusetts. Before starting his teaching career, he spent 25 years in professional practice as a consultant, an EPA regulator, and as the President and CEO of several consulting firms doing site assessment for the presence and remediation of oil and hazardous material releases. He continued his consulting work while teaching to maintain currency in his field and to bring current concepts into the classroom.

Dr. Abigail J. Charest is Associate Professor of Civil Engineering at Wentworth Institute of Technology (WIT) in Boston, Massachusetts. She is currently the Blittersdorf Endowed Professor and utilizes that professorship to address topics of sustainability in the curriculum. She is an avid researcher, experimenter, and innovator in the laboratory. She has also served as the lead faculty member in the redevelopment of a graduate program in civil engineering at Wentworth and incorporated significant research and experimentation opportunities into that program. She received her doctorate from the Worcester Polytechnic Institute, and prior to entering into academia, she worked in the field of environmental consulting in the New England Area. During this time, she received her Professional Engineering license in the Commonwealth of Massachusetts.

1 Introduction

Civil engineering is a compilation of a multitude of specialties, all of which serve to provide the infrastructure for a modern society. Civil engineering projects include, but are not limited to, roads, bridges, tunnels, airports, shipping ports, marinas, dams and levees, railroads, sewer systems, water systems, water and wastewater treatment plants, solid waste disposal facilities, and the foundations for vertical structures. These projects are designed, built, and maintained through the interdisciplinary relationships of several specialties, such as structural engineering, geotechnical engineering, transportation engineering, material science, soil mechanics, hydraulics, hydrology, water resources engineering, environmental engineering, and sustainability engineering. Civil engineering projects are necessary for a safe and healthy built environment. This means that civil engineering projects are built to last, such as the Roman aqueducts and the Great Wall of China. In spite of all of that, however, civil engineering relies on designing based on imprecise science and information. For example, the soil of a particular type in one location may not act the same way in another location. Materials that are long-lasting in one environment may be short-lived in another. To optimize design based on the environment and construction materials, it is often necessary to perform some test – or experiment – to determine the best material, method of construction, or other characteristic of the project that will best sustain the project into the future.

Experiment design is as much an art as a practice. Elegant experiments exist, but not nearly as often as mundane exercises of practical judgment superimposed on some physical phenomenon. Indeed, most experiments do not work, at least not the first time. If they did, there would be limited need to do them. It is the failure of experiments and the iteration of design from which the most is often learned, and this new knowledge allows engineers to narrow the scope of subsequent experiments to better define, describe, or demonstrate the outcomes the initial experiment was designed to demonstrate. It is a noble goal, of course, to minimize the number of iterative experiments needed to resolve an issue or question, but it should not be considered a bad thing if more than one or two tries are needed to get something complex right. The more complex the issue, of course, the more difficult it will be to get the experiments right on the first try.

This is a book about designing the experiments needed to effectively practice civil engineering. More specifically, it is about how to *design* an experiment for the university laboratory, and in some cases, for use in remote areas of the world, to verify theories and prove or disprove concepts. It is also a book of complete and detailed experiments. Experiments that have been designed to demonstrate specific engineering phenomenon. The experiments in this book are for teaching various physical traits of common engineering basics so that the novice engineer can begin to see how the world works and begin to develop necessary skills for the effective design of new experiments.

DOI: 10.1201/9781003346685-1

Chapter 2 describes what goes into designing a successful experiment and how to do it. Chapter 3 describes the most effective ways to sample media under different sample purity requirements. Chapter 4 examines how to establish the expected outcomes from an experiment; how to collect data for effective evaluation of actual results; how to interpret the data generated by an experiment, including what to do when the data do not coincide with the expected outcome; and considerations of uncertainty in the experimental outcomes, including a discussion of how probability plays into evaluating data. Chapter 5 provides the model design methodology and format that is used to develop the completed experiments in the last chapters. Chapter 6 provides a suggested template for a laboratory report that mirrors the data and information in the experiment outline with actual results, discussions, recommendations, and conclusions. Chapter 7 evaluates some considerations for preparing the outcome data for effective presentation to others. Chapter 8 discusses the differences between the experiments described in this book and the type of experimentation needed in a research project and discusses how to develop appropriate research projects. Chapters 9–13 provide a series of successful experiments designed to show engineering students some of the basics of civil engineering.

Each of the experiments in Chapters 9–13 provides an introductory section on the basic theory behind the experiment. These are not intended to be detailed theses on the underlying theory, but rather a reminder of the engineering basics underlying the experiment. It is assumed that the student has a sufficient educational background to grasp the fundamentals outlined in the experiment introductions and to understand the context for the data to be developed during the experiment.

The experiments in Chapters 9–13 are divided into some of the various specialties of civil engineering. Chapter 9 provides eight material and structural analysis experiments. Chapter 10 provides nine concrete testing experiments. Chapter 11 provides ten soil testing experiments. Chapter 12 provides eight environmental assessment experiments. Chapter 13 provides seven fluid mechanics and hydraulic engineering experiments. All of these chapters include experiments of varying difficulty for consideration by students with a variety of educational backgrounds and experiences.

2 How to Design an Engineering Experiment

The fundamental design of an experiment contains several distinct design elements. Those include the question to be answered; the variables involved; how the variables will be adjusted; the potential interferences that can occur; how the investigator intends to minimize or avoid the effects of those interferences, or to account for them in the experimental data; and what theoretical outcomes are expected. Once the data are generated, how those data are interpreted and how the results are presented will go a long way to validating the outcomes.

Note that, as suggested in Chapter 1, not all experiments succeed. If they did, there would be no need to do an experiment because the outcome could be accurately predicted. It is important, therefore, to recognize that failure *is* an acceptable component of investigation. That recognition will minimize the tendency to interpret data in a way that supports the expected outcome and to reject data that do not support that outcome. *If the data are not what is expected, it should be assumed that the data are correct and that the theory is wrong or that there was an error in the experiment design or conduct.* The investigator then needs to try to figure out why the theory or the experiment was wrong and how to redo the experiment to account for the new thinking.

Certainly, equipment will occasionally fail, people will do things in a manner inconsistent with the planned protocol of the experiment, reagents will become contaminated, and all sorts of other things will go wrong with experiments. The data that are generated are *always* correct for the experiment that was done. If those data do not reflect the expected outcome, either the theory is wrong or the experiment incorporated some unknown flaw. Either way, it is imperative to accept the data as real and to adjust the experiment or the theory, or both, to incorporate what was learned from the unexpected outcome.

2.1 DEFINING THE QUESTION TO BE ANSWERED

It is a foundational concept in engineering that in order to solve a problem, it is first necessary to define the problem accurately and completely. Much of what is done in experimentation is aimed at defining the basic problem. That implies that the initial question posed may not be the final question the investigator is attempting to answer. Sometimes an intermediate question is required to define the underlying details of the ultimate question.

In either case, the question posed needs to be as clear and concise as possible. Ambiguous questions lead to ambiguous results. For example, "Can the maximum compacted density of soil be predicted from the effective size of the soil particles?" might be an interesting question, but there are too many variables in

DOI: 10.1201/9781003346685-2

the question that are not defined. For example, what effect does the moisture content have on the outcome? How does the uniformity coefficient of the soil affect the outcome? What effect does the percent fines (smaller than a 200 sieve) in the soil have? Does the type of soil (sand, gravel, loam, mixtures, etc.) change the outcome? What other soil grain characteristics (round, smooth, granular, sharp edges, etc.) impact the experiment? These examples, and probably a few more, will affect the maximum density that is attainable, regardless of the effective size, and all must be accounted for in some appropriate manner for the answer to the basic question to be useful.

A clearer, more concise question might be "Can the maximum compacted density of river run sand with a uniformity coefficient less than 7 and a moisture content of less than 40 % be predicted from the effective size of the soil particles?" This describes the sand used in most sand filters designed for water and wastewater treatment at the time of construction (the sand becomes saturated and achieves 100 % moisture content during use), and the answer to this question may provide insights into how to design the filter supporting structures and the resulting mass of sand to be placed inside the filter. By adjusting the effective size of various samples of sand meeting the uniformity coefficient criterion and drying them to a moisture content less than 40 %, then conducting a maximum density test on the samples, many of those questions can be answered. Note that, as usual, there are some uncontrolled variables that may be at play here that must be considered, discussed, and accounted for as well. For example, river run sand generally tends to be rounded and smooth particles due to the grinding action of the moving water. If the sand selected is young sand, or taken from an area subject to frequent flooding and intermittent stream flow, the rounding may not be as complete as desired. If the samples are not taken from the same location, the data may not be consistent. If the samples are taken from only one location but filter sand is from a multitude of locations, the outcome data may not be universally applicable. All of these variables need to be considered and accounted for in the interpretation of the outcome data.

2.2 DEFINING THE VARIABLES INVOLVED IN A QUESTION

Very few things in engineering are simple. Engineering questions are driven by physics, chemistry, biology, mathematics, mechanics, geology, and a host of other "ologies". Those subjects all form engineering basics to which engineers routinely turn to find answers. How those basics interact with each other in any specific instance or circumstance is not always clear, but can be significant. Defining the variables for an experiment, determining whether those variables are to be controlled, and how to account for those not being controlled is important.

A note about variables. There are generally three kinds of variables involved in experiments, and knowing which kind is being discussed is important. *Independent variables* are those that the experimenter controls. *Dependent variables* are those that respond to the changes made to the independent variables. *Controlled variables* are those that are kept constant throughout the experiment. The ideal experiment will be designed so that only one independent variable is changed at a time and the results in all the dependent variables can be observed. Occasionally, more than one variable may need to be changed at the same time, but that should be avoided to the maximum degree possible.

Note, too, that an important characteristic of a variable is that it can be reliably measured. Time, temperature, velocity, mass, effective grain size, uniformity coefficient, and similar characteristics are measurable and are good variables. Emotions, feelings, opinions, judgments, and similar characteristics are not measurable and therefore are not suitable experimental variables.

In the question posed regarding the maximum soil density as a function of effective grain size, there are many potential variables. Five variables have been identified as key to the selected question: the source of the sand will be river run deposits, the sand will have a variable effective size, the uniformity coefficient will be less than 7, the sand will have a moisture content less than 40 % at the time of testing, and the maximum compacted sand density is the variable outcome. All five of the variables have been defined as measurable variables in the statement of the question. The source of the sand is selectable, the water content of the sand is measurable, the effective size is measurable, the uniformity coefficient is measurable, and the maximum compacted density is measurable. The effective size is the independent variable and the maximum soil density is the dependent variable. All of the others are considered controlled variables. Once this experiment is complete, however, these controlled variables must be further evaluated, possibly by further experimentation, and the results used to modify or verify the outcomes from the current experiment. The question statement would require dismissal of data generated if the sand was from a different source or did not meet the uniformity characteristics and moisture content at the time of testing.

It is noted, however, that a lot of variables can exist in a natural setting that have not been identified or considered in the question statement. For example, natural water bodies typically contain a wide variety of dissolved minerals and organics that can result in various types of acidic corrosion (surface pitting) or inorganic accretion build-up on sand particles and intermittent flow streams do not always round the edges of the sand particles as well as a continuously flowing stream. Organics may also infiltrate to a variable degree into various sand deposits, and those may interfere with the experimental results at some undefined concentration. The temperature of the sand at the time of testing could affect the maximum attainable density. These kinds of variables are neither controlled nor measured by this experiment but can significantly influence the engineer's recommendations for the project. Other experiments may need to be conducted to determine the effects of those considerations if the basic outcome data appear to be inconsistent for some otherwise unknown reason.

2.3 MEASURING AND CONTROLLING THE VARIABLES

There are several things that could go wrong with any experiment. Minimizing negative effects takes some planning. In the filter sand experiment example, the experimenter is not fully in control of any of the controlled variables. Even the independent variables that are the basis for the question are subject to fluctuation, and the test for the maximum soil density is fraught with uncontrollable variables. The question presupposes a constant environment with the measurement of only five specific variables over time. Note that because only those five variables are being measured and there are so many other variables that are not being measured, the data generated would apply only to the specific sand sample being tested. This experiment would not generate truly universally applicable data, but may identify an applicable relationship that is a good guide for other sand sources and characteristics that are sufficiently comparable to provide useful guidance to other engineering projects.

There is usually more than one way to do everything in engineering, and there are two general ways to conduct this experiment. The first is to identify a source of river run sand that is reasonably close to the stated characteristics, sift and sort the sand to create the desired characteristics of uniformity coefficient and effective grain size, ideally at one end of the range over which it is reasonable to adjust the effective grain size, dry the sand, place a constant amount into a plastic bucket and conduct a maximum compaction density test. The effective size will then be varied and the particle sizes adjusted accordingly to maintain the uniformity coefficient, and the density test run again. That procedure would be repeated as many times as desired and the data plotted on a curve of effective grain size and maximum compacted density.

A second way to conduct this experiment would be to create a sand source from non-river run sand, tumble the grains in a laboratory tumbler until the particles are smooth, then run the experiments as described earlier. This could require the tumbling of a lot of sand to the same degree of roundness or the results may be very inconsistent and unreliable. The tumbled sand would need to be sifted and adjusted to ensure proper uniformity coefficient with an effective size of varying value.

Regardless of the experiment design employed, there are several factors that would need to be measured or controlled during this experiment, including the soil moisture content, the uniformity coefficient, the effective grain size, and the maximum compacted soil density.

Controlling the experiment for potential effects of unmeasured parameters is more difficult. Over a short period of time, it may be assumed that the parameters, such as soil temperature, do not change dramatically and that their effects are negligible. Over longer periods of time, such as over a year of winter and summer temperatures, this assumption might well be false. Over a very long time frame, such as several years, the effects of the unmeasured variables may average out and become negligible again, even if they are significant in the magnitude of their variability. The experimenter could, of course, also run the series of experiments under various conditions of soil temperature to see what effect temperature has on the outcomes. The experimenter needs to weigh the costs of the long-term data-collection effort with the benefits of short-term data collection and decide whether the short-term data will best represent long-term outcomes.

Note that this does not mean that a researcher or experimenter can assume or state that the time frame is sufficiently short or long as to render unmeasured impacts negligible. A short time period in natural environments is typically measured in minutes, usually fewer than 5 minutes. A long time frame is typically measured in years, often decades. Most analyses are done using a time frame that is not within either set of guidelines, and those guidelines are not relevant to all types of engineering analyses.

2.4 EVALUATION OF POTENTIAL INTERFERENCES THAT CAN OCCUR

Unmeasured interferences need to be considered in the analysis of data. Since they are unmeasured, they are, by definition, unknown and cannot be numerically applied to the data. In order to rationally consider their potential impacts on the data, it is first necessary to determine what those interferences are and what possible impacts they could have on the experimental data. It is necessary to ask the pertinent question: "What could possibly go wrong?" A lot of things, it turns out.

Tabulation of potential interferences is a useful exercise to start to focus the mind on what could go wrong. Table 2.1 is a possible table of potential interferences for the example problem. This table is expanded shortly to show how these potential interferences may be considered and managed.

This table does not represent a comprehensive list of all possible interferences related to the example experiment but is sufficient to indicate the range of potential problems that the investigator should consider when designing an experiment.

2.5 MINIMIZING, AVOIDING, OR ACCOUNTING FOR THE EFFECTS OF INTERFERENCES

Given that unmeasured interferences are likely to create issues with any experiment, it is useful to give considerable thought to how those effects on the data will be determined or measured. The easiest way to account for those effects, of course, is to actually measure them during the experiment. That may not always be economically or practically possible. Nevertheless, adding some measuring devices, such as a pH probe at the source site to determine the acidity of the water at that location and therefore a need to microscopically examine the soil particles for pitting or accretion, or burning a sample of the source sand in a muffle furnace at 500 °C for a few hours to determine the organics content, could prove very useful. Without such efforts to measure the variables, educated guesses are the best that can be had. It is usually far easier to obtain extraneous data during the experiment than to try to back track and account for those factors later. Note that there are a lot of unintended consequences that can occur from trying to mitigate identified interferences and those need to be considered if experiment results are inconsistent with expectations.

In many cases, consideration of how to manage unmeasured interferences is usefully tabulated in an expansion of Table 2.1, as shown in Table 2.2. When the number of interferences is expected to be small (fewer than 5 to 10, for example) and easily managed, the use of tables like these is suggested. A literature search may be conducted on any of the interferences and possible ways to mitigate them identified. It is best not to reinvent the wheel, but to learn from others.

TABLE 2.1
Potential Interferences for Example Experiment

Interference	Possible Cause
Sand is not uniform for all tests	Source is a seasonally flooded plain or intermittent stream
Chemical sorption or accretion onto sand particles	Acid rain, climate change, or other random events
Compacted soil density test data are inconsistent	Different technicians run the tests on different samples and their techniques are not identical
Pitting of sand particles that affects measured moisture content	Changes to water chemistry over time that creates an acidic water environment
Organic contamination of the sand	Source location is subject to surface runoff from agricultural sites or other organic sources

TABLE 2.2
Potential Interferences for Example Experiment

Interference	Possible Cause	Management Plan
Sand is not uniform for all tests	Source is a seasonally flooded plain or intermittent stream	Collect enough sand to do all tests and mix well in the lab before starting test sample collection
Chemical sorption or accretion onto sand particles	Acid rain, climate change, or other random events	Check pH of water at source and microscopically examine representative soil particles if appropriate
Compacted soil density test data are inconsistent	Different technicians run the tests on different samples and their techniques are not identical	Standardize procedures for all testing and have the same person conduct the tests when possible
Pitting of sand particles that affects measured moisture content	Changes to water chemistry over time that creates an acidic water environment	Particles will tend to stick together at the source location under these conditions. Examine source material in situ before collecting samples
Organic contamination of the sand	Source location is subject to surface runoff from agricultural sites or other organic sources	Examine source material in situ before collecting samples, for example, organic material is generally of a darker color than source sand

2.6 SENSITIVITY ANALYSES OF EXPERIMENTAL DATA

When looking at a series of independent variables and their combined effect on one or more dependent variables, complications can arise. Most independent variables interact with each other as well as with the dependent variables being measured.

It is not always clear how a particular change in one variable is affecting the other independent variables (which are then acting as dependent variables) and what those changes are doing to the dependent variable being measured. To try to get a handle on that issue, a *sensitivity analysis* is conducted.

To conduct a sensitivity analysis, it is first necessary to identify the things that are causing changes in the dependent variables. These are generally going to be the previously identified independent variables but could include a few other variables as well. Any characteristic of the system in which the variables under consideration reside, which could affect the dependent variable, should be viewed as a possible factor to be included in the sensitivity analysis. Then a set of standards must be established by which the sensitivity of the data will be measured.

For the example problem described in this chapter, the independent variables will be accepted as the only causative factors for change in the dependent variable. That is not an entirely valid assumption, but since nothing else is being measured, and the initial intent is only to determine the sensitivity of the dependent variable to three of the four measured independent variables, that assumption will work in this case. The assumptions upon which this sensitivity analysis is based are the following:

1. Percent change in maximum compacted soil density as a function of a percentage change in effective particle size
2. Percent change in maximum compacted soil density as a function of a percentage change in uniformity coefficient
3. Percent change in maximum compacted soil density as a function of a percentage change in soil moisture content

To conduct the analysis, each independent variable is mathematically changed *while all the other independent variables are held constant.* The change in the dependent variable as a function of the measured change in the varied independent variable will yield a sensitivity curve. It is possible for the dependent variable to change linearly with a constant change in the specific independent variable. It is also possible for any other form of change to occur, such as a parametric curve, a nonlinear change, and so forth. Until the controlled changes are made and the data plotted, the effects will not be known.

In this example, the changes in maximum compacted soil density as a function of effective particle size and uniformity coefficient are not well-established physical properties of soils. The sensitivity analysis, however, is based on percentage of parameter change, not on the absolute change. Therefore, it is necessary to calculate the percentage change represented by each parameter as changes to the effective size are made during the experiment. The percent change is based on the difference between the first effective size used and each different effective size tested. Additionally, this is also true of the other two parameters used for this analysis. The consistency of the terms of measurement, such as percentage change, is important when comparing outcomes.

The uniformity coefficient is likely to change as the effective size is changed. Calculating the percentage change in the uniformity coefficient related to each change in the effective size is what gets plotted against the percent change in maximum compacted density. Similarly, the percentage change in moisture content, if any, is plotted against the percentage change in compacted soil density. There should be no change in the moisture content during this experiment, but if one occurs it will be important to evaluate the sensitivity of the outcome data to those changes.

Generally the percent change in the effective size, the uniformity coefficient and the moisture content are plotted on the y-axis and the percentage change in compacted density is plotted on the x-axis. When all the plots have been completed, overlaying them on a single graph will yield a graphic representation of how the maximum soil density is affected by the change in each of the three independent variables being measured. Ideally, each of the three lines will be flat, meaning that the percent changes are linearly correlated and a singular linear equation to predict the maximum soil density at any set of those three variables may be reliably developed. If that is not the case, which is more common, then a nonlinear equation will need to be developed to better predict the impact of the various independent variables on the singular dependent variable of interest. Moreover, the relative effect of each of the three measured independent variables will be clear, and if one is significantly different from the others it will indicate a need to more closely and carefully monitor that parameter during the testing.

It is instructive to note that all data may be assessed as statistics. Statistics are subject to analysis by statistical analysis techniques. This generally includes the various analytical tools described in Tables 4.4 and 4.5 in Chapter 4. The derivation and use of those analytical tools is not done here for two significant reasons. The first is that they can be complex to understand and there are dozens of excellent texts available for review, should a reader choose to pursue them.

The second reason is that all of those tools assume that all data are random. On the other hand, there is nothing truly random about nature. All things in nature are connected in some form or fashion to everything else in nature. Therefore, while the application of statistical analysis tools that rely on randomness is often done with measured *environmental* data, it is not appropriate for most *experimental* data since the changes in the dependent variable are presumed to be a direct result of changes in the independent variable, and thus, not random.

2.7 SAFETY CONSIDERATIONS IN EXPERIMENT DESIGN

Safety is paramount in the design of experiments, just as it is in the conduct of professional civil engineering in general. It is, therefore, important to incorporate appropriate safety considerations into every experiment design. All college and university laboratories, private laboratories, and rooms or spaces in which experiments are carried out contain a wide variety of physical, chemical, and biological agents that can seriously harm or kill inattentive experimenters.

Acids, alkalis, and other caustic substances can cause serious burns to the skin and blindness in the event of eye contact. Some organic chemicals can cause toxic reactions to anyone, but particularly to those with sensitivities to allergens. For example, volatile organic compounds (VOCs) can cause respiratory distress when inhaled, cancer in the long run, and death at high concentrations. Biological agents used to grow or enhance the growth of bacteria and other organisms can be toxic to humans. Most of the bacteria and other organisms grown in laboratories are harmful to humans, which is why they are being researched, and those effects can lead to gastrointestinal symptoms, dehydration, and death. Samples studied in civil engineering laboratories can include industrial wastes, human sewage and surface water contaminated with human sewage, animal wastes, and the potentially toxic chemicals used to treat them. In soil mechanics and geotechnical labs, certain organic and inorganic chemicals are likely to be present that can cause similar symptoms and illnesses. In addition, dust particles in the air in soils labs can cause serious lung and throat diseases and conditions.

Common considerations:

- There is no reason to believe at any time that the tables and counters in any laboratory are germ free. The principal route of exposure to harmful organisms is through ingestion and touching a contaminated surface, then touching the face or mouth, even just to scratch an itch, can cause accidental ingestion or subcutaneous infection.
- Soil-drying ovens and hotplates can achieve temperatures from 20 °C (68 °F) to 500 °C (932 °F). It should be assumed that they are very hot at all times and that severe burns can result immediately on contact from accidentally touching the sides or edges when inserting or removing samples.
- Personal protective equipment consisting of safety glasses, latex or vinyl gloves (or heat-resistant gloves when working with ovens, incubators, or hotplates), and lab coats should be required for all people entering a research laboratory or experiment space, including non-participatory observers.
- No fewer than two persons should be present at all times in any space in which an experiment is being conducted or data are being collected or read.
- There should be a direct way to contact emergency response personnel available, such as a dedicated telephone line to the public safety office, inside the laboratory space, to summon aid when needed.
- No work should be allowed to proceed without those safety protections in place for everyone present. Accidents do happen in laboratories, and preplanning for safety will minimize the risks of harm when accidents occur.

In addition, when accidents happen, it is important not to be cavalier about the consequences. Even what appears to be a minor scrape or burn can result in the introduction of serious toxic agents to the blood stream with significant health consequences. *All* accidents and injuries, however minor or major, must be reported to the laboratory supervisor immediately and recorded in a logbook. The log should contain the date and time of the incident; a list of those present at that time; a description of what happened; and the immediate effects of any actual, suspected, or potential exposure. Open wounds, puncture wounds, and second- or third-degree burns require medical attention immediately. Any eye exposure requires immediate eye washing and medical attention.

Safety is everyone's responsibility. There should always be at least two people in a laboratory or experiment space at all times that experiments are being conducted, and there should be a direct way to contact emergency response personnel available, such as a dedicated telephone line to the public safety office inside the laboratory space, to summon aid when needed. It is not possible to be *too* safe, and safety plans are an important part of every experiment design.

2.8 ETHICS CONSIDERATIONS IN EXPERIMENT DESIGN

It is important to consider ethics throughout one's life, academic pursuits, and professional career. While it may seem like a simple concept in theory, the application of ethics is complicated. To quote Disney's *Frozen* (2013), "people make bad choices when they are mad, or scared or stressed". Engineers are guided by many things, including personal ethics and morality, professional ethics, ethical standards of the community, and ethical standards promulgated by various professional organizations and regulatory agencies.

It is noted that some texts tend to use the terms "morality" and "ethics" interchangeably, but they are actually very different concepts, although both are equally important in the practice of professional engineering.

According to *Webster's New World College Dictionary, Third Edition* (1996):

> Morality pertains to behavior measured by prevailing standards of the rightness of judgment; i.e.: the difference between right and wrong as determined by the society or community in which the professional is practicing. Ethics approaches behavior from a philosophical standpoint; it stresses more objectively defined, but essentially idealistic standards of right and wrong, such as those practiced by lawyers, doctors, engineers, and businesspeople.

Individuals can, and must, decide for themselves how to differentiate the concepts of morality and ethics. Professional ethics and personal ethics may be distinguished in the following way:

> Personal ethics are standards of conduct adopted by an individual usually based on early religious training; family values, or personal reflection, conscious or unconscious, on what is right or wrong behavior. Professional ethics are standards set by members of a profession to define how those members will act. Morality relates to a set of standards established by a culture or society that defines right and wrong for that culture or society – which is why what is seen as reprehensible activity in one country may be seen as normal behavior in another (such as bribery, prostitution, various forms of gambling, bull fighting, etc.)

Professional ethics are generally defined by "codes of ethics" published by various professional or regulatory organizations. These organizations might include, for example, the American Society of Civil Engineers (ASCE), the Boston Society of Civil Engineers (BSCE), the Licensed Site Professional Association (LSPA), the American Society for Engineering Education, the Water Environment Federation (WEF), and various regional and local chapters of each national organization. Each organization publishes, and expects its members to adhere to, a specific set of professional standards typically identified as a "code of ethics".

Ethics are everyone's business, and reporting suspected cases of unethical behavior to appropriate regulatory agencies or professional organizations is an important part of professional practice.

2.8.1 Ethics in Experimentation – Citations

The execution of ethics in an academic and laboratory setting can be nuanced. This is because a lot of the experiments that are done in a college laboratory have already been designed by others. Use of those designs is the intent of the designers, and it is not necessary or expected that a separate license or prior permission is required to do so. It is important to note where a design came from, however, and to give due consideration to the original designer – even if the original design is modified in some fashion to adapt to available equipment or specific experiment outcome objective.

Similarly, if someone assists with the design of an experiment, such as by providing a key comment that triggers a unique approach to controlling an experiment, for example, it is necessary to note the contribution of that person, regardless of whether they ultimately participated in the final design.

Note that using the work of others without their consent and acknowledgment is called *plagiarism.* It is unethical and often illegal. Professionals can lose their license to practice for doing it. Note, too, that self-plagiarism is also unprofessional. When we write something or create something and then reuse it, without rewriting or acknowledging the original source in which it was used, we are *self-plagiarizing.* While not as egregious, perhaps, as plagiarizing someone else, it is important to recognize and cite where the material was first published.

2.8.2 Ethics in Experimentation – Data

Once the experiment has been conducted and all the data are accumulated, it is not uncommon to find several data points that do not seem to fit the model of the rest of the data. These are typically referred to as "outliers": data points that are outside the normal, expected range of the data.

How those outliers are handled is an important ethical consideration for the experimenter. If the outliers are simply ignored, the results presented and the interpretations made could be very wrong. If they are included but are, in fact, incorrect data for some reason, they can skew the outcomes inappropriately. It is important, then, to realize that outliers might be valid data for which an explanation is needed, or they could be inaccurate data that should be excluded. It is necessary to do an analysis of those data to determine which is the correct interpretation.

In Chapter 4 there is a discussion regarding the handling of statistical data. Part of that discussion includes the misuse of data that are real, the deliberate failure to report data that do not support a predetermined outcome, and falsification or fabrication of data to make a data set appear more compelling than it is. All of those actions are unethical.

All data are assumed to be correct until proven otherwise, beyond any doubt. They are always correct for the experiment that was actually conducted, even though that experiment conducted may not have been the intended one due to error or inaccurate equipment. It is imperative, then, to verify why the data are outliers and then whether they should be considered.

In Chapter 4 there is also a discussion of how to handle outliers when making presentations. The analyses need to be run both with and without the outliers to determine how much of an effect those data points could have on the outcomes. When presenting the data, all the data, even the outliers, need to be shown, and then a rational explanation provided for the exclusion of any data determined to be invalid. Failure to show all the data is unethical.

It is important to realize that everyone runs into time constraints on occasion. Time slips away, and it is too late to run the experiment before a deadline to turn in a laboratory report, or someone forgets to take readings at one or more specified time intervals, or something else interferes with the reading or development of specific data. The impulse is to make up data that seem to fit what we expected the outcome to be and hope that nobody notices the ruse. This is called *falsification of data*. It is unethical, and practicing professionals have been sentenced to jail for doing it. It is also a good way to fail a course and be dismissed from a college or to destroy a promising career in civil engineering before it even starts.

3 Sampling Source Media

Engineering experiments often involve some form of material testing. Those materials may be natural, such as water, soil, or air, or they may be manufactured materials. This chapter examines the accepted ways to sample the various forms of the selected materials to provide the most reliable outcomes.

It is noted that not all experiments require absolute purity of the media or materials being sampled. It often happens that the outcomes of an experiment are never going to be precisely correct because the media are constantly changing and their parameters are highly variable. The characteristics of in situ soil, for example, are not uniform even over relatively small areas around specific locations. The procedure for sampling the soil, then, is more a matter of making certain that the value of the parameters within the sample to be tested are as consistent with the average values over the entire study area as possible. Similarly, soil does not change as rapidly as liquid in terms of chemical or mineral composition during sampling events. The key with soil and other solids is primarily to avoid contamination of the sample from outside sources such that the test data represent only the soil or solids conditions at the sampling location at the time of sampling and nothing else. Soil gases, however, are very difficult to maintain even over short periods, and careful consideration of how to avoid agitation of samples during sampling and how to stabilize the gas content is important in sampling those soil conditions where soil gases are an important component being tested.

3.1 SAMPLING SOIL OR OTHER SOLIDS

Sampling soils or other solid and semisolid materials requires a somewhat different approach from that used for liquid samples. This type of medium includes all soil types, sludges, lees, sediments, and other solid or semisolid materials. Small samples of these materials are generally collected from larger samples that are well mixed to ensure that when a 1 g aliquot is later tested in the laboratory, that very small aliquot adequately represents the soil characteristics of the larger mass.

When compositing a sample from a larger mass, several samples of approximately equal size are collected from various locations throughout the mass of material to be tested. Care needs to be taken to ensure that all components of the mass are represented in approximately the same concentration in the collected samples as they exist in the original mass. The number of samples collected in this manner depends on the mass or area to be tested. Generally, one sample per cubic yard (0.75 m^3) of material will be adequate. Those samples are then placed on a clean plastic tarp or in a clean mixer and thoroughly mixed so that any sample collected from the mixed pile will likely be a fair representation of the mass as a whole. A sufficient sample for testing is then collected from the composite mix and returned to the laboratory for analysis. Grain size analysis may require a larger sample of several cubic feet, while chemical analysis may better be represented by a smaller, 8–16 oz (250–500 mL) sample from the composite mix. Chemical analyses are done on a 1 g aliquot from the sample jar,

DOI: 10.1201/9781003346685-3

and commercial laboratories seldom remix samples prior to testing unless specifically instructed to do so (and paid for that time and effort).

Note that soil being tested for volatile organic content cannot be sampled and mixed in the manner described for general soil sampling. Doing so would dramatically reduce the volatile component concentrations and destroy the accuracy of any future tests. In that case, a small, one milligram sample is collected from each of the proposed sampling locations within the mass and carefully placed into a sampling jar or VOA vial. The composite sample is not mixed at that time, but the composite sample is fixed with a preservative, typically methanol in a 1:1 ratio of methanol volume per unit volume of soil, to extract the petroleum components. The methanol is then tested for the concentration of the extracted petroleum components, with the methanol components being subtracted from the results, rather than the soil being tested directly. A Headspace Analysis (MA DEP) test is performed on-site when volatile petroleum hydrocarbons are expected to provide an indication of the concentration of those contaminants in the sampled media.

Table A.2, in the Appendix, provides several useful columns of information relative to the sampling and testing of soils and other solids. This table contains consolidated data based on published guidelines from various analytical laboratories. Verification of data with the testing lab prior to sampling is strongly recommended. In most cases, undergraduate students in university laboratories do not need to be quite so precise in their sampling and handling methods. However, the discussion that follows does provide guidance on how routine samples are collected and handled for consistent laboratory results. The following data are provided in Table A.2.

- Column 1 – a selected list of parameters for which solid samples are frequently tested.
- Column 2 – the suggested sample testing methodologies defined by U.S. Environmental Protection Agency (EPA) test protocol numbers with which all commercial laboratories are familiar and which are described in various Standard Methods of Analysis manuals. Note that not all laboratories use the same test procedure for each compound or chemical and that it is important to know which test method is being used for consistency of results.
- Column 3 – the type of recommended sample containers and sample volumes to be collected.
- Column 4 – the recommended preservatives to be used to prevent or minimize changes over short time periods after sampling or preservation.
- Column 5 – the accepted transport and holding conditions recognized as essential for minimizing changes over short time frames.
- Column 6 – the accepted holding times for samples that have been preserved as recommended in the previous columns. Data from a sample that is not tested within the recommended holding times should be regarded as suspect.

3.2 SAMPLING CONCRETE

The collection of a representative sample of freshly mixed concrete presents some unique challenges, particularly when the concrete is being delivered directly to a job site. When collecting samples, whether from a delivery truck at a job site or from a small mixer in a university laboratory, the elapsed time should not exceed 15 minutes between obtaining the first and final portions of the composite sample. Once transported to where the test is to be conducted, samples should be combined and remixed with a shovel the minimum amount necessary to ensure uniformity. Tests for slump, temperature, and air content should be done within 5 minutes after obtaining the final portion of the composite sample and completed as quickly as possible. The molding of specimens for strength tests should be done within 15 minutes after fabricating the composite sample. These tests should also be conducted as rapidly as possible while being protected from sun, wind, and other sources of rapid evaporation or contamination.

Samples collected for strength tests require a minimum volume of 28 L (1 ft^3). Smaller samples may be used for routine air content, temperature, and slump tests. The required size of those samples will be dictated by the maximum aggregate size.

Every precaution that will assist in obtaining samples that are truly representative of the nature and condition of concrete sampled must be taken during concrete sampling. Sampling at a job site should normally be performed as the concrete is delivered from the mixer to the conveying vehicle used to transport the concrete to the forms. When sampling from a laboratory or other small, stationary mixer, the concrete is sampled by collecting two or more portions taken at regularly spaced intervals during discharge of the middle portion of the batch and combined into one composite sample for testing purposes. Portions of a composite sample should not be collected from the very first or last part of the batch discharge.

The sampling portions are collected by passing a receptacle completely through the discharge stream, or by completely diverting the discharge into a sample container. If the concrete discharge stream is too rapid to divert the complete discharge stream, the concrete should be discharged into a container or transportation unit sufficiently large to accommodate the entire batch and then the composite portion collected from that batch. The intent is to provide samples that are representative of widely separated portions, but not the beginning and end of the load.

Note that ASTM Standard Practice for Sampling Freshly Mixed Concrete, C172/C172M – 17, provides far more detailed instructions for the sampling of fresh concrete from a variety of variable mixing sources. That reference should be consulted for job site sample collection, but the directions outlined here, which are consistent with that reference, are adequate for laboratory mixers.

3.3 SAMPLING WATER AND OTHER LIQUIDS

The sampling of liquids for experimentation presents some interesting challenges for civil engineers. Liquids tend to be subject to component concentration changes based on surface pressures, temperatures, and short time frames during which reactions of internal components can occur. It is imperative, then, to ensure that samples are collected in such a way as to minimize the opportunity for such changes to occur. It is also useful in many cases to add a preservative to a sample to reduce chemical reactions within the sample between the time of sampling and the time of testing.

Liquids are generally amenable to in situ testing of various parameters using portable testing equipment. Dissolved oxygen, temperature, pH, suspended solids concentration, electrical conductivity, and some chemical concentrations, such as ammonia, ammonium, nitrite, nitrate, chlorides, and phosphate, can be measured in real time by inserting probes into the liquid, whether flowing or still, to generate accurate, real-time data. When possible, this is the preferred method of sampling and testing liquids.

When in situ testing of liquids is not possible for any reason, samplers must consider ways to minimize changes in the composition of the chemicals contained in the soil between the sampling and the testing.

Table A.1, in the Appendix, provides several useful columns of information relative to the sampling and testing of water and other liquids. This table contains consolidated data based on published guidelines from various analytical laboratories. Verification of data with the testing lab prior to sampling is strongly recommended. In most cases, undergraduate students in university laboratories do not need to be quite so precise in their sampling and handling methods. However, the discussion that follows does provide guidance on how routine samples are collected and handled for consistent laboratory results. The following data are provided in Table A.1.

- Column 1 – a selected list of parameters for which liquid samples are frequently tested.
- Column 2 – the suggested sample testing methodologies defined by EPA test protocol numbers with which all commercial laboratories are familiar and which are described in various Standard Methods of Analysis manuals. Note that not all laboratories use the same test procedure for each compound or chemical and that it is important to know which test method is being used for consistency of results.
- Column 3 – the type of recommended sample containers and sample volumes to be collected.
- Column 4 – the recommended preservatives to be used to prevent or minimize changes over short time periods after sampling or preservation.
- Column 5 – the accepted transport and holding conditions recognized as essential for minimizing changes over short time frames.
- Column 6 – the accepted holding times for samples that have been preserved as recommended in the previous columns. Data from a sample that is not tested within the recommended holding times should be regarded as suspect.

Where the table calls for the addition of a preservative, it is usually considered appropriate to ask the laboratory providing the sample jars to add the preservative in advance. The sample jar then needs to be carefully filled, using a nonturbulent filling method, to avoid overfilling and losing any of the preservative. It is common to use a secondary container to collect the sample and to fill the transportation container slowly from the collected sample. Note that when grease and oil components are being measured, the use of a secondary container is strongly discouraged due to the potential for various components of the oil and grease mixture to adhere differently to the walls of the secondary container, thereby changing the concentrations in the tested sample.

Nonturbulent sampling means that the sample is collected slowly in a jar or vial by submerging the sampling container into the liquid and allowing it to slowly fill when not using a sample container with preservative pre-entered. This usually implies laying the jar into the liquid sideways, slowly lowering it into the liquid, then turning it upright, slowly removing it from the liquid, and immediately screwing a tight lid on top. If the sample requires a preservative, the preservative may be added by inserting an eye dropper or pipette with the appropriate volume of preservative to the bottom of the filled sample jar and slowly injecting the preservative, allowing the excess liquid to spill out of the top of the jar. The cap is then screwed on tightly as soon as the preservative has been added. There should be no air trapped above the sample inside the sample jar – a condition referred to as leaving no headspace. If a sampling container does have preservative pre-added, it is usually best to use a secondary container to collect the sample and to then fill the transport container slowly from the secondary container such that a small meniscus appears above the lip of the container to prevent air entrapment when the lid is immediately screwed on tightly. As noted, when grease and oil components are being measured, the use of a secondary container is strongly discouraged due to the potential for various components of the oil and grease mixture to adhere differently to the walls of the secondary container, thereby changing the concentrations in the tested sample.

It is not uncommon to require samples of liquids to be collected from various depths at the same sampling location. Inserting a sampling jar at the surface and lowering it to the desired depth before pulling it out and then capping it generally causes cross-contamination of the sample from the shallower depth to the deeper one. To avoid this, sampling devices are available commercially that allow a sealed container to be lowered to the desired depth, then opened at the selected depth. The container is resealed in situ, at depth, after the container is filled, and then retrieved. This helps to mitigate the cross-contamination issue.

When sampling a monitoring well, the sample is collected in a bailer, typically with a small ball valve at the bottom. The valve ball is lifted with a short tube inserted into the bottom of the bailer after it has been extracted from the well. The tube is inserted slowly enough to avoid rapid discharge of the fluid into the sample jar or vial, and the sample is allowed to run down the side of the container as the container is slowly turned right side up. A small meniscus on top of the liquid should be seen as the cap is tightly placed on the container.

Most tests of liquid samples require only a very small aliquot of sample – typically 1 mL or less. These are typically collected in a 40 mL volatile organic analysis (VOA) vial rather than a larger jar. That allows for preservatives to be used in small quantities when things like petroleum hydrocarbons are being tested, for example, but requires more care when filling the vials to avoid issues such as loss of preservatives, nonturbulence when filling, and the presence of a meniscus on top of the liquid before applying the cap to avoid an air pocket on top of the liquid. Smaller sample jars also conserve space in field coolers and minimize the waste disposal problem at the laboratory. Thus, smaller jars are preferred when they are reasonable.

A special scanning test for volatile petroleum hydrocarbons in water is the Headspace Analysis (see Attachment II in the Massachusetts DEP Guidance Manual at: www.mass.gov/doc/wsc-94-400-interim-remediation-waste-management-policy-for-petroleum-contaminated-soils/download). This test is performed on-site when volatile petroleum hydrocarbons are expected to be found in order to provide an indication of the concentration of those contaminants in the sampled media. An 8 or 16 oz "mason-type" jar of sample is collected half full and immediately sealed with aluminum foil followed by a screw-on lid without the metal insert. The sample is then "vigorously" shaken for 15 seconds and then allowed to rest for about 10 minutes for the headspace concentrations to develop. The sample is then vigorously shaken a second time, immediately before testing. When the ambient temperatures are below 0 °C (32 °F), headspace development should be done within a heated vehicle or building.

At the end of the 10 minute development period, the probe of a photoionization detector (PID), a flame ionization detector (FID), or a similar testing device is inserted through the foil liner into the interior of the headspace to a depth of about half the depth of the headspace, avoiding direct contact with the sample. The highest reading on the PID or FID device that shows immediately after the probe is inserted (it will decline quickly from the peak reading) is taken as an indication of the approximate (*very* approximate) concentration of volatile petroleum hydrocarbons present in the headspace over the sample. Normally, if that initial reading indicates a likely concentration of 10 ppm or higher, a sample is required to be collected for laboratory analysis.

3.4 SAMPLING AIR AND OTHER GASES

There are several ways to sample air and other gaseous media besides taking a deep breath and hoping that nothing bad happens. These methods span the continuum from scanning tests that will indicate the presence of a target compound and provide a colorimetric indication of the concentration, to precise, time-based, composite samplers from which samples can later be collected for very precise measurements in the parts per trillion (ppt) range.

Among the least expensive (but also one of the least accurate) methods is a Draeger Tube, a Tenax Tube, or an equivalent system. These devices are thin, clear, glass tubes filled with a specific proprietary resin that is selected to capture the target compound for that tube and to change color in the process. The length of the tube resin that changes color, or sometimes the brilliance of the color, will indicate the approximate concentration of that compound in the tested air. Markings along the side of the tube provide an easy mechanism for determining the approximate concentration. The air is introduced to the tube by breaking off a seal at both ends, inserting one end into a small displacement pump, drawing the handle of the pump to pull a preset volume of air through the tube, allowing a specified amount of time to expire for the reactions to take place, and reading the results from observing the changes in the resin. These are single-use devices and after use they are disposed as hazardous waste. They are available, however, for the testing of air for a very wide range of chemicals and petroleum hydrocarbons. These tests are most appropriate as a scanning tool to indicate the probable need for more precise testing.

When testing air for particulates, and certain specific compounds such as oxygen (O_2), carbon dioxide (CO_2), hydrogen sulfide (H_2S), and particulates, including specific sizes of particulates such as $PM_{2.5}$ and PM_{10} particulates, a small pump is employed to draw composite samples over a predetermined time and a set volumetric draw rate. For chemical testing, the air is drawn through a proprietary observation chamber for chemical constituents, which typically uses a specific light source to ionize the target chemical and record the concentration in parts per million (ppm) or parts per billion (ppb) of the volume passed through the device. For particulates, a proprietary filter is used to capture the particulates. The clean filter is weighed before being attached to the pump and weighed again at the end of the preset sampling period. The difference is the mass of the target particulates captured for the volume of air pulled through the filter. Note that this means that a $PM_{2.5}$ filter, for example, will capture everything larger than 2.5 μm and that it will be necessary to subtract out all unwanted sizes by simultaneously filtering the air through a prefilter that removes everything larger than the target filter size.

There are a limited number of samplers available for specific air contaminants; however, those of significant concern in certain industries are well developed. H_2S, O_2, and CO_2 samplers, for example, are widely used in the hazardous material (HazMat) responder industry and in those professions that require entrance to confined or potentially hazardous spaces, such as mines, pits, sewers, and subsurface access holes.

For sampling air in order to conduct very accurate testing, using things such as gas chromatographs, for example, it is necessary to use a gas sampling bag or a SilcoCan. These are typically stainless-steel canisters of various volume capacities that are evacuated to a very high vacuum, connected to a special valve arrangement that can be set to pass a specific volume of air per specified unit of time or to cycle on and off for preset sampling times at preset time intervals. That allows flexibility for the sampler to be able to sample a one-time sample or a time-based composite sample over any time frame up to, typically, 24 hours. The unit then shuts down completely, and the sampler can collect the canister at his or her leisure. The canister is then taken to a laboratory, and the contents are released directly into a gas chromatograph or similar electronic device for testing of various known or unknown compounds in the sampled air.

An alternative to the canister is the use of gas sampling bags. These are typically neoprene bags of standard fixed volume. A specialized valve is attached to a port on the bag, similar to the valves and ports on a SilcoCan. These valves are also programmable for collection options and will allow an attached pump to force a preset volume of air into the bag at preset intervals. At the prescribed end of the test period, the valves shut off completely and the sampler may collect the bag at a convenient future time. The issue with this procedure is that any target compound that could react with the neoprene will not be measured in useable concentrations and occasionally an unknown compound will react with the bag and add or subtract from a target compound even when the reaction of the target compound with the bag is not direct.

Canisters and bags have the advantage of collecting large samples of air and being subjected to electronic analysis can yield data on a wide range of compounds not specifically targeted. The location of a peak on the timeline of the resin used in the gas chromatograph may indicate a specific compound and concentration not expected, or at least indicate the need to test for a few compounds that elute in the same temperature and time range using different resins where problems are potentially observed.

Canisters are quite expensive to buy and difficult to clean and maintain. Accordingly, most samples collected using a canister are contracted out to a specialized air-quality testing firm that specializes in this equipment and can guarantee the cleanliness of the canisters, the reliability of the valves and programming systems, and the accuracy of the test data.

3.5 SAMPLING HEALTH AND SAFETY PLANS

Health and Safety Plans (HASPs) will vary depending on the site and the materials being tested. The purpose of a HASP is to ensure maximum protection for those doing the sampling and the environment. For work on designated or suspected hazardous waste disposal or release sites, HASPs should follow guidelines provided by OSHA at:

www.osha.gov/Publications/OSHA3114/OSHA-3114-hazwoper.pdf.

At a minimum, HASPs for hazardous waste disposal or release sites should include instructions and guidelines regarding the following items:

- Names, positions, and contact information for key on-site sampling personnel and for health and safety personnel on-site or available for immediate contact
- Site- or incident-specific risk assessment addressing sample collection activities
- Training requirements
- Personal protective equipment (PPE) on-site and usage requirements
- Medical screening requirements (maintain confidential documents properly and securely)
- Site or incident control
- Emergency response plan, containing off-site emergency contact information such as local hazardous materials response teams or additional trained rescue personnel (29 CFR 1910.38)
- Entry and egress procedures
- Spill containment
- Decontamination procedures

Most sites are not going to be hazardous waste disposal or release sites. Accordingly, the extent of sampling is not going to pose sufficient risk to justify the extensive procedures outlined in the cited OSHA standard. In those cases, the HASP needs to address the following items:

- Sampling location
- Sampling plan (what media will be sampled, a map of where to collect samples, a list of sample containers needed, including coolers with ice, sampling tools needed, cleaning supplies, preservatives to be used, and PPE needed)
- Personnel assigned to the site to do the sampling (at least two people are required for all field sampling events)
- A listing of the risks known or suspected to be present on the site, including things like slippery slopes, poisonous plants, bees and hornets, spills of

acidic preservatives on a person or to the environment, and any other risks known or perceived
- A plan for minimizing potential exposure to identified hazards
- A plan for what to do if exposed to a hazard on-site
- PPE to be used (generally vinyl or rubber gloves and eye protection at all sites and with respirators only if volatile or toxic gases are expected to be present)
- Plan for disposal of contaminated PPE and hazardous wastes at the end of sampling

A HASP is not intended to be a burden or a hinderance to effective sampling, but rather a reasonable evaluation of risks posed, risk management, and exposure response so that safety is properly considered before the team visits the sampling location.

In the laboratory, the same safety concerns encountered in the field are expected to be encountered during testing. A Safety Plan is included as part of every experiment design to ensure that the experimenter has thought through the risks inherent in the work at hand, how to minimize or eliminate those risks, how to protect against exposure, and what to do in the event of accidental exposure.

4 Expected Outcomes and Interpretation of Data

4.1 EXPECTED OUTCOMES

Experiments are conducted to determine outcomes, whether predicted or expected, when there are changes in the conditions and then to determine whether the outcomes are reasonable or accurate. It may be reasonably predicted, for example, from the sample experiment discussed so far that the maximum compacted density of river run sand will be directly correlated to the effective size of the sand. However, there is very little experimental data to either demonstrate or refute this hypothesis. It is useful, therefore, to think through from the beginning and to hypothesize what could cause the outcome data to vary. It is not clear from the outset, for example, how any accretions or acidic pitting could alter the expected effect on soil compactability associated with altering the effective size.

Sometimes, it may not be possible to predict potential outcomes from an experiment. That is rare, however, and it is a situation that should be avoided whenever possible. Going forward without any idea of the expected outcome can lead to dangerous experimental conditions for which adequate safety precautions have not been taken. No data are worth dying for or being severely injured over.

More normally, a question of "What would happen if. . .?" has been posed with the desired outcome being a positive one (even if that implies reducing the effects of a negative outcome). A discussion of the likely effects will normally ensue between knowledgeable persons and an expected outcome will be determined. An experiment is then designed to verify, *or refute*, the expected outcome. The reason it is necessary to understand, recognize, and manage this process is that a designer will almost always be predisposed to develop a design to prove his or her point. That design may then actually, although not necessarily deliberately or consciously, eliminate any data that may disprove the expected outcome or minimize its value, even if the basic theory was correct. The fallacy not found in the data will be discovered by others trying to replicate the data, and the original experimenter will end up being professionally embarrassed. Therefore, the experiment design must assume a stated objective and expected result before the design is done. This will minimize or eliminate the interferences discussed earlier, lead to a cleaner outcome data set, and minimize the potential for flawed experiment design to be shown later by others.

4.2 INTERPRETATION OF THE DATA

Most data are fundamentally useless. As independent numbers or values, most data are meaningless. All data, then, require analysis and interpretation before they become useful. In the example experiment, the experimenter may only be interested

DOI: 10.1201/9781003346685-4

in the variation in maximum compacted density of river run sand meeting the specified uniformity coefficient and measuring the density variation only as a function of the effective size. Then, the measurement of those two parameters over an extended array of effective grain size variations should provide sufficient data to predict an outcome, such as the average compacted soil density over that range of effective sizes. Whatever unmeasured parameters were present and how they affected the maximum compacted density are unimportant to the resulting value. By plotting the effective size against compacted density from the collected data, a reasonably accurate prediction could be made for the compacted density of any river run sand meeting the specified uniformity coefficient and a measured effective size of a sand sample from the sampling location used for the experiment.

A more generalized form of the basic question, as posed at the beginning of Chapter 2, requires that the compacted density of any river run sand meeting the uniformity coefficient can be predicted by measuring the effective size of the sand at that location. When the basic question is so broadened, the simplicity of the experimental data used in the first instance is no longer valid. Now the question involves a consideration of all of the unmeasured parameters previously discussed. It is incumbent upon the experimenter, then, to measure as many of those otherwise unmeasured parameters as possible and to relate the change in compacted soil density by adjusting for the effects of the other parameters. A mathematical formula can then be established in which the concentration or presence of other factors can be input to generate an expected maximum compacted soil density based on the effective size of the grains for a given sample for which those other parameters are also known or estimable.

Note that "estimable" is an important concept here. Even where data have been measured over an extended period of time, they are going to be variable. No matter what value is then chosen to represent the parameter at any given moment, it is likely to be wrong. That introduces a parameter of uncertainty into the mix. Uncertainty can be useful if a broad range of values for the outcome is desired since that is what uncertainty forces the experimenter to accept. More commonly, however, the more precise the answer can be, and therefore the least uncertainty in the data, the better the experimenter will feel about the results.

Uncertainty is a function of statistics, and statistics are a function of probability. To understand the risks of uncertainty, it is necessary to understand a few fundamental facts about probability and statistics. While not a treatise on probability or statistics, the following discussion should help ease the way to evaluating data on a rational basis.

4.3 UNCERTAINTY CONSIDERATIONS

The future is not known. Yet engineering experiments are done all the time as though the future was accurately known and predictable and those experimental outcomes will always be the same for the same experiment. Clearly, there is a disconnect here. That disconnect is called *uncertainty*. In fact, what the future will hold and how unmeasured parameters of experiments will change over time are not known, and it is necessary to consider future uncertainty in the analysis of experimental outcomes.

The best way to do that is to begin by making a careful estimate of all the important variables involved in the decision. Then the concept of predicting a range of possible outcomes should be carefully considered. Finally, what happens when the probabilities of the various outcomes are known, or can be reliably estimated, should be incorporated into the experimental design process.

Estimates and their use in experimental design require evaluation of the future consequences of alternatives. The accuracy of these estimates can, and often do, have significant consequences on the decisions made, particularly when the estimates are not made carefully enough at the outset.

4.4 ESTIMATING FUTURE PARAMETER VALUES

It is usually more realistic to describe the future value of parameters with a range of possible values, rather than a single estimated future value. Such a range could include an optimistic estimate, the most likely estimate, and a pessimistic estimate, for example. Then the outcome data analysis can determine whether the decision is sensitive to the range of projected values and, if so, by how much.

A table can be created with all the variables listed followed by their individual optimistic, most likely, and pessimistic values. If there are only a couple of such variables, it is easy enough to calculate from the analysis using all the values to see which is most sensitive. When there are a lot of variables, however, this becomes more problematic, and, because each of the variables is likely affected in some way by all the others, the total range of possibilities gets out of hand quickly.

This can be addressed by using an average or mean value for each parameter to conduct the analysis. The equation for this mean value weights the "most likely" value four times as heavily as each of the other two estimates, as follows:

$$\text{Mean Value} = \frac{\left[(\text{optimistic value}) + 4(\text{most likely value}) + (\text{pessimistic value})\right]}{6}$$

4.5 PROBABILITY

Statistical analysis of environmental data is an important tool for understanding the implications of the data over time. It is important to understand the relationship between changes in data over both long time periods and short time periods. Parameter values are not static, and they can change rapidly. They tend to go up and down regularly, and it is the trend over time that matters most. The causes of short-term changes are important but not as critical as the long-term changes, in most cases. Statistical analyses can smooth out the seasonal and noise variability in the data to indicate to the engineer whether the long-term trends are favorable.

Statistical data, however, are seldom believed or taken at face value by an audience. Even a sophisticated and technically savvy audience will look at statistical data askance and be wary of interpretations made on the basis of those data. Therefore, it behooves the engineer to be as precise and careful as possible with the presentation of statistical data if the engineer is to be believed in the end. People "know" that engineers and mathematicians "lie" with statistics "all the time" and that any outcome derived from statistics must be viewed with severe skepticism.

4.5.1 Probability as Basis of Statistics

One of the least well-understood concepts of statistical analysis is that statistics are based on probabilities, not absolutes. Factual data are developed from which to predict future events and outcomes, but the data set is never complete, and it is a historic data set, not a future data set. Thus, any event predicted by the data for the future will necessarily be subject to doubt and some real probability that the prediction will be wrong. This uncertainty is a paramount parameter of probability analyses and, therefore, also a paramount parameter of statistical analysis.

Probability plays a significant role in the management of risk and uncertainty in engineering data analysis. There are certain logical or mathematical rules for probability. If an outcome can never happen, then the probability is 0, or 0 %. If an outcome will certainly happen, then the probability is 1, or 100 %. This means that probabilities can neither be less than 0 nor more than 1. They must fall within the range of 0 to 1.

Probabilities are defined such that the sum of the probabilities for all possible outcomes is 1, or 100 %. Summing the probabilities that the value of a specific parameter will remain the same, increase in the future, or decrease in the future, for example, shows that each of these three possible outcomes could have a probability of exactly one out of three, or 0.3333, for a sum of 1.0 for all possible outcomes. They could also each have a range of possible values, but the sum of the three values must always add to exactly 1.0.

For example, a drilling program for a potable water well can have three possible outcomes: a dry hole; water, but not in useable quantities; or water in a useable quantity. The probabilities of these three outcomes must always sum to 100 % since one of these outcomes must occur.

In most probability books, many different probability, or *frequency*, distributions are presented. These distributions describe the results of large numbers of trials yielding large populations of data from which probabilities can be quite easily derived. In engineering experimentation, however, it is more common to use two to five outcomes to determine probabilities, even though two to five outcomes can only approximate the total range of possibilities. This is done for two reasons. First, the data are often estimated from expert judgment, not actual experimentation, so using numerous estimated data values would suggest a false sense of accuracy or precision. Second, each outcome requires additional analysis, and the resources are seldom available to provide those analyses. The use of two to five trials, then, has become a standard, accepted trade-off between accuracy and efficiency.

There are lots of different probabilities that can be developed. If there is more than one variable present, then the probabilities of any two or more occurring at the same time are called *joint probabilities*.

For example, if a fair coin is flipped and a fair die rolled at the same time, the probability of a head coming up for the coin toss is 1/2. The probability of a 4 coming up when the die is rolled is 1/6. The joint probability, when flipping the coin and rolling the die, of getting both a head on the coin and a 4 on the die is the product of the two probabilities, or 1/2 × 1/6 or 1/12. Notice that there are 12 possible outcomes between the head or tail of the coin and any of the six numbers on the die. Therefore, each outcome has a probability of 1/12. The sum of all the possibilities is still equal to exactly 12 × 1/12 = 1 or 100 %. Note here that the order in which the results are determined is not important and that there is only one unique set of data possible for each data point.

This simple example clearly demonstrates the concept of determining the probability of an outcome based on *permutations*, rather than *combinations* of the options. The difference between *combinations* and *permutations* is the order in which the outcomes are selected or occur. With *permutations* the order in which the outcomes are selected or occur matters, whereas with *combinations* it does not matter. For example, if a lock "combination" were made up of four numbers, 1–4, there are 12 possible arrangements that could be made from those four numbers. If the correct arrangement, or order, is 2143, but 1243 is entered, the lock will not open because the numbers entered are in the wrong order – in essence, the *permutation* selected is incorrect. With "combination locks", the correct sequence is actually a permutation, not a combination, in probability terms.

The equation for determining the number of possible permutations among a set of values is:

$$P = \frac{n!}{(n-r)!}$$

where:
P = the number of possible permutations
n = the number of things to choose from
! = the factorial function, which means that the answer is a multiplication of each value in the series together (3! = 3 × 2 × 1, for example)
r = the number of things to be selected

This equation assumes no repetition of the selected numbers, but the order of selection *does* matter.

Therefore, the probability of winning a lottery run with permutation rules (which is NOT the way lotteries are run) and which requires the gambler to select five numbers from a pool of 69, followed by one number from a pool of 26, would be calculated as:

$$P = \frac{69!}{(69-5)!}$$

or

$$P = \frac{(69)(68)(67)(66)(65)(64!)}{(64!)}$$

Which means that P = 1,348,621,560

In this case, however, there is also a need to select a single number from a pool of 26 numbers, so the previous quotient needs to be multiplied by 26 to yield a total number of permutations of

$$P = (1{,}348{,}621{,}560)(26)$$

Where, P = 35,064,160,560

And the odds of winning that lottery are 1 in 35,064,160,560.

Calculating the odds in this fashion implies that the order in which the numbers are selected by the gambler must be identical to the order in which the numbers are selected by the lottery computer. That means that this calculation infers that there are 35,064,160,560 possible combinations of the five numbers *when the order in which the numbers are selected is important.*

Lottery games do not rely on selecting the numbers in any specific order, as shown by the order in which they are drawn when watching that event occur live on television. Therefore, the six numbers selected by the gambler can be selected in a different order by the lottery computer and the gambler will still win. When the order of selecting the numbers does not matter, the calculation becomes a combination problem, not a permutation problem, and is done in the following manner.

$$C = \frac{n!}{\mathrm{r}!(\mathrm{n}-\mathrm{r})!}$$

Where:
C = the number of possible combinations
n = the number of things to choose from (in the current lottery problem that would be 69)
! = the factorial function
r = the number of things to be selected (in the current lottery problem that would be 5).

This equation also assumes no repetition of the selected numbers and that the order of selection does *not* matter. The reason the 26 is used as a separate multiplier is that this value includes the same numbers to select from as the first 26 of the 69 from which the first five values are selected. The "replacement" of the value to select the last digit requires a separate multiplication.

Using this equation, the odds of winning the example lottery, a combination problem, rather than a permutation problem, are calculated as:

$$C = \frac{69!}{(5)!(64)!}$$

$$C = \frac{(1{,}348{,}621{,}560)}{(120)}$$

Where, C = 11,238,513

Multiplying that sum by 26 for the number of options for the last number yields:

$$P = (11{,}238{,}513)(26)$$

Where, P = 292,201,338

And the chances of winning this lottery are increased to a mere 1 in 292,201,338. Still not very good, but way better than 35 billion to one against! Those are the odds of winning that would be published by the lottery commission involved in the lottery used for this example. That is also why lottery jackpot winners are rare relative to the number of players.

This concept of how probability influences experimental data starts to become important when the interferences become less well defined but are important to the evaluation of all data. The probability of things happening assumes a random selection of outcomes. When outcomes stop being random, the probabilities change rapidly and often unpredictably.

4.5.2 Estimating Probabilities in Experimentation

The probabilities of interest during experimentation are almost always associated with negative events: the probability that something bad is going to happen. It is usually very difficult, if not impossible, to accurately measure the probability of potential adverse events occurring. It can be instructive to examine past experiments of similar kinds to see what went wrong there and to carefully consider the adverse events occurring with the prior experiment to see what could go wrong with a current experiment. Looking then at what caused the adverse event, and what could have been done to prevent it, can be useful in reducing the risk of that event occurring anew.

A useful process for developing an estimate of the probability of an adverse event can be the use of average probabilities. Consider, for example, the potential failure of a specific probe being used to generate data. A specific probability is unlikely to be available to reflect the potential for this failure. Begin, then, with a *range* of possibilities for this event. Consider the most optimistic probability, the most likely probability, and the most pessimistic probability.

Once the three probabilities have been estimated, an average probability is calculated from the following equation previously proposed.

$$P_{avg} = \frac{\left[(\text{optimistic estimate}) + (4)(\text{most likely estimate}) + (\text{most pessimistic estimate})\right]}{6}$$

The resulting estimated probability is weighted toward the most likely value but tempered with the reality that either of the other two probabilities could also occur.

For example, if the most optimistic estimate of the probability of failure of the probe is determined to be 1/100, the most likely estimate is 1/50, and the most pessimistic estimate is 1/20. Then the calculation would be:

$$P_{ave} = \frac{\left[\left(\frac{1}{100}\right) + (4)\left(\frac{1}{50}\right) + \left(\frac{1}{20}\right)\right]}{6}$$

$$P_{avg} = \frac{\left[\left(\frac{1}{100}\right) + \left(\frac{4}{50}\right) + \left(\frac{1}{20}\right)\right]}{6}$$

$$P_{avg} = \frac{\left[\left(\frac{1}{100}\right) + \left(\frac{8}{50}\right) + \left(\frac{5}{100}\right)\right]}{6}$$

$$P_{avg} = \frac{\left[\frac{14}{100}\right]}{6}$$

Where, $P_{avg} = \frac{2.3}{100}$ or 2.3 %

Notice that the resulting estimate (2.3 %) in this case is higher than the most likely estimate of 2 %. That is because the difference between the most optimistic estimate and the most likely estimate is 1 %, while the difference between the most pessimistic estimate and the most likely estimate is 5 %. That smaller difference skews the average estimate toward the most optimistic estimate. The average in this case is worse than the most optimistic and slightly better than the most likely, but much better than the most pessimistic.

4.5.3 Misuse of Probability Data

Consider the following concepts and cases in which statistical data were poorly used, or even misused for nefarious purposes. The basic problem facing the statistical analyst is that any three or more data points can be analyzed by statistical methods. But the fewer the number of data points, the louder the noise, or misinformation, that is provided and the shakier the analytical outcome becomes. That causes poor or improper interpretation of the data and that poor interpretation yields incorrect analyses. The number of data points needed for effective prediction into the future depends on what the data represent and the use to which the analyses are intended to be put.

The following examples demonstrate this phenomenon:

Premise: The premise of an analysis presented at a technical seminar was that a new engineering course, based on project-based service learning ideals, provided added value to the overall engineering program at the presenter's university and that specific design skills, and a broader set of skills for graduates to use to help them thrive within a global environment, were being successfully imparted to the students through this course.

Details: To document the validity of the premise, a survey instrument was sent to all program alumni. The survey was internally tested, first, to eliminate bias and ambiguity; 1,500 electronic requests were sent out and 1,000 other alumni were sent a postcard with an electronic link to the survey. In addition, to encourage participation, a premium was offered to those who responded, regardless of their feelings or responses. There were approximately 600 respondents.

Findings and Conclusions Drawn: A very high percentage of the respondents found the course to be very helpful to them, individually, and agreed that the results would improve their project management skills going forward. The majority also agreed that this course would provide significant added value to the education offered by this program.

Problem: A total of 600 responses out of 2,500 survey requests is a reasonably large response pool, and a 23 % response rate would normally give some fairly reliable data. However, in this case, the pool of respondents was self-selected, probably already felt good about the university (or, conversely, had really bad feelings about the university), and the data were going to have a bias, probably a positive bias, right from the start.

It is further noted that any survey that offers a premium for responding, regardless of the value of that premium, is already biased by the self-selection of the respondents and the generally improved feelings about the surveyors and the survey. Any analyses, then, are going to be unreliable, at best.

Similarly, a professor at a different school conducted a survey on the following basis:

Premise: A specific leadership and service learning module in the first year of study positively impacted the success of students in subsequent years and increased interest in pursuing leadership careers.

Details: The subject leadership module was well designed and implemented as an elective course for all students who wished to take it, and the course included excellent leadership and service components. The first class of students who took the course, and a similar group of students in the same major and cohort who did not take that course, were surveyed at the end of the course to determine the validity of the premise. There were 87 respondents – 24 % of those surveyed who took the leadership module responded (54 of the 87 respondents), and 18 % of those surveyed who did not take the module responded (33 of the 87 respondents).

Findings and Conclusions Drawn: Those who did take the course found it very helpful, while those who did not take the course saw limited value in it. It was concluded, therefore, that the premise was correct, and it was predicted that this module would improve the leadership skills of all who took it in the future.

Problem: The course module was voluntary and taking the survey was voluntary. Those who took the course were already predisposed to leadership, and those who did not take the course were predisposed to not be interested in leadership. In addition, while the response rate is fine for statistical analysis, the actual numbers are so small as to be meaningless when projected over a much larger cohort and much further into the future. The data say only that the module may help those who want to lead and that it would likely do nothing for those who do not already want to lead. Any interpretation beyond that is speculative, at best.

4.5.4 Number of Data Points Needed

Finally, it is important to note that the number of data points used in statistical analyses is important. Consider the following case of the traveler and the train:

> A traveler who routinely rode a train into the city for work decided one day to accept a ride from a friend who happened to be going into the city on that day. Upon arrival at his office, on time, he noted a report on the radio that the train he normally rode, and the one on which he would have ridden that morning had he not accepted his friend's offer, had broken down and was long delayed in arriving at the city. He had never experienced a train delay prior to that date and none since.

The statistics say that any time this rider accepts a ride from a friend, the train is going to break down – with 100 % certainty – since every time he had taken a ride with a friend the train had, indeed, broken down, and every time he had ridden the train it had arrived on time. That is, of course, a false premise because the data set is nowhere near sufficient to justify either the premise or the conclusions and there is no conceivable cause and effect relationship between the two events.

4.5.5 Selective Use of Data

The selective use of data implies the deliberate omission of data that do not support the hypothesis being tested. For example, if 152 data points were collected but only 75 supported the hypothesis being tested, selective use of data would indicate that the 77 points that did not support the hypothesis would be ignored. Regardless of how the data are then displayed, ignoring the data that do not support the hypothesis is unethical and wrong.

Consider the case of air temperatures collected at precise intervals every month for 10 consecutive years and then averaged over those months to provide ten data points for each month of the year; in essence, 10 years of average monthly data. Suppose a hypothesis that average air temperatures at the test site are declining, not increasing, as a result of presumed climate change effects. Assume, further, that the data are tabulated as shown in Table 4.1. This is a hypothetical example, so none of the data are real, but they do illustrate the principle involved.

The change in average annual temperature over the stated time frame is shown in Figure 4.1 on the basis of the fictitious data from Table 4.1.

TABLE 4.1
Average Air Temperatures by Month and Year (Degrees F) (Fictitious Data)

Yr Mo	1996	1997	1998	1999	2000	2001	2002	2003	2004	2005	10-Year Average	Change in Value
Jan	23.7	23.5	21.0	23.7	22.1	21.3	23.6	24.4	21.9	21.8	22.7	−1.0
Feb	27.5	26.7	26.6	26.7	27.5	28.9	27.6	29.4	32.3	31.6	28.5	1.0
Mar	38.7	38.2	39.4	37.5	39.5	37.7	36.8	35.2	37.3	37.4	37.8	−0.9
Apr	53.6	52.7	50.9	48.3	52.7	58.9	60.1	59.4	58.9	61.3	55.7	2.1
May	58.6	57.5	57.0	57.2	58.1	59.3	58.3	57.2	59.3	58.2	58.1	−0.5
Jun	66.1	66.9	66.4	66.1	65.9	67.8	66.3	65.2	67.7	66.8	66.5	0.4
Jul	72.6	71.2	70.6	69.9	70.5	70.4	71.2	71.9	74.2	72.7	71.5	−1.1
Aug	72.1	73.8	71.6	73.9	75.2	71.6	72.3	72.4	76.8	73.1	73.3	1.2
Sep	67.5	66.5	69.5	69.3	69.0	66.3	67.8	65.9	61.7	65.8	66.9	−0.6
Oct	50.1	51.7	50.8	50.9	50.1	50.3	50.4	50.3	50.1	50.8	50.6	0.5
Nov	32.6	33.7	31.9	30.7	32.6	31.6	30.5	32.9	30.9	31.8	31.9	−0.7
Dec	26.0	25.9	24.8	27.8	26.8	26.6	27.6	25.8	24.7	26.4	26.2	0.2
Avg. for the year	49.1	49.0	48.4	48.5	49.2	49.2	49.4	49.2	49.7	49.8	49.1	0.0

Source: Hopcroft, Francis J. *Engineering Economics for Environmental Engineers*. New York, NY: Momentum Press, LLC, 2016.

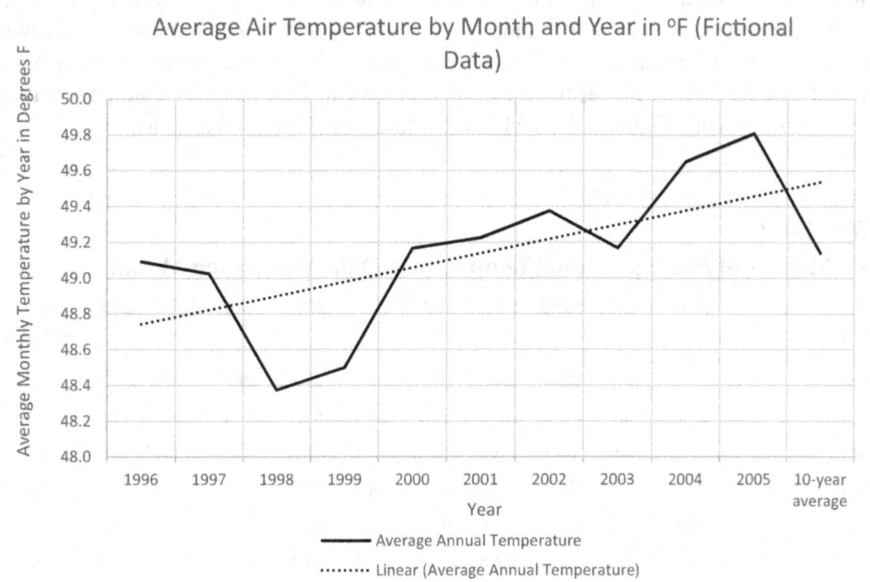

FIGURE 4.1 Average annual temperature over 10-year period (fictitious data from Table 4.1).

Source: Hopcroft, Francis J. *Engineering Economics for Environmental Engineers*. New York: Momentum Press, LLC, 2016.

Looking at the data in Table 4.1, it can be argued that the average annual temperature during the entire 10-year period did not change at all at this hypothetical location. The average temperature for the year 1996, shown at the bottom of the second column, is identical to the average of all the years at the end of the 10-year period, as shown at the bottom of the 12th column, and reinforced by the zero-change calculated in the lowest right-hand cell. That would, however, be a false reading of the data.

Notice is called to the linear trend line in the graph in Figure 4.1. There is an initial decline in average annual temperatures from 1996 to 1998. Thereafter, however, the average annual temperatures increase almost every year relative to the previous year, with the exception of 2002–2003. At the end of 2005, the average annual temperature is actually 0.7 °F higher than that recorded in 1996 and 1.4 °F higher than the lowest point recorded in 1998. Including the 10-year average number as the final data point on the chart skews the data so that the casual reader may imply a conclusion that is not valid. Figure 4.2 shows this same chart without the 10-year average number, and the trend line increase is more dramatic – and truthful.

A person wanting to promote the concept that global warming has made a big difference in this community by raising the average annual temperatures might show Figure 4.2 without the average value in the calculations and neglect to note the effect of averaging the temperatures over the 10-year data period. That person could also show the data for only every second month, as shown in Table 4.2.

From these 6 months of data, it can be argued that the average monthly and annual temperatures at this location rose about 1.2 °F over the 10-year period and that climate change effects are real. This conclusion is further demonstrated by Figure 4.2.

A person who is a climate change denier might use the other 6 months, as shown in Table 4.3, and argue that the average annual temperatures at that location have actually declined 0.5 °F over the same 10-year period. That position would be further demonstrated with Figure 4.3 on which the data from Table 4.3 are plotted.

TABLE 4.2
Six Months of Average Annual Temperature Data, Every Even Month

Yr Mo	1996	1997	1998	1999	2000	2001	2002	2003	2004	2005	Change in Value
Feb	27.5	26.7	26.6	26.7	27.5	28.9	27.6	29.4	32.3	31.6	4.1
Apr	53.6	52.7	50.9	48.3	52.7	58.9	60.1	59.4	58.9	61.3	7.7
Jun	66.1	66.9	66.4	66.1	65.9	67.8	66.3	65.2	67.7	66.8	0.7
Aug	72.1	73.8	71.6	73.9	75.2	71.6	72.3	72.4	76.8	73.1	1.0
Oct	50.1	51.7	50.8	50.9	50.1	50.3	50.4	50.3	50.1	50.8	0.7
Dec	26.0	25.9	24.8	27.8	26.8	26.6	27.6	25.8	24.7	26.4	0.4
Avg. for the year	24.6	24.8	24.3	24.5	24.9	25.3	25.4	25.2	25.9	25.8	1.2

Source: Hopcroft, Francis J. *Engineering Economics for Environmental Engineers*. New York, NY: Momentum Press, LLC, 2016.

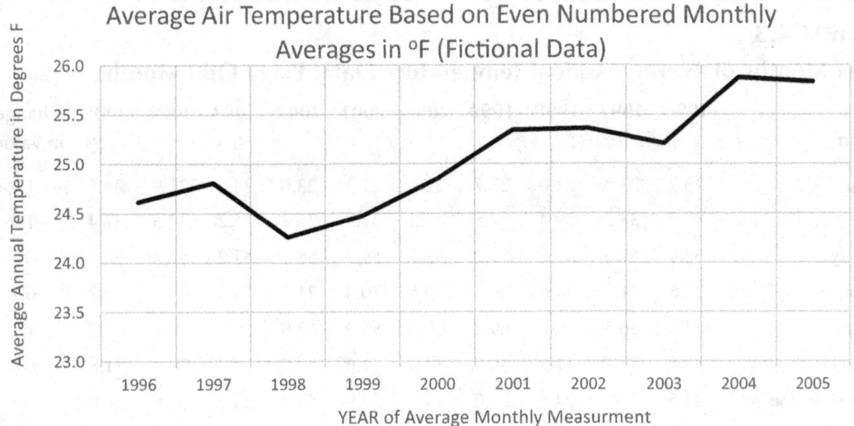

FIGURE 4.2 Average annual temperatures based on data from Table 4.2 (even months Feb. through Dec.).

Source: Hopcroft, Francis J. *Engineering Economics for Environmental Engineers*. New York, NY: Momentum Press, LLC, 2016.

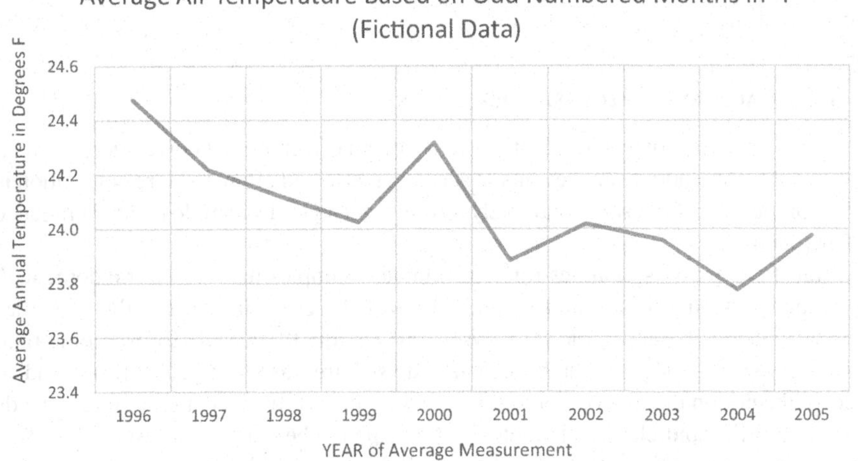

FIGURE 4.3 Average annual temperature over a 10-year period (odd months, Jan. through Nov.).

Source: Hopcroft, Francis J. *Engineering Economics for Environmental Engineers*. New York, NY: Momentum Press, LLC, 2016.

Clearly, the climate change proponent and the climate change denier cannot both be correct at the same time, but both have selected those data that support their position and ignored those which oppose their position. Both approaches are unethical and wrong. Note that these temperature values have been deliberately selected to illustrate the points being made here and that data separation is seldom this neat and

TABLE 4.3
Six Months of Average Annual Temperature Data, Every Odd Month

Yr Mo	1996	1997	1998	1999	2000	2001	2002	2003	2004	2005	Change in Value
Jan	23.7	23.5	21.0	23.7	22.1	21.3	23.6	24.4	21.9	21.8	−1.9
Mar	38.7	38.2	39.4	37.5	39.5	37.7	36.8	35.2	37.3	37.4	−1.3
May	58.6	57.5	57.0	57.2	58.1	59.3	58.3	57.2	59.3	58.2	−0.4
Jul	72.6	71.2	70.6	69.9	70.5	70.4	71.2	71.9	74.2	72.7	+0.1
Sep	67.5	66.5	69.5	69.3	69.0	66.3	67.8	65.9	61.7	65.8	−1.7
Nov	32.6	33.7	31.9	30.7	32.6	31.6	30.5	32.9	30.9	31.8	−0.8
Avg. for the year	24.5	24.2	24.1	24.0	24.3	23.9	24.0	24.0	23.8	24.0	−0.5

Source: Hopcroft, Francis J. *Engineering Economics for Environmental Engineers*. New York, NY: Momentum Press, LLC, 2016.

clean. Careful selection of data, however, without a clear presentation of the known data supporting an alternative view, can be used to skew any analysis unfairly and unethically. A presenter only has to be caught at this once to ruin a reputation for a very long time.

4.5.6 Cause and Effect vs. Correlation

Notice, too, that statistics do not denote cause and effect, only a correlation between two presumably independent events. Ascribing cause and effect to mere correlation is an improper use of the statistical analysis tools and a good way to lose the confidence of any audience.

Statistical analysis assumes random samples. Samples that are not random, such as responses from self-selected individuals, are inherently biased, and the outcomes and conclusions based on those analyses will be equally biased. To overcome that, it is necessary to accept that statistical analyses are based on probabilities and to accept the notion that the outcomes are rife with uncertainty. That uncertainty needs to be carefully and clearly characterized as part of the statistics presentation. It is important to recognize and to show the audience that the data are only *perceptions* of reality and *not actual reality*.

To minimize the perception that the engineer or mathematician is not being truthful when presenting statistical data, it is also useful to first understand some additional basic concepts regarding why people do not believe statistics. An approach pioneered by Edward R. Tufte, in his classic book on *The Visual Display of Quantitative Information*,[1] was to display several data sets that are statistically identical, but graphically and numerically very different. See, for example, the following data sets, which are a variation of Anscombe's Quartet used by Tufte.

Anscombe's Quartet was published by Francis Anscombe in 1973[2] to demonstrate the necessity of visualizing statistical data in order to properly understand them.

It consists of four sets of paired numbers, *X*/*Y* values, that are statistically equivalent within reasonable limits of common measures of statistical equivalence. A data set that comprises a variation of Anscombe's Quartet is shown in Table 4.4. The parameters used to evaluate equivalence among the data sets, their values for the original Anscombe's Quartet, and their values for the variation used here are shown in Table 4.5.

Without going through the proof of the statistical analyses here, they were all calculated in Excel by proper formulation, suffice it to be said that these four data sets are statistically identical based on the analytical tools utilized and the values shown in Table 4.4. Most of the accuracy statements in the third column of Table 4.5 come from an anonymous short essay titled "Anscombe's Quartet", published May 16, 2011, at www.vernier.com/innovate/anscombes-quartet/that describes the original Anscombe's Quartet. Other values shown in that column came from an Excel analysis of the presentation of this quartet by Edward Tufte in his book titled *The Visual Display of Quantitative Information*. The accuracy values shown in Column 5 of Table 4.5 were calculated in Excel on the basis of the values shown in Table 4.4.

TABLE 4.4
Data Sets for Variation of Anscombe's Quartet

	Variation of Anscombe's Quartet							
SET	**X1**	**Y1**	**X2**	**Y2**	**X3**	**Y3**	**X4**	**Y4**
	12.0	8.05	12.0	6.38	11.70	8.77	20.0	9.40
	8.0	6.81	8.0	8.51	8.05	8.04	9.0	9.00
	14.0	7.35	14.0	8.43	14.05	8.03	9.0	8.70
	7.0	6.85	7.0	7.63	6.81	7.21	9.0	8.58
	13.0	6.75	13.0	7.31	12.80	8.53	9.0	7.68
	15.0	9.71	15.0	9.35	15.29	6.69	9.0	7.78
	9.0	6.59	9.0	9.15	9.22	8.50	9.0	7.40
	5.0	6.12	5.0	5.69	5.11	5.10	9.0	7.98
	11.0	8.96	11.0	7.35	10.95	8.79	9.0	6.20
	10.0	9.16	10.0	8.55	9.99	8.67	9.0	6.01
	6.0	8.36	6.0	6.36	6.06	6.38	9.0	5.98
AVERAGE	10.00	7.70091	10.00	7.70091	10.00	7.70091	10.00	7.70091
SUM OF SQUARES	110.00	14.58549	110.00	14.59409	110.00	14.58989	110.00	14.59409
VARIANCE	11	1.458549	11	1.459409	11.00418	1.458989	11	1.459409
CORREL		0.460866		0.488184		0.483003		0.466471
R SQ		0.212398		0.238324		0.233292		0.217595
LINEST		0.167818		0.177818		0.175872		0.169909

Source: Hopcroft, Francis J. *Engineering Economics for Environmental Engineers*. New York, NY: Momentum Press, LLC, 2016.

TABLE 4.5
Statistical Equivalency Values for the Original Anscombe's Quartet and the Variation of Anscombe's Quartet Shown in Table 4.4

Property	Anscombe's Value	Accuracy	Variation Quartet Value	Variation Quartet Accuracy
Number of data points per set	11	Exact	11	Exact
Mean of x (AVERAGE in Excel)	9	Exact	10.00	To 2 decimal places and exact in three of the four sets
Mean of y (AVERAGE in Excel)	7.50005	+/− 0.0005	7.70091	Exact
Linear regression line	Y = 3 + 0.5 x	To 2 and 3 decimal places, respectively	Y= 5.97 + 0.1728	+/−0.03 and +/− 0.005, respectively
Standard error of estimate of slope	0.118	+/− 0.0001	0.1065	+/− 0.05
t	4.24	+/− 0.002	1.6183	+/− 0.06
Sum of squares, $x - \bar{X}$	110.0	Exact	110.00	To 4 decimal places except for 2 decimal places in set 3
Sum of squares, $Y - \bar{Y}$	41	To 0 decimal places	14.59	To 2 decimal places
Regression sum of squares	27.49	+/− 0.02	3.2880	+/− 0.19
Residual sum of squares of Y	13.7594	+/− 0.017	11.3021	+/− 0.186
Correlation coefficient (CORREL in Excel)	0.816	To 3 decimal places	0.5	To 1 decimal place
r^2, or Coefficient of Determination of the Linear Regression (RSQ in Excel)	0.67	To 2 decimal places	0.22795	+/− 0.01035
Sample variance of x (VARIANCE in Excel)	11	exact	11	To 3 decimal places
Sample variance of y	4.125	plus/minus 0.003	1.459	To 3 decimal places

Source: Hopcroft, Francis J. *Engineering Economics for Environmental Engineers*. New York, NY: Momentum Press, LLC, 2016.

It can be seen directly from the data in Table 4.4 that the numbers in these four sets of data are very different. It may, in fact, be difficult to discern where, or how, the sets of data were altered to make them identical – and they *were* mathematically altered to ensure that they are statistically equivalent – to the degree shown in Table 4.5.

To show these data in a way that allows an understanding of how different they are, in spite of their statistical equivalence, it is necessary to graph the four data sets separately; as shown in Figure 4.4, where it can be seen that the four data sets graph very differently. It is also clear from the graphs, and in hindsight from the data sets, where most of the outliers are that cause the statistical equivalence.

Therein lies a key consideration with the presentation of statistical data. Whether the outliers are used in the statistical analysis of the data is a matter of judgment on the part of the engineer. The decision to use or not use outliers will always be subject to challenge regardless of which decision is made. Generally, a true outlier can be ignored in most civil engineering work, but only if the presence of the outlier is acknowledged, the justification for the decision to ignore it is provided and is cogent, and the effect of including it in the analyses is also acknowledged and described. In this case, for example, ignoring the outliers would dramatically change several of the plots and destroy the claim of statistical equivalence among the data sets.

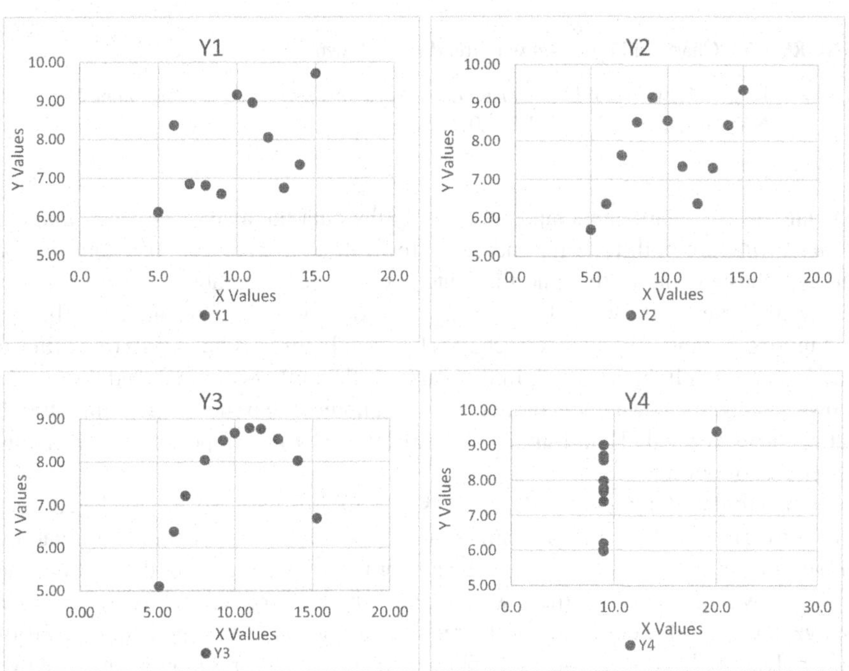

FIGURE 4.4 Charts of variation of Anscombe's Quartet.

Source: Hopcroft, Francis J. *Engineering Economics for Environmental Engineers*. New York, NY: Momentum Press, LLC, 2016.

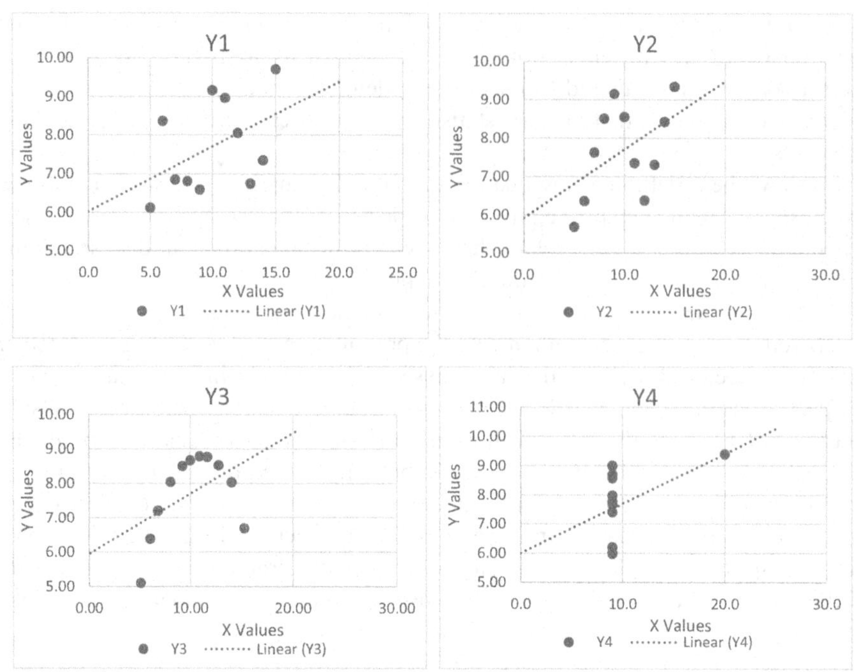

FIGURE 4.5 Charts in Figure 4.4 with trend lines shown.

Source: Hopcroft, Francis J. *Engineering Economics for Environmental Engineers.* New York, NY: Momentum Press, LLC, 2016.

The engineer or mathematician can use only the data the analyst believes, *and can demonstrate*, are both relevant and valid, regardless of whether those data support the hypothesis being tested, and that outcome will be acceptable to the majority of reviewers if the justifications for ignoring other data are cogent, clear, and valid.

Figure 4.5 shows the same four charts shown in Figure 4.4 but with a trend line, or linear regression line, provided. In this case, all the outliers are included. Figure 4.6 shows what happens when the outlier in Y4 is ignored; the trend line remains straight but becomes vertical. That dramatically indicates why it is important to understand the effects of outliers.

Notice in the aforementioned examples that all four data sets are analyzed using a linear regression line. Linear regression lines are typically used on data that are, in fact, linear. It can be seen from the plots of those data that only the first data set is in any way linear. The others would generally, and arguably more correctly, be analyzed using different forms of the regression line, such as exponential, logarithmic, polynomial (in one of several optional ways), power, or a moving average over an optional period length. Using a linear regression line with those data will yield interesting lines but provide no useful information, particularly when the objective is to predict the future with a reasonable degree of certainty. The results mean exactly nothing, which is precisely what Anscombe was trying to point out in his use of a

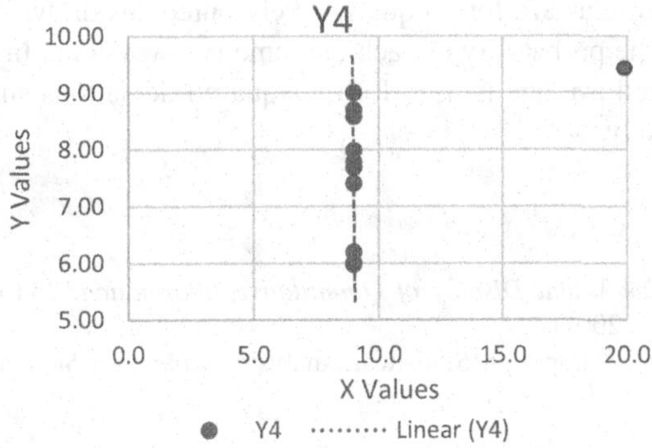

FIGURE 4.6 Y4 from Figure 4.5 with the outlier ignored; trend line is vertical at 9.0.

Source: Hopcroft, Francis J. *Engineering Economics for Environmental Engineers*. New York, NY: Momentum Press, LLC, 2016.

linear regression analysis on these types of data sets. Nevertheless, for consistency with the equivalency calculations, and for consistency with the analyses done by Anscombe, a linear regression line was used here for all the data.

4.6 RISK

Risk can be thought of as the probability of getting an outcome other than the expected outcome, usually with an emphasis on something negative happening. One common measure of risk is the probability of a loss, for example. The other common measure of risk is the standard deviation, which measures the dispersion of outcomes around an expected value. With a normal distribution, represented by a standard bell curve, 68 % of the values lie within ±1 standard deviation from the mean value, and 95 % of the values lie within ±2 standard deviations of the mean value.

Mathematically, the standard deviation is defined as the square root of the variance. The variance is defined as the weighted average of the squared difference between the outcomes of the random variable x and its mean. Thus, the larger the difference between the mean and the value, the larger the variance and the larger the standard deviation.

The equation is as follows:

$$\text{Standard deviation} = \text{Expected value}\left(x - \text{mean}\right)^2]$$

Squaring the differences between the individual outcomes and the expected (mean) value ensures that positive and negative deviations all receive positive weights. Consequently, negative values for the standard deviation are not possible. Moreover, the standard deviation equals 0 if there is only one possible outcome. Otherwise, it is positive. Note that this is not the standard deviation formula built into most calculators, just as the weighted average is not the simple average built into most calculators.

The calculator formulas are for *n* equally likely outcomes drawn from a sample of outcomes so that the probability of each outcome is always one. In engineering data analysis, a weighted average is used for the squared deviations since the outcomes are not equally likely.

NOTES

1 E.R. Tufte, *The Visual Display of Quantitative Information*, 2nd ed. (Cheshire, CT: Graphics Press, 2001).
2 F.J. Anscombe, "Graphs in Statistical Analysis," *American Statistician* 27 (February 1973): 17–21.

5 Model Design Methodology

5.1 DISCUSSION

As indicated in several of the previous chapters, the design of an effective experiment requires consideration of many peripheral issues, along with the main question to be answered or observations to be made. It is useful to organize those concepts into a uniform format in order to manage them. This chapter provides an example of a useful format for achieving this goal. This example is a recommended model and not a template, so notice that where spaces are provided for data, the size of the space is irrelevant and sufficient space should be utilized to completely and adequately provide the information requested.

5.2 MODEL FORMAT

The following pages provide the recommended format for experiment design and data collection.

The title of the experiment provides a way to easily refer to the specific experiment design for the convenience of the investigators and others who may be reviewing or approving it. The title should be descriptive but does not need to be as comprehensive as an academic research paper.

The date is shown at the top so that any changes can be identified easily. Whenever a document is modified or adjusted, insert a new date so that everyone using the document will be able to ascertain which version is being discussed.

The name and contact information for the principal investigator and any collaborators should be easily available and space is provided for that information.

The objective or question to be addressed needs to be clear and concise, following the guidance in Section 2.1 of Chapter 2 in this book.

The theory behind the experiment provides background and context to the question being addressed. It is intended to force thinking about why this experiment is being conducted, how it should be done, and the expected outcomes. Good research into this issue prior to initiation of the experiment design will ultimately lead to a much better outcome from the experiment.

The data to be collected section provides a way to focus the investigators on how the experiment will be conducted. It is axiomatic in engineering that in order to solve a problem it is first necessary to fully define the problem. Sometimes, it is just as important to determine what is not a part of the problem or beyond the scope of the experiment. How data are to be collected is best addressed after the list of desired data has been created, thereby defining the problem to be solved and how data will be collected.

A list of tools, equipment, and supplies is required to be developed to ensure that the procedures have been thoroughly thought through and that everything necessary for the experiment is available and on hand when needed.

DOI: 10.1201/9781003346685-5

Safety is a key consideration in everything an engineer does, including experimental work. In this section, the collaborators need to think through what could possibly go wrong, how to minimize that risk, and how to protect the experimenters and anyone around them in the event something does go wrong.

The next step is to think through the procedures to be followed during the experiment. That includes a discussion of what is included in each step and a plan for how each step will be performed. For example, collection of a water sample usually involves more than sticking an empty bottle into a stream and screwing the cap on after the jar is full. There is often a need to provide some preservative in the bottle before the sample is added, and therefore over-filling will wash out the preservative and destroy the sample. In addition, rapidly filling a sample container can introduce oxygen during the transfer and invalidate a subsequent DO reading. Filling a sample jar too slowly can allow volatiles to escape, thereby invalidating those analyses later. How the sample is to be collected, then, is dictated by the preservation methods and the planned analysis procedure for the sample. Chapter 3 describes the best methods used to sample various media.

Potential interferences and their resolution are key to a successful set of outcome data. Section 2.4 of this book discusses interferences in more detail and provides a mechanism for identifying and managing the effects of those interferences.

Once the data are collected, they need to be analyzed. How those data are analyzed can dramatically affect the outcomes. For example, consider the measurement of pH and dissolved oxygen. Digital meters are generally more accurate than colorimetric methods. pH meters are generally more accurate than pH strips. But Winkler titration analyses for dissolved oxygen are often more accurate than digital meters. It is useful to determine the degree of accuracy required by the experiment, the amount or variance allowed within the expected outcomes, and how often simple tests may need to be calibrated with more precise tests. The data analysis section needs to address these questions and define any requirements for continuous calibration of digital equipment proposed for use. The equipment should be specified as completely as possible, such as, for example, an "XYZ Corp. Model 62 Dissolved Oxygen meter" and an operating and calibration manual for that equipment should be appended to the experiment design sheet when available.

Once the data have been collected and analyzed, there will be a set of outcomes from the experiment. With luck, those outcomes will match the expected outcomes, although in reality they seldom do. It is important, therefore, to consider how the actual outcome data differ from the expected outcome data and to think through why there are variances. Variances can occur from a variety of causes, including poor experiment design, poor implementation, operator error using the analyzers, improper calibration of equipment, equipment failure, or an unrealistic expectation of anticipated outcomes. In any case, the actual data should be believed *unless and until* proven false. The vast majority of experimental data are real and correct for the experiment that was actually performed. Why they did not match the expected outcome is usually because the expectations were not correct or there were some interferences not anticipated. See Section 2.4 for further discussion of this issue.

Chapters 9–14 provide a data sheet for each completed experiment. Those sheets can be used as models for experiment design, data collection, data analysis, and outcome interpretation for new experiments designed by civil engineers.

5.3 EXPERIMENT DESIGN FORMAT MODEL

TITLE OF EXPERIMENT

Date: ____________________________________

Principal Investigator

Name: ____________________________________

Email: ____________________________________

Phone: ____________________________________

Collaborators

Name: ____________________________________

Email: ____________________________________

Phone: ____________________________________

Name: ____________________________________

Email: ____________________________________

Phone: ____________________________________

Name: ____________________________________

Email: ____________________________________

Phone: ____________________________________

Name: ____________________________________

Email: ____________________________________

Phone: ____________________________________

Name: ____________________________________

Email: ____________________________________

Phone: ____________________________________

Objective or Question to Be Addressed

This section defines the intended outcome of the experiment. Experiments are done for a reason, and that reason needs to be clearly defined in this section of the experiment and subsequent laboratory report.

Theory behind Experiment

The theoretical basis for any experiment informs the procedures to be used and the methods of data analysis to be used. This section of the experiment and report provides an outline of the theory behind the experiment and justifies the bases for the data collection, experiment procedures, and data analysis sections.

Potential Interferences and Interference Management Plan

This section requires the experimenter to think through the procedures to be followed and to ensure that a minimum of potential interferences are unaccounted for, or that those not addressed are not likely to skew the outcomes significantly. It may be possible to allow certain interferences if their effects can be easily quantified and the outcome data can be appropriately corrected for the interferences.

Data to Be Collected with an Explanation of How They Will Be Collected (Ex.: "Water Temperature Data for 3 Days Using a Continuous Data Logger")

This section requires the experimenter to think through the intended procedures and to develop a plan for collecting useful data during the experiment. The quality of the data to be developed in the laboratory is directly related to the quality of the samples collected in the field. Therefore, where field sampling is part of the exercise, this section must completely outline the field collection procedures to be used to protect the quality of the samples collected.

Tools, Equipment, and Supplies Required

This list should be sufficiently complete so that an experimenter taking this list of equipment into an isolation chamber on the moon could conduct the experiment without leaving the chamber or having anything else delivered to it.

Safety

This section describes the health and safety plans utilized for the collection, handling, and analysis of the samples and for the disposal of materials at the end of the experiment. Disposal procedures of the laboratory may be appended to the report to justify procedures used in this section.

Planned Experimental Procedures and Steps, in Order

In this section, the procedures to be followed need to be very clearly described and defined. The steps should be sufficiently defined and ordered such that a complete novice, who did not read the theoretical section of the experiment outline, could conduct the experiment successfully and generate the required data even if the experimenter has no idea what the collected data mean.

1.
2.
3.

Data Collection Format and Data Collection Sheets Tailored to Experiment

In this section, the experimenter will think through the most reasonable format and manner of collecting the data to facilitate analysis of the data at the conclusion of the work. A form should be prepared in advance on which all the collected data can be entered as it is developed without having to worry about where to find data points later.

Anticipated Outcomes

It is useful to know what to expect from an experiment so that the actual data can be compared to some baseline expectation. These are not always what the actual outcomes turn out to be, but the expected outcomes should be described here.

Data Analysis Plan

Here the experimenter will determine what is going to be done with the data that are collected. Questions such as what analytical processes will be used to verify the accuracy of the data, what statistical analyses are going to be performed, whether the data will be graphed in some format, what that format would look like, and similar issues are discussed.

6 Laboratory Report

6.1 CONVERSION OF THEORY TO EXPERIMENTAL OUTPUT DATA

The purpose of an experiment is to provide new information to investigators in the present and in the future. The design of the experiment is based on a theory, but the theory may not be fully correct. Most theories are based initially on a pragmatic review of unverified anecdotal observations or other forms of unproven assertions. The experiment is designed to determine the veracity of the theory based on the earlier observations and to modify the theory to the extent necessary to incorporate both the earlier observations and the experimental data.

To accomplish the goal of incorporating the new experimental data into the existing theory, it is first necessary to understand the existing theory. Indeed, the entire experiment design must be based on that understanding or there will be no relationship possible between the existing theory and the new data. Consequently, the first step is to examine existing literature to determine what the theory is and how a new experiment can be designed to verify or refute that theory. The existing theory should then be described in the experiment design so that a reader of the design can understand how and why the experiment was designed to relate to the theory.

The theoretical basis for experiments is usually found by first examining online sources to identify the basic theory. Then, the academic documents in which it may be described and in which its basis is defined can be identified. Those academic and research documents are then examined to see the finer details. Online sources, such as webpages, are good for finding rough data sources and identifying possible word searches in the academic literature. Academic literature may be found online but would not be considered a "webpage" source because they are formally published as a part of a peer-reviewed journal. Webpages (that are not online sources of published journals) do not provide defensible data, in most cases, and should not be relied upon as the source of information. Often these sources are not reviewed, or fact checked. Only known academic sources should be relied upon for accurate information and data. Google, for example, provides an academic search engine (Google Scholar), and most libraries, certainly most university libraries, have good online academic resources available to researchers.

6.2 COMPARE AND CONTRAST EXPERIMENTAL OUTPUT DATA TO EXISTING LITERATURE

Scientific and engineering theories are devised from the output of various experiments. Some of those experiments may be serendipitous outcomes from unrelated research that ultimately tweaks an idea that requires further investigation. Other

DOI: 10.1201/9781003346685-6

times the idea comes first, and some experimentation is done to verify or refute the expected outcome or basis for the idea. It happens, however, that there are very few questions that have never been asked previously in science or engineering. Sometimes the truth has not yet been observed or recognized but provides an opportunity for innovative research. Other times it is recognized but not proven and someone decides to better understand a specific phenomenon or observed effect. The key takeaway from all of that is that there is almost always *some* literature that relates to the new theory or practice to provide context to experimental work.

That is important. Reinventing wheels really is a waste of time. Modifying wheels for different purposes can be useful, but the concept of a wheel is well defined and accepted now and need not be further pursued. That means that it is not only useful but also necessary to do some research on new ideas to see who else may have come up with a similar idea in the past and what they did about it. Research conducted by others also provides context and validity to new work. Often, people start to investigate things, lose interest for a variety of reasons, and then do not pursue the end objective. The idea may have merit, however, if evaluated slightly differently. Knowing what the first person or first few people did who investigated the question will save a lot of time, eliminate failures, and lead to greater value in the final outcome.

It is also true that not all data are properly recognized or evaluated. Sometimes a literature review will uncover a failed experiment because the investigator was looking for the wrong outcome, or the outcome was not what was expected or desired, so it was ignored. The new investigator may see the true meaning of those data and be able to expand the notions involved to great advantage rather quickly.

Regardless of the reasons for the work described in the prior literature, anyone reviewing the new work will want to know about the prior work and how the new work expands on that prior work. The prior literature may also suggest some new ways of looking at a new experiment or provide updated methods for current experiments. Those who are familiar with the concepts being investigated will likely be familiar with at least the most important prior work and will discount any new work that does not at least consider the prior efforts and outcomes. Consequently, any report of new experimentation must include a section on how the new work relates to prior studies of the same and related topics.

6.3 SUGGESTED LABORATORY REPORT TEMPLATE

The following template follows the outline of the recommended experiment design process with the addition of sections for the presentation of actual data collected, interpretation of those data, conclusions, and recommendations. Be mindful that this is only a recommended format and others (your professors) may have different report format requirements. All data included in this template should be provided somewhere in the experiment laboratory report, however.

TITLE OF EXPERIMENT

Laboratory Report

Date: ________________________________

Principal Investigator

Name: ________________________________

Email: ________________________________

Phone: ________________________________

Collaborators

Name: ________________________________

Email: ________________________________

Phone: ________________________________

Name: ________________________________

Email: ________________________________

Phone: ________________________________

Name: ________________________________

Email: ________________________________

Phone: ________________________________

Name: ________________________________

Email: ________________________________

Phone: ________________________________

Name: ________________________________

Email: ________________________________

Phone: ________________________________

Objective or Question to Be Addressed

This section defines the intended outcome of the experiment. Experiments are done for a reason, and that reason needs to be clearly defined in this section of the laboratory report.

Theory behind Experiment

The theoretical basis for any experiment informs the procedures to be used and the methods of data analysis to be used. This section of the report provides an outline of the theory behind the experiment and justifies the bases for the data collection, experiment procedures, and data analysis sections.

Potential Interferences and Interference Management Plan

This section is used to consider what might go wrong with the experiment procedure and to describe what the researcher did to minimize or otherwise avoid the effects of any potential interferences identified.

Data to Be Collected with an Explanation of How They Will Be Collected (Ex.: "Water Temperature Data for 3 Days Using a Continuous Data Logger")

This section provides a detailed description of the data-collection procedures used during the experiment. The quality of the data developed in the laboratory is directly related to the quality of the samples collected in the field. Therefore, where field sampling was part of the exercise, this section must completely outline the field collection procedures used to protect the quality of the samples collected.

Tools, Equipment, and Supplies Required

The list of equipment actually used should be sufficiently complete so that an experimenter taking this list of equipment into an isolation chamber on the moon could conduct the experiment without leaving the chamber or having anything else delivered to it.

Safety

This section describes the health and safety plans utilized for the collection, handling, and analysis of the samples and for the disposal of materials at the end of the experiment. Disposal procedures of the laboratory may be appended to the report to justify procedures used in this section.

Planned Experimental Procedures and Steps, in Order

In this section, the procedures actually followed need to be very carefully and very clearly described and defined. The steps should be sufficiently defined and ordered such that a complete novice, who did not read the theoretical section of the experiment outline, could successfully conduct the experiment following the stated procedures exactly and generate identical outcome data even if the experimenter has no idea what the collected data mean.

1.
2.
3.

Data Collection Format and Data Collection Sheets Tailored to Experiment

In this section the experimenter will provide a description of the data collection form used; the timing of data collection; the sample preparation requirements used, if any; and any other relevant consideration related to collection of the samples tested.

Data Analysis Plan

Here, the experimenter will describe what was actually done with the data that were collected. Questions such as what analytical processes were used to verify the accuracy of the data, what statistical analyses were performed, whether the data were graphed and what that format looks like, and similar issues. Consider that data and results are different from one another. Data may include a multitude of quantitative results, while results come from the interpretation of the data. Experimenters should be mindful of various ways to show results without including all of their data.

Actual Outcomes Observed with Discussion of Variances from Anticipated Outcomes

If the actual outcomes are not consistent with the expected outcomes, it will be necessary to determine why that is so. Any variation from the expected outcomes must be described and explained in this section with documentation of how and why the outcomes varied. This section should also include a discussion of prior work on the same and related topics that influenced the expected outcomes from the current experiments.

Conclusions and Recommendations

Here the investigators will draw conclusions from the actual outcome data and any variations from the expected outcome data. Those conclusions will be described and documented. Any recommendations for further testing resulting from this experiment will be outlined and described here.

Appendices (If Any)

Appendices will include such things as the filled-in data collection sheets, special operating procedures for specific analytical equipment used, or descriptions of unusual or special statistical analyses used and detailed explanations, documentation, and justification for special data collection or analysis techniques utilized.

7 Effective Presentation of the Data to Others[1]

7.1 INTRODUCTION

One of the more important things that engineers are called upon to do as professionals is to present information of a technical nature to nontechnical audiences. The ability to communicate professionally can make or break a career. This is true for all professionals but particularly important for engineers due to the complexities of design.

These two concepts – technical data and nontechnical audiences – when taken together, seem both onerous and ridiculous (but they are not). The ways that engineers present themselves, the way they interact with clients, and the way they interact with the general public will dictate how those audiences perceive the accuracy, clarity, and substance of their presentations. Engineers are the key to everything people do, from roads and bridges to clean drinking water, including the equipment and the outcomes. Nothing lasting will be built or manufactured without quality engineering.

Engineers are called upon to assess problems to be solved, collect data during the analysis, analyze data accurately, present various interpretations of those data, discuss alternative interpretations of the data, and determine the recommended alternatives.

The audience to whom the engineer is presenting data and analyses may or may not be technically competent, or a mix of stakeholders may have various degrees of competence in the technical areas under discussion. The presentation needs to be tailored to meet the expectations and competence of the entire audience at hand. The technically competent audience does not generally need to hear about the theory behind the science; only an identification of the methodology used, the results of the analyses, and the recommendation. The nontechnical audience needs to understand the basic concepts behind the science used and the justification for the use of that specific analytical technique before they try to understand the results of the analyses. The mixed audience needs some understanding of the science, but not as much as a nontechnical audience. The key is to ensure understanding and to keep everyone involved with the discussion throughout the presentation.

7.2 PRESENTATION OF THE PRESENTER

Presentations occur, by definition, before an audience. How that audience perceives the presenter can dramatically affect the success of the presentation. If the presenter is wearing clothes that are too flashy, too sexy, too dirty, too crude, too out-of-style, or just uncoordinated, the audience attention will be grabbed by the appearance of the presenter and not by the message. If the presentation is delivered in a haphazard

DOI: 10.1201/9781003346685-7

or unintelligible manner, the audience will fall asleep and not hear the message. If the presenter fidgets and moves around, does not speak clearly, or seems unduly nervous, the audience will not respect the truth of the message and will not hear the message.

It is important, therefore, for the presenter to appear before the audience in such a way as to convey professionalism, confidence, and knowledge without conveying arrogance, superiority, ignorance of local customs and mores, or being the center of attention.

7.2.1 Dress Code

There is generally no "dress code" for presenters that is universally applicable to all situations. There are, however, some important guidelines that can be followed to help minimize the appearance of a presenter in what the audience perceives as not following the unwritten, local, dress code for presenters.

Rule 1: Know the Audience

Every audience is different, and the reaction of the audience to a presenter will depend in large part on the nature of the majority of those attending. The members of some professions are quite comfortable discussing technical issues and hearing presentations on topics related to that profession while dressed in jeans and T-shirts. Members of other professions would not even consider the notion of attending a formal presentation unless dressed in business formal attire. Most audiences, and mixed audiences in particular, tend to prefer business or business casual dress for sitting through or giving a presentation.

Rule 2: Dress for the Occasion

Professional dress requires consideration of the dress codes used by the audience, as well as the code assumed by the presenter. If an engineer needs to talk to the operator of a bulldozer who is busy moving earth on a construction site, wearing a suit coat and wing-tip shoes will seldom garner the respect the engineer is seeking. Similarly, when meeting the president of a client corporation, walking into the CEO's office with muddy work boots, dirty jeans, and unwashed hands would likely destroy on sight any concept of respect the CEO may otherwise have offered. While those are extreme examples, the concept is valid: dress the way the person you are meeting expects a knowledgeable professional to dress.

Rule 3: When in Doubt, Dress in Removable Layers

Years ago, an employee of the U.S. EPA attended a public hearing on a project proposed for partial federal funding as part of his job. It was a hot summer afternoon that day and the meeting was held in a small room in a municipal building on a coastal island off New England. As the EPA representative on the project, the EPA employee's presence at the hearing was more a formality and no presentation was expected to be made. He was there solely to observe the hearing and to see what issues the public may have with the project that he would have to later address. Consequently, his attire consisted of chino pants, a short-sleeved white shirt, and a conservative necktie. He remained at the rear of the room for the entire hearing and waited while the audience left the room at the end of the hearing, having said not a single word during the entire event.

As the audience was leaving, one person suddenly stopped in front of the employee and said, "You must be from Boston!" Having no reason not to do so, the employee agreed, and asked how this person knew that. The respondent replied, "You're too well dressed to be a local." At that point, the tie was quickly removed and put away for the duration.

Even though this person was not making a presentation at this hearing, his dress for the day was, first, a violation of Rule 2, but second, a perfect example of why Rule 3 is applicable.

7.2.2 The Meaning of the Dress Code Concepts

In the article "What Is Business Attire?" Susan M. Heathfield (www.thebalancecareers.com/what-is-business-attire-1918075) four categories of business attire are described: Traditional Business, Smart Casual Business, Business Casual, and Casual. The category of Business Formal has been added to that list here, but the rest of the definitions are very close to those described by Heathfield.

There are subtle differentiations between the five categories of attire as described here. The successful engineer will pay attention to what the norm is for the firm he or she works for and will adjust his or her attire appropriately to match the norms for the firm.

7.2.2.1 Business Formal Attire

There are very few circumstances under which business formal attire should be necessary for a practicing engineer. These would include such events as a "black-tie" dinner; typically, where tickets cost several hundred to several thousands of dollars each and the event is intended to raise significant numbers of dollars for a charitable cause or other reason. In that case, people in the audience will likely be wearing tuxedos or evening gowns. A presenter speaking to that audience needs to do the same. Formal attire implies tuxedos and evening gowns. In general, the darker the attire, the more formal the outfit.

7.2.2.2 Traditional Business Attire

Normal and routine business activities are generally carried out in business attire. This will include office meetings, luncheon meetings, negotiations, and similar activities that make businesses run. While the following is "traditional" for the authors, traditions change over time and the following suggestions need to be modified to adapt to the times and the culture of the organization.

This normally means a formal suit, tie/scarf, business shirt, tailored sports jacket and a business shirt/blouse, leather dress shoes, and appropriate conservative leather accessories, such as briefcases, portfolios, and diaries. Accessories like watches, jewelry, makeup, and perfume should be subtle.

People who want respect from an audience need to dress for respect in order to start the process of earning continued respect.

7.2.2.3 Smart Casual Business Attire

The concept of smart casual business attire is to step down just a notch from traditional business attire. The clothing list is similar, but with some subtle differences.

This means a sports jacket/sweater with a tie/scarf, dress pants/skirt, button-down or traditional business shirt or nice turtleneck, dress shoes, and accessories as described in traditional business attire.

7.2.2.4 Business Casual Attire

Business casual attire implies a step down from smart casual business attire and implies that ties and jackets are not expected, but more comfortable, neat, and professional-looking sports clothing is anticipated. Note, too, the lack of a need for a briefcase or folder.

This category means khakis or chino pants, dress, or Dockers-type pants; shirts/blouses or golf-type shirts with collars; sweaters; vests; occasionally an informal jacket and tie; and closed-toe shoes and accessories.

7.2.2.5 Casual Business Attire

This is the least formal category of attire for business activities and is typical of what might be worn to a company picnic, boating trip, or other form of informal corporate gathering.

This category includes casual pants and jeans, shirts with collars or not, blouses, tops, sweaters, vests, sweatshirts, and casual shoes, including sandals and athletic wear.

7.3 ACTIONS AT THE PODIUM

A well-designed presentation by a properly dressed and groomed presenter can still end in disaster if the presenter does not act appropriately on stage. The presenter needs to present the message without becoming the message. The presenter should most effectively blend into the background and not be noticed while the audience is focused on the message.

7.3.1 Speaking

A speaker has a message to deliver. That message needs to be delivered in a slow enough cadence for everyone in the room to absorb the message and understand it. A rapid delivery is simply not effective in any public speaking situation. Ignoring this dictum leaves the speaker appearing to not understand his or her own message and leaves the audience bewildered about what the message really is. In addition, speaking softly, even with a microphone, leaves the audience straining to hear the words, rather than focusing on what they mean. Speaking in the lingo or language of a profession is not helpful to a general audience unless the words are continuously explained in nontechnical terms (which dilutes the message and further confuses the audience). If the audience is essentially all members of the same profession, the use of professional language and lingo is appropriate, but not for a general audience.

Speakers who use expressions such as "um", "er", or other non-words to fill in gaps in their thinking destroy their credibility with an audience very quickly. It is painfully difficult to listen for very long to a speaker who cannot put together a simple sentence without an interruption with a non-word while the speaker thinks about the next real word to say. This habit of repeating non-words to fill speaking gaps can be overcome with practice. It is necessary for a speaker to listen carefully to the way he or she talks and to consciously stop when a gap is needed, saying nothing, if necessary, until the right word is found to proceed with. This takes time and practice, but with a conscious effort, this very distracting habit can be overcome.

7.3.2 The Use of a Written Script

Most speakers are not able to properly memorize a 30–60-minute presentation and then deliver that presentation without notes from a stage. There are those who can do that with uncanny regularity, and they can only be admired for that skill. For the rest of us, however, it is important to recognize the admiration we have for those who do not need notes and the disdain we have for those who stand before a podium and read a long dissertation, seldom even looking up to the see whether the audience is even still there.

Having a set of notes on the podium is helpful and should not be avoided simply to avoid them. It is helpful to prepare notes in 14- or 18-point font, using all capital letters, and to underline or highlight key statements or elements so that the presenter can occasionally give a quick *glance* at the notes to stay on track. When using PowerPoint or other slide show aides, there is typically a notes box attached to each slide on the presentation laptop that does not show on the big screen. Placing key comments or reminder notes there can also help a presenter stay on track.

The key is to rehearse the presentation a sufficient number of times so that the message is clear in the mind of the presenter before the presentation begins and so that the presentation can be made more like a conversation that a reading. Standing behind a podium and reading a speech is seldom going to work well for the presenter.

7.3.3 The Use of a Microphone

That message needs to be delivered loudly enough to be heard clearly by all in attendance, and not so loudly as to offend the ears of anyone listening. In almost all cases where the room is larger than four rows of listeners, the use of a microphone is necessary, regardless of how well the speaker thinks he or she can be heard without one. One need only sit through a question-and-answer period at any public presentation to note the difficulty hearing most questions from the audience, regardless of whether the questioner is in front of the listener or behind. Where microphones are provided to the audience, however, everyone hears both the question and the answer.

7.3.4 Movements on Stage

Moving around on the stage is distracting to the audience and standing still can focus the attention of the audience on the rigidity of the speaker and not the message. This dichotomy presents a challenge to the speaker. It is necessary to move to appear real, but essential not to overdo it. Standing rigidly behind a podium is tiring for the speaker, so it is necessary to move a bit.

Fidgeting is distracting, so simply walking around slowly and in a prescribed manner behind the podium is a good idea. Hands need to be up on the top of the podium, in most cases. Hands in pockets need something to do, so they start to find things to play with, like keys, coins, and other pocket contents. Those things typically create noises and movements that distract the audience. Keeping hands out of pockets eliminates these distractions.

Hand movements can be positive or negative. It is often considered cultural for some people to "talk with their hands". People who have grown up in such cultures, whether a home culture or a country culture, tend to express ideas with hand movements as much as with words. This is not necessarily a bad thing when it is an ingrained habit that helps the speaker move smoothly through a presentation – so long as it is not overdone. As stated many times already, the message is the key point the audience needs to remember, not the presenter or the presenter's activities on stage.

In addition, it is common for inexperienced speakers to move to the big screen to point out or emphasize something on a slide on screen shot to the audience. This action necessitates moving away from the podium – and therefore away from the microphone, in most cases – and leaves the audience straining to hear what the speaker is saying rather than understanding the words being said. There are pointers built into virtually every slide or presentation program that can be used for this purpose without the speaker ever even looking at the big screen. There are electronic laser pointers that work in smaller rooms (the dot on the screen is too small for larger rooms), and an assistant can be used, where necessary, to point out things on the screen while the principal presenter discusses the meaning of those things being pointed out. The speaker should never leave the podium far enough to defeat the microphone. If the speaker is wearing a mobile microphone, it is easier to move to the screen and still be heard clearly, but rude to turn a back to the audience in order to point out something on the screen.

7.3.5 Eye Contact

Much has been made over the years over the concept of eye contact with an audience. There is a reason for that; it is extremely helpful in keeping the entire audience engaged. When people speak to one or two other people in a casual setting, it is easy to maintain eye contact with all the people engaged in the conversation. In an audience of 20 or 30 people, that becomes more difficult. A habit worth developing is for the speaker to continuously move his or her eyes around the room, focusing for one or two seconds at a time on a different person in the audience. In a small audience, this is very effective at maintaining control of the conversation. In a large room, where the audience may comprised hundreds, or even several thousand people, even seeing all the people is difficult and maintaining eye contact in the manner just described is impossible. In those cases, however, the presentation is almost always televised on large screens with at least a closed-circuit television system within the room and sometimes is live streamed outside the room. That allows the speaker to focus on the cameras, not on the audience and to thereby focus on everyone at once.

It is noted that a favorite trick often promoted to help overcome nervousness on the part of a speaker is for the speaker to focus on a spot at the rear of the room, perhaps moving his or her eyes around the room every few seconds, but never looking directly at anyone in the audience. The idea is to avoid the need to see the individual audience members and to thereby eliminate the one-on-one characteristic of individual conversations with people the speaker does not know. The first or second time a person makes a public speech or presentation, the use of such a technique may be all that maintains enough confidence to get through the moment. It is a habit that needs to be overcome as quickly as possible, however, for long-term success in public speaking. Nervousness is normal in public speakers. It needs to be accepted, embraced, and managed. Recognition of the notion that the speaker knows more than the audience, at least about what the speaker is about to say, can help with this. The audience is generally not there to harass or dispute the speaker, but rather to hear, learn from, and understand what the speaker has to offer.

7.3.6 Question-and-Answer Sessions

Almost all public presentations include a question-and-answer period at the end. These are helpful to the audience understanding of things the presenter has said and to fill in background gaps the presenter has left out. How these sessions are managed can be instrumental in the overall effectiveness of the presentation.

Whenever possible, the audience should be provided with a roving microphone to use when asking questions. Most audience members do not think they need a microphone, of course, because they are not comfortable talking to "the whole world" when they unrealistically really only want to have a one-on-one question-and-answer session with the presenter. The presenter must insist on the questioner using the microphone, however, to ensure that everyone in the audience can hear the question. When a roving microphone is not available, the presenter must be careful to fully paraphrase or repeat the question for the benefit of everyone listening before trying to respond to it. Unless everyone in attendance accurately hears all the questions and all the answers, the end of the presentation becomes tattered and the whole effect of an otherwise excellent presentation can be destroyed.

7.4 WHEN THERE IS MORE THAN ONE PRESENTER ACTING AS A TEAM

7.4.1 Introduction

When there is more than one person making a presentation as a team, it is often difficult to divide the time properly between the presenters without losing content or short-changing the last speaker. The more people on the team who are presenting part of the presentation, the more difficult it becomes to keep the presentation on track.

The problems inherent in a team presentation include the time allotted for the entire presentation, the way that time is divided among the speakers, how each speaker utilizes his or her time, and how the speakers transition from one to the other and, when necessary, back again. In most cases, all the speakers will inherently believe that their portion of the presentation is the most important and that if they run long, the rest of the team, presenting what the first speaker believes is less important material, will just have to adjust. That is a classic recipe for disaster.

7.4.2 Allocation of Time

Anything that is not important should not be in the presentation in the first place and, therefore, anyone who thinks their work is paramount to that of the entire team should not be on the team or presenting anything. This does NOT mean that everyone on the team should be allotted the same amount of time to speak. The audience does not care whether everyone speaks for the same amount of time or whether one or two presenters speak for the majority of the time. In some cases, it will be clear that some of the material requires a longer time to present than other material. In those cases, a longer time to present the material needs to be allotted to that speaker or to those speakers who will be addressing that issue or material. Note that it becomes even more important for a speaker allotted extra time to be very frugal with that extra time and to not run over the allotted amount.

Where a presentation has been prepared by a large group of people (more than three, for example), it is often best to select no more than one or two members of the group to make the presentation of the group effort. The rest of the group members should be present, when possible, to answer questions that may come up regarding areas in which they were the lead investigator or writer, but not everyone on the team needs to participate in the presentation for the presentation to be effective. Indeed, it is often more effective for one or two people to make a presentation than a larger group, particularly when the time allotted for the entire presentation is short.

7.4.3 Time Utilization

How a presenter uses his or her allotted time can make a big difference in how well the presentation as a whole is received by the audience. There needs to be agreement among the presenters on how this will be done. The most common presentation method is utilization of a PowerPoint or similar program as a crutch to guide the speakers and the audience. Occasionally, flip charts are useful to present things that can only be seen well on a larger format than is reasonably possible on a presentation screen, or when interactive activities are required by the nature of the presentation. These presentation media can be mixed, so long as they do not conflict with or otherwise interrupt the smooth transitions between speakers.

In all cases, it is useful to have a guide sitting in the front row who can, and will, surreptitiously flash a sign indicating how much longer each speaker has so that the speakers can adjust the speed of their presentation to the remaining time allotted to them.

When allocating time, it is also important, in most cases, to allow sufficient time for questions and answers at the end of the presentation. One person on the team needs to be the lead on this segment and allocate the question to the person most familiar with the concept being questioned. That is done by the gatekeeper simply turning to the appropriate team member and saying something like "Tom?" Tom will then address the question succinctly and quickly so that the gatekeeper can then select another question for consideration. The time allotted to each answer cannot be determined in advance, of course, since the specific questions are not known, but each answer MUST, by agreement among the team members, be limited to not more than one minute, in most cases (timed by the timer and a sign flashed with 30 seconds and 15 seconds left). If a longer explanation is needed, the questioner should be asked to remain at the close of the presentation for a much longer and clearer discussion one-on-one with the appropriate presenter.

7.4.4 Transitions between Speakers

One of the most difficult things to accomplish smoothly in a group presentation is the transition from one speaker to the next and, when appropriate, back again. The most commonly used transition is for the first speaker to say something like "And now Tom will address . . ." Tom then steps forward and does his spiel. Tom then says something similar to turn it over to whoever is next, and so on.

This is incredibly disruptive to the flow of the narrative and an excellent way to lose the audience by the third speaker.

It is much cleaner if everyone knows the entire narrative and chimes in exactly when, and only when, they are supposed to do so. The presentation may be thought of as a play. Watching a play on stage, nobody would expect one actor to interject what is about to be said by the next actor or otherwise introduce that actor. Indeed, that would be so disruptive as to turn the audience off and most would leave before the second act. There is no reason for a group of presenters to do anything different from a group of actors. Certainly, the actors are reciting a continuous dialogue and the presenters are reciting a continuous story, but the concept is the same. When the first actor (presenter) reaches that point at which the second presenter (actor) is going to take over, the first presenter must simply stop with a good segue line for the next speaker, and the next speaker must immediately pick up the segue and move forward. With a little practice, this can result in transitions that the audience barely notices; and that should be the goal of the transitions.

7.5 PRESENTING THE DATA

How the engineer presents the data at the outset can have a significant, and expensive, impact on the decision made by the client. This section describes some of the more common issues and errors made when presenting technical data, particularly to a nontechnical audience, and how to overcome those conceptual issues.

There are a variety of ways to present data. Tables are useful for organizing and presenting large sets of numerical data, if the tables are printed, but not so useful when presented in a slide presentation to a large audience because the data points will necessarily be too small for the audience to see or interpret.

Presenting data, particularly large or very large data sets, can be much more effective when presented as graphs and charts of the trends in the results, rather than the actual data. The concept is to *present as much information as possible, in the smallest amount of space, with maximum clarity*,[2] for both the technical audience and the nontechnical audience. This is the difference between data and results. The presenter wants to aggregate all the data into meaningful and presentable results.

The use of PowerPoint and similar programs for the presentation of data is a common practice, but one that is fraught with pitfalls. Very few slideshows end up wowing the audience. Presentations that do wow an audience present the material in a way that supports the discussion but is not the main delivery tool. The engineer who writes out the entire presentation on slides and ends up basically reading the slides to the audience may indeed afflict the audience with "death by PowerPoint".

The first common mistake made by presenters is a lack of preparation. The slides are there to keep the audience focused and to help guide the presenter through the topics. Anyone who thinks they can pick up a set of slides prepared by someone else and use them to make a cogent presentation is kidding themselves. The person who does not prepare the slides will have no idea why particular information is provided or how it relates to the topic at hand. They will also have no idea what the next slide is going to be or what information is provided later in the show. This will be painfully obvious to the audience and destroy the credibility of the presenter. It takes a lot of practice to get it right, but the effort is always worth it. Every slide needs to be there to help the audience *focus* on the narrative, not to *tell* the narrative.

The second problem that often occurs is that the slides are too busy or complicated, overloaded with bullets and lists, lacking in focus, or filled with poor-quality images. Slides are a visual aid to help the presenter and the audience stay focused on the message, not an entertainment for the audience. Visual simplicity, a clear and meaningful message, and quality visuals are essential to achieving those objectives. A general 6/6 standard is recommended, meaning no more than six bullets on a slide with no more than six words for each bullet. Slides should use a minimum of words, set large enough for the hard of seeing to read from a distance (25–30 points minimum), to avoid having the audience try to read the words on the screen instead of hearing the presenter.

The third problem is using images that fail to convey the intended message. In fact, the audience should never actually notice the image but immediately perceive the message when looking at the image. Clip art images do get attention but seldom convey the proper message, particularly in a professional setting. Professional, stock images generally work best for this purpose, although personally photographed images (not snapshots) have their place in specific presentations. It is true that, in general, simple graphics and diagrams illustrate or highlight information better than text.

A fourth problem has to do with the use of color on slides. Most presentation software has built-in color palettes that can be used to color the words, the background, or the graphics. Unfortunately, the colors that show up best on the computer screen are typically those that show up the worst on the projection screen. In general, colors tend to become washed out and faded when projected from the computer screen to the presentation screen. Yellows, light blues, pinks, and grays, for example, tend to disappear when projected. Yellows and pastels tend to fade the most because they are already so weak, even on the computer screen. It is also best (when possible) to practice the presentation using the presentation screen to observe how the colors actually appear during the presentation.

Even bright colors may appear faded and often difficult to see, particularly when the lighting in the presentation room is not adequately dimmed and the background color of the slide does not provide adequate contrast. Changing the background color can allow some otherwise weak colors to pop but will also cause other colors that are usually bright to fade. In general, black and red colors show up well on the presentation screen. Outlined letters and wide letters should be avoided unless used very sparingly to highlight specific comments. A deep blue or dark green color will show up well on both screens but beware of the fact that writing on a deep blue or dark green with black lettering will look great on the computer screen but be unreadable on the presentation screen. Using white letters on the blue or green background, or other dark backgrounds, can work well.

A fifth issue is font size selection and font type. Selecting a font and size other than the defaults preprogrammed into the slideshow can accent a presentation if done carefully and judiciously. Font size is a matter of personal preference. However, if the presenter prefers to allow the audience the opportunity to read the slide, the font size needs to be large enough for that to happen. In general, a font size of at least 28 should be used for the main message on any slide, with 24- or even 20-point fonts for less important information. Anything smaller than 20 points should be avoided.

Transitions should be avoided, particularly those that are distracting, such as cutesy, fading, twirling, and snapping in and out. Sounds can be more distracting if used throughout the presentation. An occasional sound clip or short video can enhance presentations if they pertain clearly to the message. Otherwise, they should be avoided. If videos are used, they are often web-based links embedded into the slideshow. They only work where there is sufficient bandwidth and an internet connection available at the site of the presentation. Videos often do not work smoothly in a presentation. Advance verification of this availability is important before adding videos.

NOTES

1 Adapted in part from Francis J. Hopcroft, *Presenting Technical Data to a Non-Technical Audience* (Momentum Press, 2019).

2 E.R. Tufte, *The Visual Display of Quantitative Information*, 2nd ed. (Cheshire, CT: Graphics Press, 2001).

8 Designing Research Experiment Projects

Experimenting and assessing data obtained from standard methods are important parts of engineering. To further the profession, the next steps for an engineer would be to innovate and research topics of interest. Research projects in civil engineering generally involve more experimentation than interpretation and understanding of undergraduate engineering concepts. A colleague recently described the various level of academic research as:

- Bachelor's level experimentation looks at a question that has an answer, which is derived from standard experimental protocols.
- Master's level experimentation considers a question that may not have an answer that is fully resolved with a single experiment.
- Doctoral level experimentation involves the unknown, both in terms of the actual question to be asked and the answer to be obtained.

The following describes two areas of research that are used to demonstrate the idea of further exploring the concepts presented in the previous chapters. The research areas are examples of currently unresolved questions. Both explore ways to challenge the accepted protocols for testing in order to accelerate the processes or otherwise improve the test outcomes.

8.1 RESEARCH TO ASSESS SOFT ROADBED SUPPORT SOILS

The first example considers the fact that roadbeds built over soft bedding materials do not historically last. Most roadbeds are built over gravel or compacted gravel bases, within a layout that has been excavated to an appropriate depth to avoid soft soils, then compacted and built up again to the desired grade using highly designed and carefully placed gravels and stone drainage layers. Occasionally, however, it is not realistically possible to excavate entire thick layers of soft material due to the sheer depth of the material, a normal very high water table, intermittent flooding, and other factors. Such was the case not long ago in Germany where a new section of the Autobahn was opened, but within 5 years had totally failed to the point where it had to be closed and rebuilt.

Investigations revealed that there was a peat layer a few feet below where the original compaction of native soil had occurred and that this layer was periodically subject to various degrees of inundation and saturation. Over time, the swelling and recompression of that layer, and seepage of the compacted material horizontally into the softer material over time, caused stress cracking in the foundation of the roadbed and the roadbed failed. A better way to stabilize soft soils at great depths in areas

DOI: 10.1201/9781003346685-8

such as those described here is needed. The method must be relatively easy to utilize, be effective at great depths (perhaps as deep as 100 feet or more), and be relatively low cost. Relativity in this context implies relative to the cost of excavation of 100 feet or more of soft material and replacing it with stable material while also preventing side slippage of the stable soils into the unconsolidated soft material at depth, resulting in the future failure of the stable material.

The researcher on a project of this nature needs to define the nature of economically viable materials that could stabilize the softer materials without changing their natural flow characteristics or otherwise negatively affecting the local environment, how to insert that material into the soft layers, how to verify that the concept will work in practice (as opposed to working well at a lab scale, but failing at full scale), and then conduct experiments on prospective materials to verify all those requirements.

8.2 RESEARCH TO DEVELOP A STRONGER ASPHALT PAVING MATERIAL

A second example of this type of experimentation involves the development of a stronger asphalt paving material. One of the biggest banes of motorists is the existence of potholes that develop in asphalt roadways every spring, particularly in the northern portions of the country where freezing and thawing cycles occur regularly and frequently. The fundamental cause of the potholes has to do with several factors.

The first is that most asphalt roads are overlays of previously paved roads. The surface layer may be scraped away, ground up, and reused, but the second and third layers (and sometimes even more than that) are seldom removed. A binder coat and a surface coat are then installed over the scarified layer to restore the original elevation of the road surface. The base and subbase are seldom rebuilt due to cost considerations. A new roadway can cost between $1,500 and $2,000 per mile to pave, while potholes cost about $500 per pothole of average size to repair.

The second problem then arises as the new surface layer is subjected to sunlight and ultraviolet radiation every day for an average of 12 hours a day or more, resulting in high temperatures that semi-melt the asphalt binder and cause expansion of the emulsion, followed by rapid cooling in the evening, resulting in stress cracking of the surface. This can be seen in the form of "alligatoring", or large areas of the surface that are covered with thin hairline cracks over wide areas. During the fall and winter, when rain and snow melt get into those cracks and freeze, the cracks are expanded and the asphalt is moved about on the lower layers, breaking the bonds between the two layers.

Moreover, the sliding and moving stresses on the surface layer, exacerbated by traffic (particularly traffic heavier than what the road was designed for) can cause cracking and water infiltration into the binder course, which reacts the same way as the thin surface course to freezing and thawing. Occasionally, the original scarified layer can experience similar failures, even when a bonding agent is used between the original paving and the new.

Potholes develop because small pieces of the asphalt pop out during the freeze cycle, increasing the rate of deterioration in a small area. As vehicles pass over the now damaged zone, greater and greater damage is accelerated and a large mass of material is rapidly removed from tire contact and bumping from vehicles hitting the pothole.

Research in this area will require identification and examination of various forms of additive that could be added to the asphalt mix to minimize the effects of sunlight, heat, and ultraviolet radiation on the structural integrity of the asphalt, mixing samples in the lab and subjecting them to various degrees of sunlight, heat, ultraviolet radiation, freezing, thawing and moisture – a rather large array of variables that can be managed as independent or dependent and all of which may impact each of the others.

8.3 RESEARCH RECOMMENDATIONS

Research questions the known and explores the unknown. The experiments to examine an unknown need to be modified from the basic concepts generally espoused in this book to provide an open-ended question and to adjust the experiments to adapt to the outcomes developed from standard experimental protocols. It is generally recognized that experimental data are consistent only when the procedures are identical for identical aliquots of the same material or source. Ensuring good recordkeeping during the experiments is essential to later verification of the desired protocols. These steps will allow a researcher to expand upon known protocols and define methods to question standard industry procedures.

9 Material and Structural Analysis Experiments

This chapter provides completed designs for example experiments that can be used to demonstrate specific engineering phenomena in a university laboratory or field collection site. They may be used as presented or modified by the user to adapt to other desired outcomes. Sections on actual outcomes and data interpretation are necessarily left blank in these examples. They should be completed by the investigators upon collection of the data.

Note that where an experiment indicates that it is consistent with a specified ASTM method or other standard, this does not mean that the procedure is identical to, nor does it include all the details of, the stated standard method. Outcomes that require strict compliance with the stated standards must use that standard method to conduct the test. Therefore, the experiment procedures outlined in this book are not a substitute for proper compliance with the stated standards, nor are they intended to be.

The experiments in this chapter are:

9.1 Paper Beam Experiment
9.2 Effects of Live Load on Beam Stresses
9.3 Thermal Expansion of Metals
9.4 Distribution of Tensile Forces in Structural Members
9.5 Density and Specific Gravity (Relative Density) of Wood Using the Volume by Direct Measurement Procedure
9.6 Density and Specific Gravity (Relative Density) of Wood Using the Volume by Water Immersion Procedure
9.7 Density and Specific Gravity of Wood Chips
9.8 Direct Moisture Content Measurement of Wood by Primary Drying Oven

DOI: 10.1201/9781003346685-9

EXPERIMENT 9.1

PAPER BEAM EXPERIMENT

Date: ______________________________

Principal Investigator:

Name: ______________________________

Email: ______________________________

Phone: ______________________________

Collaborators:

Name: ______________________________

Email: ______________________________

Phone: ______________________________

Name: ______________________________

Email: ______________________________

Phone: ______________________________

Name: ______________________________

Email: ______________________________

Phone: ______________________________

Name: ______________________________

Email: ______________________________

Phone: ______________________________

Objective or Question to Be Addressed

Experiment design is a subset of engineering design. Consequently, there is merit to beginning design activities at an early stage of engineering education. This experiment requires the student to design a beam using only paper of predetermined dimensions and characteristics. This allows students to explore the various ways such a beam can be constructed and maximize its structural integrity defined by the maximum load it can carry under specified conditions.

Theory behind Experiment

Structural strength arises from shape, material, and unit configurations. In general, triangles are the strongest structural shape, which is why almost all trusses are based on a series of triangles. Round shapes that are subjected to equal forces from all sides do much better than other shapes subjected to similar loading. Most beams fail through bending stresses at the middle of a span rather than from shear failure at the columns.

The objective of this experiment is for the student to be creative in developing appropriate designs for a beam and to then test the designs by loading them to failure. Those designs that capture the right mix of material, shape, and structure will perform the best, while others will fail easily. Note that failure of a beam design does not imply failure of the experiment. Indeed, engineers can learn as much from a failure as they can from a success. The key is to study the failure to understand why the design failed and to then create a new design that overcomes the weaknesses of the failed design.

Design is an iterative function, and it is important to recognize the need to test various concepts before determining that an optimum design has been achieved. This experiment provides for multiple designs to be deliberately tested to failure and to then analyze the failure to determine how to overcome that failure on the next iteration.

Potential Interferences and Interference Management Plan

Few designs fail because they were intended to fail. There are circumstances in which one part of a design is deliberately constructed to fail before any other parts do, and this prevents catastrophic failures when designs are expected to be stressed to potential failure when used as intended. Shear bolts on a snowblower that allow the blades to be stopped by ice before the stress on the motor blows up the motor are a good example. The axle continues to turn, protecting the motor, while the blades stop because a weak bolt shears. The bolt is easily replaced at low cost and the machine is ready to work again within a matter of minutes.

There are no known interferences that can occur with this experiment except failure to carefully construct the designs made.

Data to Be Collected with an Explanation of How They Will Be Collected (Ex.: "Water Temperature Data for Three Days Using a Continuous Data Logger")

Designs are to be built in this experiment and subjected to a standard loading protocol. The reaction of each design to the load applied will be recorded and the load at failure will be noted. The failed design will be studied and analyzed to determine how and why it failed, those observations will be noted and recorded. A revised design will be done and the device constructed and tested as before. Each new design will incorporate a rationale for how the design differs from previous designs and why those changes are expected to overcome the previous failures.

Using the attached data sheet, record at a minimum the following information:

- Name of all experimenters
- Date of the test
- Beam descriptions
- Failure loads

Tools, Equipment, and Supplies Required

- Beams will be made from not more than two pieces of 8 ½ × 11 inch (21.6 × 27.9 cm), 20 lb bond, paper.
  - The paper may be formed and altered in any way desired, but not more than that provided by the two pieces may be used. Each iteration of the beam will use a new set of two pieces of the paper. The paper may be bonded to itself, only, using not more than 3 in (7.6 cm) of standard "Scotch" tape or 0.5 oz (14.2 g) of white glue. The beam may have any cross-sectional area or shape desired but must be 10 inch (25.4 cm) long.
- Testing device should consist of a set of two anchors – one for each end of the beam.
  - The beam should rest in a rectangular or circular base on the anchor and be clamped tightly enough to keep it from moving, but not so tightly as to cause deformation, at one end and be free at the other end. The clamping device needs to be able to be adjusted in width to accommodate any functional design of the beam. Two blocks of wood that are attachable with C-clamps work well, as would a sliding piece that can be locked down with a screw. The anchors need to be set such that 1 inch (2.54 cm) of the beam is locked in place at one end, the second end is held vertically and from moving side to side, but is not prevented from being pulled toward the center as the beam fails. The unsupported free span will be 8 inch (20.3 cm) for every test. Placing the two anchors at the edge of separate tables or benches with an opening of 8 inch (20.3 cm) of open space between them is recommended.
- Load strap 1 inch (2.54 cm) wide is required to be placed at the center of the beam and hung over the top of the beam.
  - The ends of the strap will be connected to a pouch or other container of convenient size that will hold about 22 lb (10 kg) of sand. The strap material needs to be sufficient to also hold that mass of sand, plus the weight of the strap and the container.
- Pouch or container as described in the previous paragraph to hold the sand creating the load on the beam.

- 22 lb (10 kg) of fine sand to be used to create the load.
  - Fine aggregate suitable for concrete mixing is a good material to use. In the absence of fine sand, table salt or similar crystalline material may also be used. The material is not important so long as it can be slowly poured into the pouch or container and easily removed again after each test.

Safety

When the beam fails, it may do so catastrophically. A system or plan to catch the falling strap and container, which could be up to 22.5 lb (4.6 kg), should be provided. This may be an enclosed box or padded container from which spilled sand can be easily and completely recovered prior to weighing of the mass loading on the beam at failure. Trying to catch that load by hand is NOT recommended.

Falling or bouncing sand containers and pouring of the sand into the pouch or container may allow fine particles to become airborne. Eye protection is recommended during this experiment.

As with all experiments in a laboratory, the use of gloves, eye protection, and a lab coat are recommended.

Planned Experimental Procedures and Steps, in Order

1. Put on the appropriate personal protective equipment and gather all required tools and materials in a convenient location.
2. Design a beam using only the materials specified in the material list.
3. Construct the beam as designed and estimate the maximum load-carrying capacity of the beam.
4. Erect the testing device according to the instructions in the tools and equipment section of the experiment.
5. Place the completed beam into the holding clamps and secure it at *one* end.
6. Test the beam by slowly pouring sand into the pouch or can until the beam fails.
7. Record the mass of the load at failure (the weight of the sand, the container, and the strap). Failure will be described as structural failure (bending, twisting, tearing, or collapse) of the structure such that it no longer supports the weight placed on it or becomes so deformed that it no longer maintains its original shape.
8. Study the way the beam failed and determine why the beam failed in the way that it did.
9. Redesign the beam in such a way that the new design eliminates the failure mode of the first beam. Describe how that failure mode is overcome by the new design and why the new design will not fail, at least in the same way.
10. Build and test a new design using the same loading mechanism as used for the first trial.
11. Repeat Steps 5–10 for at least four iterations, or until the maximum load on the beam exceeds 22 lb (10 kg).

Data Collection Format and Data Collection Sheets Tailored to Experiment

See attached data collection sheet.

Data Analysis Plan

The failed beams from this experiment will be examined in the lab and a determination made as to how they failed (through bending, twisting, tearing, or otherwise) and at what load. An attempt will then be made to redesign the beam to overcome the identified failure mode, being cognizant of the need to not create a new failure mode, if possible. At least four iterations of the beam will be designed.

Expected Outcomes

It is expected that all the beams will fail. This is a good outcome because of the ability to diagnose the failures and to learn from them. The lessons learned are then applied to subsequent designs and the process of learning continues with each failure.

DATA COLLECTION SHEET PAPER BEAM EXPERIMENT	
Lead Experimenter	
Date of Experiment	
LOAD AT FAILURE	
Description of Beam 1	
Beam 1 Failure Load	
Description of Beam 2	
Beam 2 Failure Load	
Description of Beam 3	
Beam 3 Failure Load	
Description of Beam 4	
Beam 4 Failure Load	
Description of Beam 5	
Beam 5 Failure Load	

EXPERIMENT 9.2

EFFECTS OF LIVE LOAD ON WOOD BEAM BENDING STRESSES

Date: ______________________________

Principal Investigator:

Name: ______________________________

Email: ______________________________

Phone: ______________________________

Collaborators:

Name: ______________________________

Email: ______________________________

Phone: ______________________________

Name: ______________________________

Email: ______________________________

Phone: ______________________________

Name: ______________________________

Email: ______________________________

Phone: ______________________________

Name: ______________________________

Email: ______________________________

Phone: ______________________________

Objective or Question to Be Addressed

As a live load moves along a span, the stresses in the structure can change dramatically, depending on the relative magnitude of the load versus the strain resistance of the beam. This experiment is designed to demonstrate the way those stresses change as a function of the load movement across the span.

Theory behind Experiment

Stresses manifest in two general forms: shear and moment. The shear stresses tend to cause the beam to fail vertically through the beam cross-section, while the moment stresses are the bending forces that cause the top of the beam to compress and the bottom of the beam to want to pull apart. Typically, a beam will undergo some degree of elastic bending, or deflection, before the fibers or molecules at the bottom begin to pull apart or the ones at the top begin to crush. The amount of elastic bending that occurs, which is reversible as the stress is reduced, versus how much permanent bending will occur, which is not reversible, depends, again, on the load stress and the stress resistance of the span material.

A stress diagram of a beam cross-section will indicate how much stress each molecule at the center of the stress diagram is undergoing during any moment in the loading scenario. By moving a mass across a span, it is possible to measure the change in the loading and to construct therefrom, a reasonable stress diagram for the beam being used. Clearly, a steel beam of a given cross-section will sustain much higher stress without deflection than a wooden beam of identical cross-section. In all cases, the compression stress at the top of the beam will exactly balance the tension stresses at the bottom of the beam at all times. The maximum compression stress and the maximum tension stress occur at the very top and bottom edges with a point of neutral stress somewhere in between – but seldom at the center.

Wood, for example, is generally much stronger in compression than in tension. Consequently, loading a wood beam to failure will always result in the beam pulling apart at the bottom before it crushes at the top. A paper beam, on the other hand, will generally collapse at the top from compression before pulling apart at the bottom from tension because the paper is stronger in tension than in compression. The cross-sectional shape and the uniformity of the internal structure, physical and chemical, of both types of beams will also affect the way the beam reacts to stresses.

This experiment uses a wooden beam and a reasonably heavy load to create stresses internally to the beam. The effects of those stresses, both in loading at the ends and deflection at the midpoint, where deflection will be the greatest, are measured and graphed to create a stress diagram for the beam. Experiment 9.1 demonstrates a more simplified version of this experiment using paper beams of student design.

Note that there are several different characteristics of wood strength that could be measured. They include:

- Compressive stress perpendicular to the grain – This is the maximum stress sustained by a compression perpendicular to the grain. Compressive stress perpendicular to the grain is what is being applied in this experiment.

- Compressive strength parallel to the grain (crushing strength) – This is the maximum stress sustained in compression parallel-to-grain by a test sample having a ratio of length to the lesser of width or depth of no more than 11. It is one of the two properties plotted on the stress diagram.
- Tensile strength parallel to grain – This is the maximum tensile stress sustained in a direction parallel to the grain. Wood is seldom exposed to this type of stress, and there is limited test data available on average values. This is, however, the second property plotted on the stress diagram.
- Stiffness (modulus of elasticity) – This property indicates how much the wood will deflect when a load is applied to it perpendicular to the grain. That property is measured with this experiment.
- Modulus of rupture in bending (bending strength) – This indicates the maximum load-carrying capacity of a wood framing member in bending and is proportional to the maximum moment borne by a test specimen.
- Shear strength parallel to grain – The ability to resist internal slipping of one part upon another along the grain.
- Tensile strength perpendicular to grain – This is the resistance of wood to forces acting across the grain that tend to split a member.

Potential Interferences and Interference Management Plan

There are various things that can cause interference in the outcome to this experiment. The first is that the changing end loads are being measured on the flat surface of a scale. As the beam is loaded, some deflection is expected in the center. That can cause the load on the scale to become slightly angular and no longer as accurate as intended. Placing the ends of the beam on a narrow, raised pivot point on top of the scale can reduce this effect to negligible impact for this experiment.

The exact locations at which the load is sequentially placed along the top of the beam will affect the accuracy of the stress diagram developed from the experiment. Careful measurement and placement of the load will overcome this issue.

Not all wood is absolutely uniform in cross-sectional density or fiber strength. A board with a knot in it will usually fail at the knot in this experiment before it fails anywhere else and usually at a much lower stress than a smooth, straight piece of wood of the same kind. Selecting the beam carefully, such that it is of uniform thickness and uniform dimensions throughout its length, without knots or surface damage, will ensure a smooth and accurate stress diagram from this experiment. Variations in the diagram that are developed may be analyzed for the probability for weak spots in the beam material and adjusted accordingly.

Data to Be Collected with an Explanation of How They Will Be Collected (Ex.: "Water Temperature Data for Three Days Using a Continuous Data Logger")

The data to be collected in this experiment will be mass measurement from two calibrated scales and deflection at one location using a fixed ruler, marked in suitable increments, usually tenths of an inch or millimeters.

Using the attached data sheet, record at a minimum the following information:

- Name of all experimenters
- Date of the test
- Mass measurements
- Deflection measurements

Tools, Equipment, and Supplies Required

- 2 calibrated scales with capacity exceeding the weight of the beam, plus the total mass expected to be placed on the beam (at least 110 lb [50 kg]) and an accuracy of 0.1 lb (0.05 kg)
- 1 ruler or measuring device fixed in position next to, but not attached to, the edge of the beam at the center
- Several pieces of dimensioned lumber of any convenient size, but typically including the following:
  - A nominal 1 × 4, a nominal 2 × 4, and a nominal 4 × 4
  - Note that dimensional lumber is smaller than the stated nominal sizes
  - A 1 × 4 is typically 0.75 × 3.75 inch (19 × 95.25 mm) in cross-section
  - A 2 × 4 is typically 1.75 × 3.75 inch (44.45 × 95.25 mm)
  - A 4 × 4 is typically 3.75 × 3.75 inch (95.25 × 95.25 mm)
- The length should be as long as possible, but not so long as to cause significant deflection at the center prior to loading.
  - A 1 × 4 should typically be less than 5 ft (1.5 m), while the other pieces may generally be up to 7–9 ft (2.1–2.7 m) long.
  - Note that these odd numbered lengths will allow the support of the beam to be a linear point 6 inch (15.2 cm) in from each end, while still leaving a long enough length to generate meaningful data.
  - The actual length of the beam is not important, but knowing the actual length between the supports is very important.
- Two 1 × 1 inch nominal dimensions, with actual dimensions of 0.75 × 0.75 inch (1.9 × 1.9 cm) pieces of wood, ideally oak or other hardwood, or a similar piece of plastic or metal
- A series of weights of known mass, such as 10 lb (4.5 kg), 20 lb (9 kg), 30 lb (13.6 kg), 50 lb (22.7 kg), and 100 lb (45.4 kg)

Safety

Some of the weights being used in this experiment are heavy. Lifting and moving masses of those magnitudes can cause strains in muscles, and dropping one on a toe or finger can result in broken bones or lacerations. Care is needed when handling those loads. Consideration should be given to using multiple smaller weights for the experiment rather than single larger ones.

In the event of a toe or finger being injured by the weights, seek medical attention quickly to ensure that no breaks have occurred since rapid setting of breaks is essential to long-term recovery.

Wood may produce splinters, particularly as it approaches failure. Splinters can cause infection if allowed to penetrate the skin. Care is needed when handling lumber to avoid puncture wounds from splinters. In the event of a splinter entering the skin, seek medical attention to avoid infection or to remove deep splinters.

Heavy work gloves, eye protection, and a lab coat are recommended for this laboratory experiment.

Planned Experimental Procedures and Steps, in Order

1. Put on personal protective equipment and gather all required equipment and materials in a convenient location.
2. Place two calibrated scales, designated Scale A and Scale B, at a suitable distance apart on a flat and level platform such that the centers of the scale platforms are approximately 1 ft (0.3 m) closer than the length of the piece of lumber to be tested.
3. Place a 1 × 1 inch nominal dimensions (actual dimensions of 0.75 × 0.75 inch [1.9 × 1.9 cm]) wood (ideally oak or other hardwood, or a similar piece of plastic or metal) perpendicular to the length of the beam to be tested on each scale. The beam will rest on these two pieces to balance the load on the scales.
4. Place a piece of 1 × 2 inch nominal lumber across the two support pieces on the scales such that the beam is centered on the two supports. Place the lumber flat with the 1 inch dimension vertical and the 2 inch dimension horizontal.
5. Carefully measure and mark the locations on the top of the beam that represent the length of the beam, *between the supports*, at each tenth of the length. It is useful to locate the beam on the supports such that the distance to each mark is an even number of inches or centimeters in order to facilitate the development of the stress diagram later.
6. Place a ruler or other measuring device perpendicular to, but not touching, the edge of the beam and carefully note, to 0.01 inch (0.025 cm), the height of the beam from the floor or base of the measuring device.
7. Tare the two scales to zero.
8. Place a 10 lb (4.5 kg) weight on top of the beam directly over the center of the support on one scale and record the mass showing on each scale, plus the deflection, if any, at the center of the beam (subtract the current reading from the original reading). The scale with the weight should show 10 kg while the other scale should show 0 kg, if the weight and the beam are properly located as described.
9. Repeat Step 8 five times using 20, 30 50, and 100 lb (9, 13.6, 22.7, and 45.4 kg) masses.
10. Repeat Steps 8 and 9, moving the series of weights 1/10 the distance to either scale each time.
11. Repeat Steps 8–10 for the same piece of lumber placed with the 2 inch dimension vertical and the 1 inch dimension horizontal.

12. Repeat Steps 8–10 for the 2 × 4, then the 2 × 4 on edge to yield a 4 × 2, and the 4 × 4 pieces of dimensioned lumber, moving the two scales farther apart for these pieces of lumber.
13. Plot the data for each piece of dimensioned lumber on a graph with the load on each scale shown as a separate line with the load on the y-axis and the mass on the x-axis. The two lines should cross at the center of the graph, indicating the maximum bending stress on the beam.
14. Calculate the load at the center of the beam, the maximum deflection, and the stress diagram for each piece of lumber in accordance with the data analysis section of the experiment.

Data Collection Format and Data Collection Sheets Tailored to Experiment

See attached data collection sheet. *Note that 25 copies of the Data Collection Sheet will be needed for this experiment if all samples and orientations suggested are carried out.*

Data Analysis Plan

Calculate the expected moment at the center of the beam with each movement of the load. That moment should be equal to the mass of the moving load times the proportion of the total distance that the load is from either support (except when the load is directly over one support; then the load at the other support should be zero and so should the load at the center of the beam. Otherwise, each support should take a portion of the total load based on the proportion of the total length of the beam from that support to the load.). For example, a 50 lb load 10 % of the way along a 10 ft beam would have a bending moment of (1/10 × 10 ft) (50 lb) or 50 ft-lb at the center of the beam. That same load 2 ft along the beam would have a bending moment of (2/10 × 10 ft) (50 lb) or 100 ft-lb at the center of the beam. Note that maximum bending moment occurs when the load is directly over the center of the beam. Once the center of the beam is reached, the distance needs to be measured from the opposite scale.

The load at each scale will also vary in direct proportion to the distance from each scale. For example, if the load is directly over Scale A, the full load should be recorded there and there should be no load on the other scale. When the load is in the center of the beam, half of the load should be on each scale.

The shear stress will be highest when the load is closest to the supports. This stress will move with the load, and it will decrease as the load moves to the center, then switch to the opposite support and increase as the load approaches that support. The bending moment will quickly overwhelm the shear stress near the ends of the beam. The bending moment will be highest at the center of the beam, and it will occur when the load is also directly over the center of the beam. The shear stress will be the highest at the ends immediately adjacent to the supports, where the bending moment will approach zero.

Plot the measured deflection at the center of the beam as a function of the changing bending moment at the center of the beam. The deflection should maximize when the load is at the center of the beam. Note that the bending moment is defined in units of ft-lb or kg-m. The load times the distance from the nearest support to that load will equal the bending moment at that location.

Using Table A.10 in the Appendix, determine the compressive and tensile strength of the wood species being used (most dimensional lumber will be fir or pine).

Draw a stress diagram for each piece of wood by balancing the cross-sectional area of the lumber times the compressive strength and the cross-sectional area of the lumber times its tensile strength such that the two products are equal. Where that sloped line crosses the vertical is the point of neutral stress in the piece of lumber.

Note that the stress on the cross-sectional area in compression and the stress on the cross-sectional area in tension is assumed as a point stress one third of the way from the surface of the lumber to the neutral point, in both directions. The ratio of the tensile strength to the compressive strength is calculated from the values in Table A.10. Then that ratio is applied to the total mass placed on the center of the beam, and the higher proportion of the load is assumed to go into compressive stress and the lower proportion into tensile stress. The location of the neutral point should occur at the same proportion of the total depth of the beam when measured by the compression proportion down from the top or the tensile proportion measured up from the bottom of the beam. It will generally be above the center of the beam cross-section – usually by a relative lot.

Expected Outcomes

The outcomes from this experiment should be reasonably consistent with the data from Table A.10. If that is not the case, revisit the measurements for inconsistencies and then the calculations for the same purpose. If no obvious errors are found, the probability exists that the lumber was not uniform in cross-sectional uniformity or that an error occurred during conduct of the experiment.

Note that the deflection in the 1 × 4 should be noticeable, while the deflection in the other pieces will be less so. The deflection in the 4 × 4 may not be measurable even with 100 lb and an 8 ft beam. The stress diagram can still be constructed from the load and the cross-sectional area.

DATA COLLECTION SHEET EFFECTS OF LIVE LOAD ON BEAM STRESSES			
Lead Experimenter			
Date of Experiment			
Mass of Load			
Mass on Scale A with Load at Scale A (Step 8)		Mass on Scale A with Load at 6/10 of distance to Scale B (Step 8)	
Deflection at Center of Beam with Weight at Scale A		Deflection at Center of Beam with Load at 6/10 of Distance to Scale B	
Mass on Scale A with Load at 1/10 of distance to Scale B (Step 8)		Mass on Scale A with Load at 7/10 of distance to Scale B (Step 8)	
Deflection at Center of Beam with Load at 1/10 of Distance to Scale B		Deflection at Center of Beam with Load at 7/10 of Distance to Scale B	
Mass on Scale A with Load at 2/10 of distance to Scale B (Step 8)		Mass on Scale A with Load at 8/10 of distance to Scale B (Step 8)	
Deflection at Center of Beam with Load at 2/10 of Distance to Scale B		Deflection at Center of Beam with Load at 8/10 of Distance to Scale B	
Mass on Scale A with Load at 3/10 of distance to Scale B (Step 8)		Mass on Scale A with Load at 9/10 of distance to Scale B (Step 8)	
Deflection at Center of Beam with Load at 3/10 of Distance to Scale B		Deflection at Center of Beam with Load at 9/10 of Distance to Scale B	
Mass on Scale A with Load at 4/10 of Distance to Scale B (Step 8)		Mass on Scale A with Load at Scale B (Step 8)	
Deflection at Center of Beam with Load at 4/10 of Distance to Scale B		Deflection at Center of Beam with Weight at Scale B	
Mass on Scale A with Load at Center of Distance to Scale B (Step 8)			
Deflection at Center of Beam with Load at Center of Distance to Scale B			

EXPERIMENT 9.3

THERMAL EXPANSION OF METALS

Date: ______________________________

Principal Investigator:

Name: ______________________________

Email: ______________________________

Phone: ______________________________

Collaborators:

Name: ______________________________

Email: ______________________________

Phone: ______________________________

Name: ______________________________

Email: ______________________________

Phone: ______________________________

Name: ______________________________

Email: ______________________________

Phone: ______________________________

Name: ______________________________

Email: ______________________________

Phone: ______________________________

Objective or Question to Be Addressed

This experiment provides a method for assessing various metals and their expansion or contraction as a function of temperature.

Theory behind Experiment

Almost all materials and all metals are subject to expansion and contraction as a result of temperature changes. In most materials this change is subtle and barely noticeable at normal temperatures around the home or yard, even in the heat of summer and the depths of winter. While this size or dimensional change is not critical for most circumstances, it can become critical in some everyday applications.

For example, steam lines used for central heating in large cities and industrial complexes are generally metallic due to the extreme pressures they carry. Steel has an expansion coefficient of around 9.2×10^{-6} inch/inch/°F (1×10^{-6} mm/mm/1.047 °C). That is a very small number, but it is multiplied by a very long pipe. Assuming an average pipe section of 500 linear feet, or 6,000 linear inches (152.4 m) the change in length from "room" temperature of 72 °F (22 °C) to a superheated working temperature of 220 °F (104 °C), the change in length will be:

$$\Delta L = (9.2 \times 10^{-6}) \times (6{,}000) \times (220\text{–}72)$$
$$\Delta L = 8.17 \text{ inch } (20.75 \text{ cm})$$

If that amount of movement, even over 500 ft in length, is constrained, the pressure would be enormous. Consequently, steam pipes are designed to fit into conduits with expansion chambers placed at convenient locations with a small loop built into the pipe. The loop allows the pipe to flex without breaking.

Since every material has a unique coefficient of expansion, it is necessary to measure the expected expansion for materials being exposed to differing site or use conditions. This experiment provides a means for doing that.

Note that there are commercially available devices, known as *dilatometers*, that can be used to do these measurements. They are difficult to use, difficult to maintain, and very expensive to purchase. The device described in this experiment is much simpler, but less accurate, than a commercial device. The results, however, should be demonstrative of the issue and be able to differentiate between various metals.

Potential Interferences and Interference Management Plan

The device described in this experiment requires some construction and specialized components. All components are readily available commercially and the construction needs to be done carefully, but absolute precision is not required. When measurements are taken, however, it will be necessary to measure very carefully and precisely. The equations shown will eliminate any lack of precision in the construction, but they must be done very precisely to accurately reflect the very small values being recorded. The use of micrometers is required.

Note that the terminal blocks specified are designed for high temperature electrical connections. There are connectors built into the blocks that are not capable of sustaining the temperatures that the basic blocks can sustain. It is likely that some of those internal components will melt or run at the experiment temperatures and potentially interfere with the very fine measurements required. It is recommended that all openings in the blocks be located on a face other than the face in contact with the sample. In addition, preheating the blocks to the experimental temperatures and letting them cool before attaching them to the plate may be useful.

Data to Be Collected with an Explanation of How They Will Be Collected (Ex.: "Water Temperature Data for Three Days Using a Continuous Data Logger")

Using the attached data sheet, record at a minimum the following information:

- Name of all experimenters
- Dates of experiment
- Material used as a sample
- Temperatures of room and oven
- Distances with inside and outside micrometers
- Thermal expansion coefficient calculated

Tools, Equipment, and Supplies Required

- Thermal expansion measuring device as follows (see Figures 9.3.1 and 9.3.2. for details):
    - A 240 mm (9.5 inch) diameter high-temperature porcelain desiccator plate (Cole-Parmer 240 mm desiccator plate, Mfg. #66456, or equivalent) that can withstand temperatures of 932 °F (500 °C) is fitted with a ceramic terminal end block (Allied Electronics Terminal Block, Part # 7033855, Allied stock # 72190158, or equivalent) by cementing the block to the plate with a special high-temperature cement (Omegabond High Temperature Chemical Set Cement, type OB-600, or equivalent). A stainless-steel strip is attached to the plate with a screw at a distance of 8 inches (20.32 cm) from the face of the block. This strip extends parallel to the end of the block and a distance of 3.75 inches (9.53 cm) to the side of the centerline between the block and the strip. The screw that holds this strip in place is located 0.5 inches (1.27 cm) from the centerline of the block. A small sheet metal screw is placed through the strip, with the point of the screw facing the block, at the centerline. A second screw is placed through the strip, in the same direction, 3.5 inches (8.89 cm) from the first one. (The tips of these screws are used as measuring points later.) A second block identical to the first one is cemented to the plate parallel to the stainless-steel strip about ½ inch (1.27 cm) back toward the first block.

- Oven capable of providing sustained heat at 950 °F (500 °C) and large enough to hold the 240 mm (9.5 inch) diameter desiccator plate
- Micrometers capable of measuring inside dimensions from 0.5 inch (1.27 cm) to 3 inch (7.62 cm) to an accuracy of 0.01 inch (0.254 mm)
- Micrometers capable of measuring outside dimensions from 0.0 to 6.5 inch (16.51 cm) to an accuracy of 0.01 inch (0.254 mm)
- High-temperature heat-resistant gloves (Steel Grip model GL 1332–14, or equivalent)
  NOTE: These gloves need to be able to withstand temperatures of 1,000 °F (550 °C) for sustained periods. Normal heat resistant gloves will not be sufficient for this application.
- A heat-resistant trivet or area to place the hot thermal expansion measuring device to cool. This trivet or area must be able to withstand temperatures of up to 1,000 °F (550 °C) for an extended period.

Safety

This experiment requires the use of very high temperatures and requires the movement of a very hot, and heavy, desiccator plate, with the measuring blocks and sample, from inside the hot oven to a cooling location and then immediately locking down a small screw. The use of high-temperature heat-resistant gloves is essential for this purpose. Regular heat-resistant gloves used in a laboratory will not be sufficient for this purpose.

At the temperatures required for this experiment, casual touching of the walls or sides of the oven, or any part of the thermal expansion measuring device, during the process will result in instantaneous loss of skin tissue, the likely dropping of the device onto surfaces unable to withstand the heat, and causing further damage to the laboratory. Extreme care is required during the conduct of this experiment to avoid accidents.

In the event of a burn, seek immediate medical assistance to minimize skin and tissue damage.

Planned Experimental Procedures and Steps, in Order

1. Put on the appropriate personal protective equipment and gather all necessary tools, equipment, and supplies in a convenient location.
2. Construct the thermal expansion measuring device in accordance with the previous description and the diagrams in the appendix following the manufacturer's instructions for preparation and application of the cement. Be sure the cement extends into the holes in the desiccator plate and the holes in the terminal blocks.
3. While the cement is setting for the designated time, prepare a sample or samples of metal to be tested. Samples should be about 1–2 inch (2.5–5 cm) in height and width and as close to exactly 6 inches (7.62 cm) as reasonably possible. The ends of the sample must be perpendicular to the sides and parallel to each other.
4. Allow the sample to cool or heat to room temperature and record that temperature to the nearest 0.1 degree, T_o.
5. Carefully measure and record the exact distance, M_1 from the tip of the small screw facing the sample to the center of the locking screw/axle, to the nearest 0.01 inch (0.25 mm).

6. Carefully measure and record the exact distance, M_2, from the center of the locking screw/axle to the tip of the second small screw facing the second block, to the nearest 0.01 inch (0.25 mm).
7. Using the outside micrometers, carefully measure and record the exact length, L_s, of the sample between the centers of the two end faces to the nearest 0.01 inch (0.25 mm).
8. After the cement is properly cured, place the sample onto the thermal expansion device with one end firmly touching the face of the first block.
9. Move the stainless-steel strip so that the end of the small screw is just touching the face of the sample and tighten the screw only so much as necessary to prevent movement of the strip during the subsequent measurement.
10. Using the inside micrometers, carefully measure and record the distance from the end of the second small screw to the face of the second block, D_o, to the nearest 0.01 inch (0.25 mm).
11. Loosen the screw on the strip only as much as needed to allow the strip to move freely on the axis.
12. Place the entire device into the oven with the strip end facing out. Be sure the sample is tight against the face of the first block and the strip is tight against the face of the sample.
13. Close the oven slowly and allow it to heat to a temperature of approximately 900 °F (500 °C) and maintain that temperature for at least two hours to allow the sample to reach the temperature of the oven.
14. Record the final temperature of the oven, T_1.
15. *Slowly* open the door of the oven.
16. Before removing the device from the oven, use the heat-resistant gloves to gently tighten the screw that forms the axel of the strip only tight enough to prevent the strip from moving during subsequent handling.
17. Using the heat-resistant gloves, carefully remove the device from the oven and set it on the cooling pad.
18. Allow the sample and device to cool to room temperature (at least four hours) without disturbance. Note and record this room temperature. If it is not the same as the original room temperature, note the average between those two temperatures and use that average value as T_o.
19. After the device and sample have reached room temperature, use the inside micrometers to carefully measure and record the distance, D_1, from the tip of the small screw to the face of the second block, to the nearest 0.01 inch (0.25 mm).
20. Using the inside micrometers, carefully measure and record the distance, D_s, from the tip of the small screw to the face of the sample, to the nearest 0.01 inch (0.25 mm).
21. Using the outside micrometers, measure the length of the sample to the nearest 0.01 inch (0.25 mm). It should be the same as L_s. If these two values are not identical, calculate the average of these two values and use that value as L_s.
22. Calculate the thermal expansion coefficient of the sample using the procedure and equations from the data analysis section.

Data Collection Format and Data Collection Sheets Tailored to Experiment

See attached data collection sheet.

Data Analysis Plan

The thermal expansion of the sample will be visually apparent from the very small gap between the end of the sample and the tip of the small screw in the strip that was originally touching the face of the sample. This would be difficult to measure precisely. However, the gap will have closed between the tip of the second small screw and the face of the second block by a multiple of the gap between the end of the sample and the small screw that had been touching it.

Calculate the coefficient of expansion as follows.

First calculate the temperature change, in degrees F or C, that the sample was subjected to.

$$\Delta T = T_1 - T_o$$

Then calculate the multiplier provided by the stainless-steel strip (M_d) as follows:

$$M_d = M_1/M_2$$

Calculate the coefficient of thermal expansion, as inch/inch/°F (cm/cm/°C) as:

$$C = [(D_1)/(L_o)/(\Delta T)] \times M_d$$

Compare the calculated value of C with the measurement of D_s, if possible. Theoretically, they should be identical. That is unlikely to happen, precisely, because of the difficulty in measuring the value of D_s. The calculated value of C will be closer to the actual value for the tested material.

Expected Outcomes

The expected outcome from this experiment will depend on the material tested. Table A.7 in the Appendix provides published coefficients of expansion for various metals. The value determined by this experiment should be consistent with those shown in the table. Exact correlation to the table will depend on the exact type of sample. There are several forms of aluminum alloy and stainless-steel alloys available, for example, and each may have a slightly different coefficient of expansion.

DATA COLLECTION SHEET THERMAL EXPANSION OF METALS	
Lead Experimenter	
Date of Experiment	
Material Tested	
Temperature of Material at Start of Test (Step 4), T_o	
Distance from Tip of the Screw to Center of Locking Screw/Axle (Step 5), M_1	
Distance from Center of Locking Screw/ Axle to Tip of Screw Facing Second Block (Step 6), M_2	
Exact Length of Sample (Step 7), L_s	
Distance from Second Screw to Face Second Block, D_o	
Time of Insertion into Oven (Step 13)	
Time of Removal from Oven (Step 13)	
Temperature inside Oven (Step 14), T_1	
Temperature of Room after Cooling (Step 18)	
Time of Cooling Start	
Time of Cooling End	
Distance from Tip of Small Screw to Face of Second Block after Cooling (Step 19), D_1	
Distance from Tip of Small Screw to Face of Sample after Cooling (Step 20), D_s	
Length of the Sample (Compare to L_s) (Step 21)	
Calculated Temperature Change during Experiment, ΔT	
Calculated Multiplier Value, M_d	
Calculated Coefficient of Expansion, C	

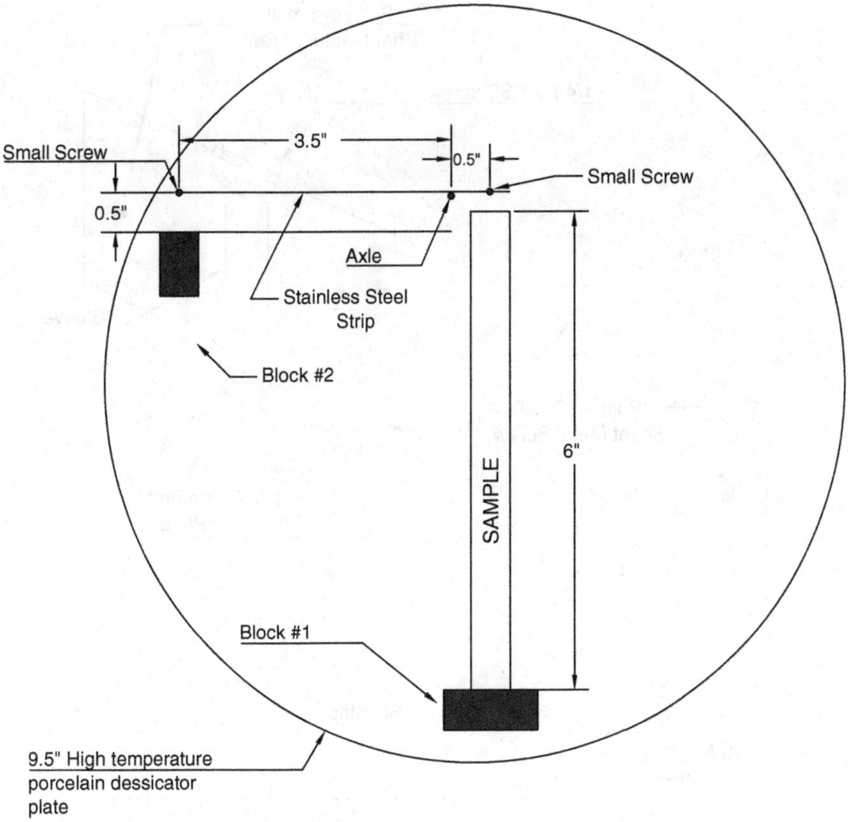

Small Screw
3.5"
0.5"
Small Screw
0.5"
Axle
Stainless Steel Strip
Block #2
SAMPLE
6"
Block #1
9.5" High temperature porcelain dessicator plate

$\frac{1}{4}$" SS sleeve cemented to the SS strip
Small steel screw each end
3.500"
0.500"
1.5" tall x $\frac{1}{8}$" thick SS strip
set small screws $\frac{3}{4}$" up from plate surface

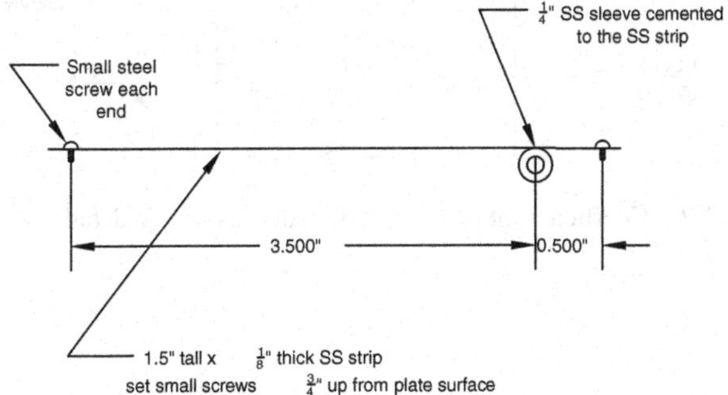

FIGURE 9.3.1 Coefficient of expansion of materials device.

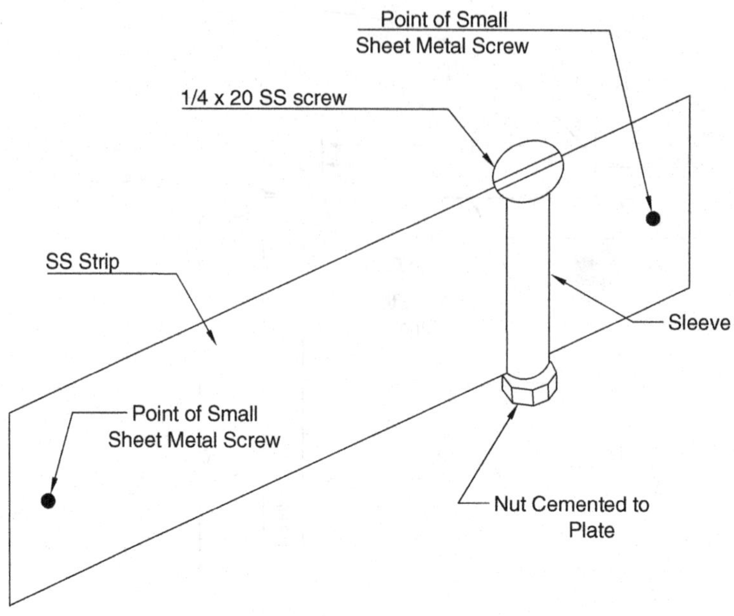

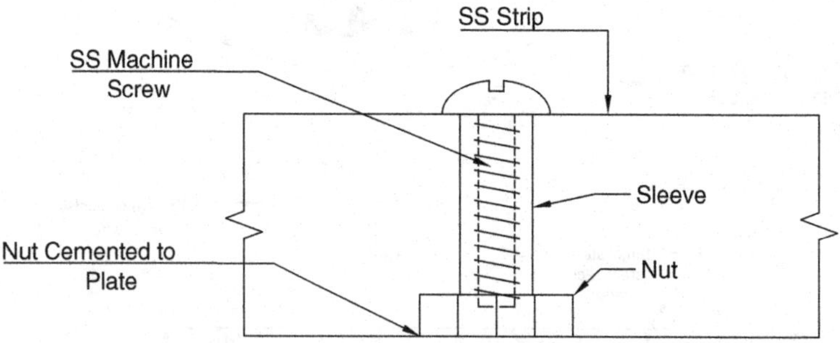

FIGURE 9.3.2 Coefficient of expansion of materials device details.

EXPERIMENT 9.4

DISTRIBUTION OF TENSION FORCES IN STRUCTURAL MEMBERS

Date: ____________________________

Principal Investigator:

Name: ____________________________

Email: ____________________________

Phone: ____________________________

Collaborators:

Name: ____________________________

Email: ____________________________

Phone: ____________________________

Name: ____________________________

Email: ____________________________

Phone: ____________________________

Name: ____________________________

Email: ____________________________

Phone: ____________________________

Name: ____________________________

Email: ____________________________

Phone: ____________________________

Name: ____________________________

Email: ____________________________

Phone: ____________________________

Objective or Question to Be Addressed

Whenever a beam or cable is pulled in the direction parallel to its length, a tension force is created inside the beam or cable. This experiment is designed to demonstrate how those forces change as the load on a structure changes.

Theory behind Experiment

Tension is defined as the drawing force acting on the body when it is hung from objects like chain, cable, string, etc. It is represented by **T** (occasionally also symbolized as $\mathbf{F}_t$). Tension is calculated from the formula:

$$T = mg +/- ma$$

Where: T = tension (N or kg-m/s^2)
g = acceleration due to gravity (9.8 m/s^2)
m = mass of the body
a = acceleration of the moving body

- If the object causing the tension is traveling upward, the tension will be $T = mg + ma$.
- If the object causing the tension is traveling downward, the tension will be $T = mg - ma$.
- If the tension is only equivalent to the weight of object causing the tension, $T = mg$.

The *tension formula* is used to find the tension force acting on or within a structural member and is usually expressed in Newtons (N).

This experiment uses a truss made of rope, not stiff members, because rope cannot transmit a compression force. The intent is to measure the tension forces in the truss members as a load is shifted around the truss.

Potential Interferences and Interference Management Plan

The devices used to measure the tensile loads are generally spring-based devices that may become stretched at some point and be less accurate. In addition, the mass of the truss itself will cause some tension within some of the members before a load is applied. Those forces need to be subtracted out of the force measured in each member resulting from the load.

Data to Be Collected with an Explanation of How They Will Be Collected (Ex.: "Water Temperature Data for Three Days Using a Continuous Data Logger")

Using the attached data sheet, record at a minimum the following information:

- Name of all experimenters
- Date of the test
- Tension forces that will be measured using fixed-in-place tension scales

Tools, Equipment, and Supplies Required

- About 50 ft of 1/8 inch (3.2 mm) diameter rope or cable (minimal stretching is most desirable)
- 11 Hanging or pull-spring scales or luggage scales with a capacity of 10 kg (22 lb) or more (AvaWeigh 22 lb Industrial Hanging Scales or Globite Luggage Weighing Scales, or equivalents, are suggested)
- Three 10 lb (4.5 kg) weights that can be hooked onto the truss at preset locations

Safety

There are no inherently dangerous components to this experiment. However, a 10 lb (4.5 kg) weight will cause injury if it is dropped onto a foot or other body part. Care is required to minimize this risk.

As with all laboratory experiments, gloves and a lab coat are recommended while conducting this experiment.

Planned Experimental Procedures and Steps, in Order

1. Put on appropriate personal protective equipment and gather all necessary tools and equipment in a convenient place.
2. Construct a truss from the rope or cable and suspend it in a frame in accordance with Figures 9.4.1 and 9.4.2.
3. The scales that are incorporated into the legs of the truss and the materials from which the truss is made add tension to the truss members independent of the loads that are going to be applied. Record the tension forces in each of the numbered scales or zero the scales if that is possible.
4. Using standard vector analysis mathematics, calculate the expected tension in each of the members with a 10 lb (4.5 kg) weight suspended at Point 1 on the truss.
5. Hang a 10 lb (4.5 kg) weight at Point 1 on the truss.
6. Observe and record the new tension forces shown on each of the member scales.
7. Determine the cause of any difference between the calculated tensions and the actual tensions.
8. Move the 10 lb (4.5 kg) weight to Point 2 on the truss and repeat Steps 4, 6, and 7.
9. Leave the 10 lb (4.5 kg) weight at Point 2 on the truss and add a second 10 lb (4.5 kg) weight to Point 1 on the truss.
10. Repeat Steps 4, 6, and 7.
11. Remove the two weights and place a 10 lb (4.5 kg) weight at the center of the lower truss member.
12. Repeat Steps 4, 6, and 7.

13. Place the 10 lb (4.5 kg) weight at a distance of 12 inches (30.5 cm) to the left of Point 1 (toward the nearest support point) on the lower truss member.
14. Repeat Steps 4, 6, and 7.

Data Collection Format and Data Collection Sheets Tailored to Experiment

See attached data collection sheet.

Data Analysis Plan

The dead weight tension caused by the mass of the materials from which the truss has been constructed will be subtracted from all readings after the truss is loaded so that only the effects of the loading will be measured. Those tension forces will be compared to the vector analyses for consistency.

Expected Outcomes

The tension readings on the scales each time the weights are moved will change. The mathematical vector analyses should coincide with the readings on all of the scales (minus the dead weight tension forces in the members prior to loading). It is expected that four of the scales will show no change because the loads are not being carried by those members. All seven of the other truss members should show varying amounts of tension as a result of the loadings.

DATA COLLECTION SHEET DISTRIBUTION OF TENSION FORCES IN TRUSS MEMBERS							
Lead Experimenter							
Date of Experiment							
Scale Number	1	2	3	4	5	6	7
Tension Force on Each Scale at Zero Load – Step 3							
Calculated Tension in Each Member with 10 lb Weight at Point 1 – Step 4							
Tension Force on Each Scale with 10 lb Weight at Point 1 – Step 6							
Calculated Tension in Each Member with 10 lb Weight at Point 2 – Step 8							
Tension Force on Each Scale with 10 lb Weight at Point 2 – Step 8							
Calculated Tension in Each Member with 10 lb Weight at Points 1 and 2 – Step 10							
Tension Force on Each Scale with 10 lb Weight at Points 1 and 2 – Step 10							
Calculated Tension in Each Member with 10 lb Weight at Center of Truss – Step 12							
Tension Force on Each Scale with 10 lb Weight at Center of Truss – Step 12							
Calculated Tension in Each Member with 10 lb Weight 12 inches to Left of Point 1 – Step 14							
Tension Force on Each Scale with 10 lb Weight 12 inches to Left of Point 1 – Step 14							

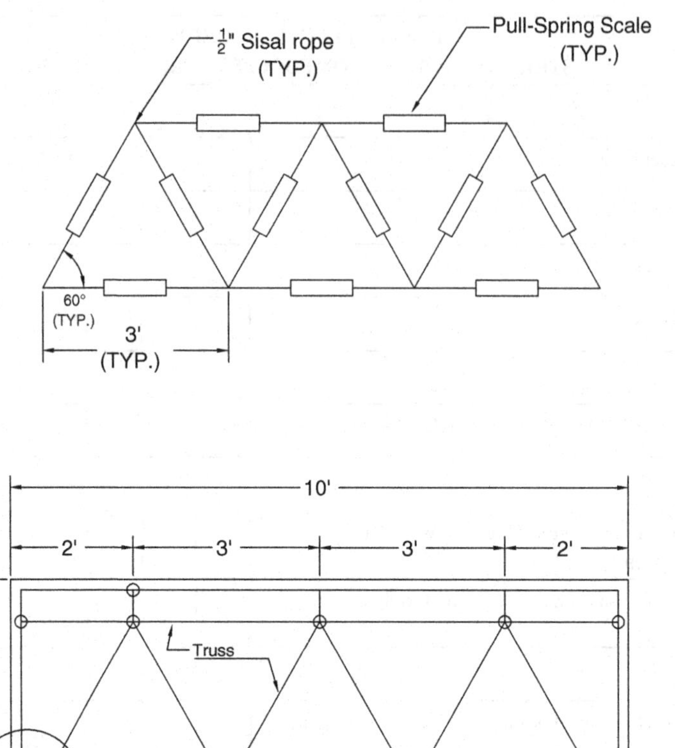

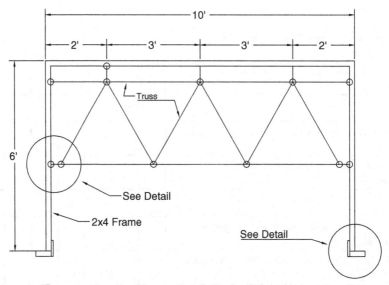

FIGURE 9.4.1 Rope truss with tension measurement system.

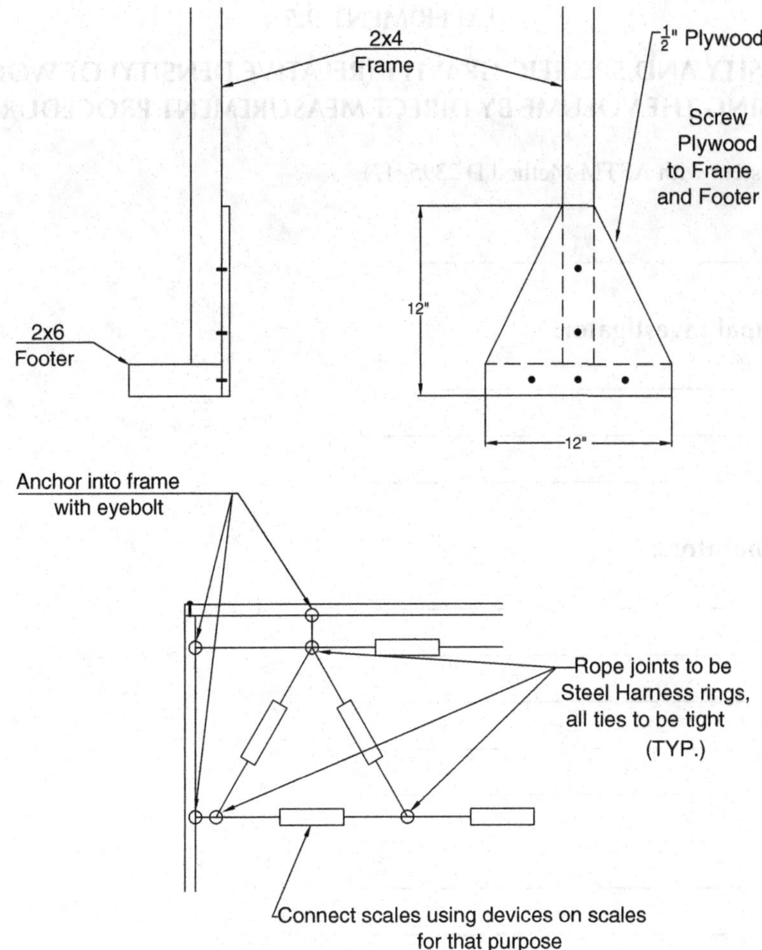

FIGURE 9.4.2 Rope truss with tension measurement system details.

EXPERIMENT 9.5

DENSITY AND SPECIFIC GRAVITY (RELATIVE DENSITY) OF WOOD USING THE VOLUME BY DIRECT MEASUREMENT PROCEDURE

(Consistent with ASTM Method D 2395–17)

Date: ______________________________

Principal Investigator:

Name: ______________________________

Email: ______________________________

Phone: ______________________________

Collaborators:

Name: ______________________________

Email: ______________________________

Phone: ______________________________

Name: ______________________________

Email: ______________________________

Phone: ______________________________

Name: ______________________________

Email: ______________________________

Phone: ______________________________

Name: ______________________________

Email: ______________________________

Phone: ______________________________

Objective or Question to Be Addressed

This experiment provides a method for determining the density and specific gravity of wood or woody materials. Several test methods are provided by ASTM Methods 2395–17 for that purpose. This experiment is based on Method A, which is used for precise measurements when the specimens are carefully prepared and regular in shape. Test Method B is used for precise measurements if the specimens are irregularly shaped; see Experiment 9.6 for that procedure. Test Method C is an approximate test method that is not covered in this book. Test Methods D and E are special methods for the determination of density or specific gravity of living trees or of in-place elements. Those procedures are not covered in this book. Test Method F is a specific procedure for wood chips: see Experiment 9.7 for that procedure. Test Method G is used to estimate the overall density or specific gravity of full-size rectangular members and is not covered in this book.

Theory behind Experiment

Density and specific gravity are commonly used to help define various useful properties of wood and wood-based materials, depending on the industry that is doing the measurements. Note that specific gravity is variable within a tree, between trees, and between species. The specific gravity of a wood cell wall is essentially constant for all species (approximately 1.53). Thus, the individual specific gravity value of a sample is an indirect measurement of the amount of wood cell wall present in the sample.

The moisture content (M) of the sample is measured to identify the basis or the determination of the density or specific gravity. See Experiment 9.8 for the determination of the moisture content by oven drying.

The dimensions of test samples must be carefully measured to yield a variance of all measurements within +/– 5 % of the average of the measurements. The mass measurement should yield a variance of +/– 0.2 % or less between several measurements. These precisions require careful attention to the shape uniformity of the sample being measured.

Samples should be dried prior to measurement to ensure uniform results of the density and specific gravity calculations. Samples used for determining the moisture content may be used for the subsequent tests if size appropriate. Drying should be done in a vented, forced convection oven that can be maintained at 103 +/– 2 °C (217 +/– 4 °F). For most wood and structural composite lumber, a sample 1 inch (25 mm) in length parallel to grain that is dried in an oven for 24 hours will achieve practical equilibrium.

Potential Interferences and Interference Management Plan

There are no know interferences likely with this experiment.

Data to Be Collected with an Explanation of How They Will Be Collected (Ex.: "Water Temperature Data for Three Days Using a Continuous Data Logger")

Using the attached data sheet, record at a minimum the following information:

- Name of all experimenters
- Sample type and source-identifying information
- Description of sample, such as all dimensions measured and volume calculations, to verify precision of measurements and all measurements made, all calculations made, and all values determined for density and specific gravity of the sample
- Equipment and methods used to determine moisture content and to dry the sample
- Method used to determine the density and specific gravity

Tools, Equipment, and Supplies Required

- Sample of material to be tested, per theory section
- Vented forced convection drying oven that can be maintained at 103 +/– 2 °C (217 +/– 4 °F)
- Heat-resistant gloves or tongs for removing the sample from the drying oven
- Meter for determining the moisture content of the wood if Experiment 9.8 is not used for this determination

Safety

The drying process utilizes a drying oven at a temperature just above the boiling temperature of water. The oven and the sample in it will be very hot and can cause burns to the skin. Heat-resistant gloves or tongs are needed to handle the sample while placing it in a hot oven or removing it from that oven.

In the event of a burn, immediately run the affected area under cold water for several minutes and contact medical personnel for assistance.

As with all laboratory experiments, eye protection, gloves, and a lab coat are recommended for this experiment.

Planned Experimental Procedures and Steps, in Order

1. Put on appropriate personal protective equipment, then gather samples and tools in a convenient location.
2. Measure and record the length, width, and thickness of the sample in a sufficient number of places to ensure a precise calculation of the volume. This is V_m.
3. Determine and record the initial mass of the sample at the time of test by weighing on a scale that measures to 0.01 g. This is m_m.
4. Dry the sample to a consistent mass equilibrium (+/– 0.2 % after several measurements) or by calculation in accordance with Experiment 9.8.
5. Determine and record the oven-dried mass of the sample by drying the sample to practical equilibrium. This is m_0.

6. Determine and record the oven-dried volume of the sample in a manner consistent with Step 2. This is V_o.
7. Determine and record the moisture content (M) of the specimen in accordance with the data analysis section.
8. Determine and record the density (ρ) and specific gravity (S) of the sample as described in the data analysis section.

Data Collection Format and Data Collection Sheets Tailored to Experiment

See attached data collection sheet.

Data Analysis Plan

If this experiment procedure or the procedure in Experiment 9.8 is used to determine the moisture content, the calculation is the following:

$$M = ((m_m - m_o)/m_o)) \times 100$$

where: M = the moisture content at the time of testing, %
m_m = initial mass
m_o = oven-dry mass

If a moisture meter is used, the oven-dry mass is estimated as follows:

$$m_o = m_m/(1+ 0.01\ M)$$

where: m_o = oven-dry mass (Step 5)
m_m = initial mass (Step 3)
M = moisture content determined by the meter

Density (ρ) is calculated from the following equations:
Density at moisture content M:

$$\rho_M = m_m/V_m$$

Oven-dry density:

$$\rho_0 = m_o/V_o$$

Basic density:

$$\rho_b = m_o/V_{max}$$

where: m_o = oven-dry mass of sample (Step 5)

V_m = volume of specimen at moisture content M (Step 2)
V_o = oven-dry volume of specimen (Step 6)
V_{max} = green volume of specimen

The specific gravity (S) of the sample is calculated using the following equations.
Specific gravity at moisture content M:

$$S_M = K_{m0}/V_M$$

Oven-dry specific gravity:

$$S_o = K_{m0}/V_0$$

Basic specific gravity:

$$S_b = K_{m0}/V_{max}$$

where: K = constant determined by the units used to measure mass and volume:
Value of K:
K = 27.680 inches³/lb when mass is in lb and volume is in inches³
= 453.59 cm³/lb when mass is in lb and volume is in cm³
= 453,590 mm³/lb when mass is in lb and volume is in mm³
= 0.061024 inches³/g when mass is in g and volume is in inches³
= 1.000 cm³/g when mass is in g and volume is in cm³
= 1,000 mm³/g when mass is in g and volume is in mm³

The relationships between specific gravity and density values are expressed as follows.
Values at moisture content M:

$$S_M = ((\rho_{M/}\rho_w (1 + 0.01\ M))$$

Oven-dry values:

$$S_o = \rho_o/\rho_w$$

Basic values:

$$S_b = \rho_b/\rho_w$$

where: ρ_w = density of water at test temperature; see Table A.5 in the appendix
M = moisture content of specimen, %

Note that when the values of density (ρ) are expressed in g/cm³, the oven-dried specific gravity (S_o) and oven-dry density (ρ_o) are numerically equal, as are the values of the basic specific gravity (S_b) and basic density (ρ_b). However, the values of specific gravity and density at moisture content M are not equal. These relationships may be useful as quality-control checks for calculations.

Expected Outcomes

The expected outcomes from this experiment will vary based on the actual moisture content of the material tested and the type of wood being tested.

DATA COLLECTION SHEET DENSITY AND SPECIFIC GRAVITY (RELATIVE DENSITY) OF WOOD USING THE VOLUME BY DIRECT MEASUREMENT PROCEDURE	
Lead Experimenter	
Date of Experiment	
Sample Wood Type	
Sample Source	
Dimensions of Sample	
Calculated Volume of Sample (Step 2). This is V_M.	
Initial Mass of Sample (Step 3). This is m_m.	
Oven-dried Mass of Sample (Step 5). This is m_o.	
Oven-dried Volume of Sample (Step 6). This is V_o.	
Calculated or Metered Moisture Content of Sample	
Calculated Density at Moisture Content M	
Calculated Oven-dried Density	
Calculated Basic Density	
Calculated Specific Gravity at Moisture Content M	
Calculated Oven-dried Specific Gravity	
Calculated Basic Specific Gravity	

EXPERIMENT 9.6

DENSITY AND SPECIFIC GRAVITY (RELATIVE DENSITY) OF WOOD USING THE VOLUME BY WATER IMMERSION PROCEDURE

(Consistent with ASTM Method D 2395–17)

Date: ______________________________

Principal Investigator:

Name: ______________________________

Email: ______________________________

Phone: ______________________________

Collaborators:

Name: ______________________________

Email: ______________________________

Phone: ______________________________

Name: ______________________________

Email: ______________________________

Phone: ______________________________

Name: ______________________________

Email: ______________________________

Phone: ______________________________

Name: ______________________________

Email: ______________________________

Phone: ______________________________

Objective or Question to Be Addressed

This experiment provides a method for determining the density and specific gravity of wood. Methods A–G (described in the following paragraphs) are provided by ASTM Methods 2395–17 for determining the density and specific gravity of wood and woody materials. This experiment covers Method B.

- Method A is used for precise measurements when the specimens are carefully prepared and regular in shape. See Experiment 9.5 for an experiment consistent with that procedure.
- Test Method B is used for precise measurements if the specimens are irregularly shaped. This experiment is consistent with that procedure B.
- Test Method C is an approximate test method that is not covered in this book.
- Test Methods D and E are special methods for the determination of density or specific gravity of living trees or of in-place elements. Those procedures are not covered in this book.
- Test Method F is used for the measurement of density and specific gravity of wood chips. See Experiment 9.7 for an experiment consistent with that procedure.
- Test Method G is used to estimate the overall density or specific gravity of full-size rectangular members and is not covered in this book.

Theory behind Experiment

Density and specific gravity are commonly used to help define various useful properties of wood and wood-based materials, the specific methods for defining these characteristics are dependent on the industry that is doing the measurements, such as the construction industry versus the paper industry. Note that specific gravity is variable within a tree, between trees, and between species. The specific gravity of a wood cell wall is essentially constant for all species (approximately 1.53). Thus, the individual specific gravity value of a sample is an indirect measurement of the amount of wood cell wall present in the sample.

The moisture content (M) of the sample is measured to identify the basis or the determination of the relationship between the density and or specific gravity.

The dimensions of test samples must be carefully measured to yield a variance of all measurements within +/– 5 % of the average of the measurements. The mass measurement should yield a variance of +/– 0.2 % or less between several measurements. These precisions require careful attention to the shape uniformity of the sample being measured.

Samples should be dried prior to measurement to ensure uniform results of the density and specific gravity calculations. Samples used for determining the moisture content may be used for the subsequent tests if size appropriate. Drying should be done in a vented, forced convection oven that can be maintained at 103 +/– 2 °C (217 +/– 4 °F). For most wood and structural composite lumber, a sample 1 inch (25 mm) in length parallel to grain that is dried in an oven for 24 hours will achieve practical equilibrium.

Potential Interferences and Interference Management Plan

This procedure is particularly suited to and designed for samples of irregular shape or having a rough surface with no knots, limb joints, or other irregularities.

Limitations on specimen size are based primarily on size of immersion tanks available. In very small size specimens, generally less than 1 cm^3 in volume, air bubbles can adhere to the specimen surface and result in substantial errors in the volume measurement and the computed density or specific gravity calculations.

Freshly cut green wood will not absorb appreciable quantities of water during the brief immersion period.

If any drying has taken place, the surface of the specimen needs to be sealed before immersion in water (see Step 3 of the procedures section) to prevent error in the volumetric displacement of the sample by an amount equal to the volume of water absorbed by the wood.

Data to Be Collected with an Explanation of How They Will Be Collected (Ex.: "Water Temperature Data for Three Days Using a Continuous Data Logger")

Using the attached data sheet, record at a minimum the following information:

- Name of all experimenters
- Sample type and source-identifying information
- Description of sample, such as all dimensions measured and volume calculations, to verify precision of measurements and all measurements made, all calculations made, and all values determined for density and specific gravity of the sample
- Equipment and methods used to determine moisture content and to dry the sample
- Method used to determine the density and specific gravity

Tools, Equipment, and Supplies Required

- Sample of material to be tested, per theory section
- Vented forced convection drying oven that can be maintained at 103 +/– 2 °C (217 +/– 4 °F)
- Heat-resistant gloves or tongs for removing the sample from the drying oven
- Water tank or graduated cylinder suitable for the selected method of immersion (see Step 3 of the procedures section)

Safety

The drying process utilizes a drying oven at a temperature just above the boiling temperature of water. The oven and the sample in it will be very hot and can cause burns to the skin. Heat-resistant gloves or tongs are needed to handle the sample while placing it in a hot oven or removing it from that oven.

In the event of a burn, immediately run the affected area under cold water for several minutes and contact medical personnel for assistance.

As with all laboratory experiments, eye protection, gloves, and a lab coat are recommended for this experiment; additionally, this lab requires the use of heat-resistant gloves.

Planned Experimental Procedures and Steps, in Order

1. Put on appropriate personal protective equipment, then gather samples and tools in a convenient location.
2. Determine the initial mass of the specimen at time of the test by weighing it on the balance. This is m_M.
3. Determine the volume of the specimen by measuring the volume or the mass of the water displaced by the specimen using one of the following four methods. This is V_M.
   NOTE: The mass of water in grams is numerically equal to its volume in cm^3. Unless the volume is being determined on a sample of green wood, the surfaces of the sample must be sealed prior to immersion to prevent water infiltration into the wood pores, skewing the resulting volume measurements. To do this, dip the partially dry or oven-dried sample in hot paraffin wax. After the wax dries, scrape off any excess wax, then weigh the specimen again and use this mass in lieu of the w_M value, in conjunction with the immersed mass for determining the sample volume, using the following options for determining the volume of the sample. Green wood samples may be briefly immersed in water unsealed without appreciable absorption that will affect volume determinations.
   a. 1) Place the sample in a tank of known volume and add sufficient water to fill the tank with the specimen being fully submerged.
      2) Remove the specimen and determine the volume of water remaining.
      3) Record the tank volume, less the volume of water remaining, as the volume of the sample.
   b. 1) Place a container holding enough water to completely submerge the sample on a suitable balance.
      2) Tare the balance to the combined mass of the container and water.
      3) Using a sharp, pointed, slender rod, place the specimen in the container so that it is just submerged completely in the water without touching the sides of the container.
      4) The equilibrium reading on the balance is equal to the mass of water displaced by the sample and that is equal to the mass of the sample.
   c. 1) Place a container holding enough water to completely submerge the specimen on a suitable balance. *The container must be large enough that immersion of the sample causes no significant change in water level.*
      2) Suspend a wire basket heavy enough to keep the sample submerged and immerse it in the water.
      3) Tare the balance to the mass of the basket when freely immersed.
      4) Weigh the sample in air.
      5) Place the sample in the basket and hold the basket and sample completely submerged without either touching the container.
      6) If the sample is lighter than water, read and record the mass of the sample as the mass reading on the balance, after the balance reaches

equilibrium, plus the mass of the sample in air as equal to the volume of water displaced.

7) If the sample is heavier than water, subtract the mass reading on the balance from the mass of the sample in air to determine the volume of water displaced.

d. 1) Immerse the sample, if it has an elongated shape, into a graduated cylinder having a cross-section only slightly larger than that of the sample.

2) Read and record the water level in the cylinder, preferably to an even graduation mark, before immersing the sample.

3) Immerse the sample and hold it submerged with a slender pointed rod if necessary and read the water level again. The difference in water level is equal to and recorded as the volume of the sample.

4. Dry the sample to a consistent mass equilibrium (+/– 0.2 % after several measurements) or by calculation in accordance with Experiment 9.8. This is m_o.
5. Determine and record the moisture content (M) of the specimen in accordance with the data analysis section.
6. Determine and record the density (ρ) and specific gravity (S) of the sample as described in the data analysis section.
7. Coat the sample with a light coat of wax; just enough to prevent absorption of water by the dry sample.
8. Determine the oven-dried volume of the sample using the procedures in Step 3. This is V_o.

Data Collection Format and Data Collection Sheets Tailored to Experiment

See attached data collection sheet.

Data Analysis Plan

If this experiment procedure or the procedure in Experiment 9.8 is used to determine the moisture content, the calculation is the following.

$$M = ((m_M - m_o)/(m_o)) \times 100$$

where: M = the moisture content at the time of testing, %
m_M = initial mass
m_o = oven-dry mass

If a moisture meter is used, the oven-dry mass is estimated as follows:

$$M_o = m_M/(1 + 0.01\ M)$$

where: m_o = oven-dry mass

m_M = initial mass
M = moisture content determined by the meter

Density (ρ) is calculated from the following equations:

Density at moisture content M:

$$\rho_M = m_M/V_M$$

Oven-dry density:

$$\rho_0 = m_o/V_o$$

Basic density:

$$\rho_b = m_o/V_{max}$$

where: m_0 = oven-dry mass of sample (Step 4)
V_M = volume of specimen at moisture content M (Step 2)
V_o = oven-dry volume of specimen (Step 8)
V_{max} = green volume of specimen.

The specific gravity (S) of the sample is calculated using the following equations.
Specific gravity at moisture content M:

$$S_M = K_{mo}/V_M$$

Oven-dry specific gravity:

$$S_o = K_{mo}/V_o$$

Basic specific gravity:

$$S_b = K_{mo}/V_{max}$$

where: K = constant determined by the units used to measure mass and volume:
Value of K:

K = 27.680 inches^3/lb when mass is in lb and volume is in inches^3
= 453.59 cm^3/lb when mass is in lb and volume is in cm^3
= 453,590 mm^3/lb when mass is in lb and volume is in mm^3
= 0.061024 inches^3/g when mass is in g and volume is in inches^3
= 1.000 cm^3/g when mass is in g and volume is in cm^3
= 1,000 mm^3/g when mass is in g and volume is in mm^3

The relationships between specific gravity and density values are expressed as follows

Values at moisture content M:

$$S_M = ((\rho_{M/}\rho_w\ (1 + 0.01\ M))$$

Oven-dry values:

$$S_o = \rho_o/\rho_w$$

Basic values:

$$S_b = \rho_b/\rho_w$$

where: ρ_w = density of water at test temperature; (see Table A.5 in the Appendix)
M = moisture content of specimen, %

Note that when the values of density (ρ) are expressed in g/cm³, the oven-dried specific gravity (S_o) and oven-dry density (ρ_o) are numerically equal, as are the values of the basic specific gravity (S_b) and basic density (ρ_b). However, the values of specific gravity and density at moisture content M are not equal. These relationships may be useful as quality-control checks for calculations.

Expected Outcomes

The expected outcomes from this experiment will vary based on the actual moisture content of the material tested and the type of wood being tested.

DATA COLLECTION SHEET DENSITY AND SPECIFIC GRAVITY (RELATIVE DENSITY) OF WOOD USING THE VOLUME BY WATER IMMERSION PROCEDURE	
Lead Experimenter	
Date of Experiment	
Type of Wood Sample	
Source of Sample	
Initial Mass of Sample (Step 2), m_M	
Volume of Sample (Step 3), V_M	
Oven-dried Mass of Sample (Step 4), m_o	
Oven-dried Volume of Sample (Step 8), V_o	
Calculated Moisture Content (Step 5), M	
Calculated Density of Sample at Moisture Content, M	
Calculated Oven-dried Density of Sample	
Calculated Basic Density of Sample	
Calculated Specific Gravity of Sample at Moisture Content, S_M	
Calculated Oven-dried Specific Gravity of Sample, So	
Calculated Basic Specific Gravity of Sample, S_b	

EXPERIMENT 9.7

DENSITY AND SPECIFIC GRAVITY (RELATIVE DENSITY) OF WOOD CHIPS

(Consistent with ASTM Method D 2395–17)

Date: ______________________________

Principal Investigator:

Name: ______________________________

Email: ______________________________

Phone: ______________________________

Collaborators:

Name: ______________________________

Email: ______________________________

Phone: ______________________________

Name: ______________________________

Email: ______________________________

Phone: ______________________________

Name: ______________________________

Email: ______________________________

Phone: ______________________________

Name: ______________________________

Email: ______________________________

Phone: ______________________________

Objective or Question to Be Addressed

This experiment provides a method for determining the density and specific gravity of wood chips. Methods A–G (described in the following paragraphs) are provided by ASTM Methods 2395–17 for determining the density and specific gravity of wood, wood chips, and woody materials. This experiment covers Method F.

- Method A is used for precise measurements when the specimens are carefully prepared and regular in shape. See Experiment 9.5 for an experiment consistent with that procedure.
- Test Method B is used for precise measurements if the specimens are irregularly shaped. See Experiment 9.6 for an experiment consistent with that procedure.
- Test Method C is an approximate test method that is not covered in this book.
- Test Methods D and E are special methods for the determination of density or specific gravity of living trees or of in-place elements. Those procedures are not covered in this book.
- Test Method F is a specific procedure for wood chips and that is the test method with which this experiment is most consistent.
- Test Method G is used to estimate the overall density or specific gravity of full-size rectangular members and is not covered in this book.

Theory behind Experiment

Density and specific gravity are commonly used to help define various useful properties of wood and wood-based materials, the specific method for defining those characteristics is dependent on the industry that is doing the measurements, such as the construction industry versus the paper industry. Note that specific gravity is variable within a tree, between trees, and between species. The specific gravity of a wood cell wall is essentially constant for all species (approximately 1.53). Thus, the individual specific gravity value of a sample is an indirect measurement of the amount of wood cell wall present in the sample.

The moisture content (M) of the sample is measured to identify the relationship between the density and specific gravity.

The dimensions of test samples must be carefully measured to yield a variance of all measurements within +/– 5 % of the average of the measurements. The mass measurement should yield a variance of +/– 0.2 % or less between several measurements. These precisions require careful attention to the shape uniformity of the sample being measured.

Samples should be dried prior to measurement to ensure uniform results of the density and specific gravity calculations. Samples used for determining the moisture content may be used for the subsequent tests if size appropriate. Drying should be done in a vented, forced convection oven that can be maintained at 103 +/– 2 °C (217 +/– 4 °F). For most wood and structural composite lumber, a sample 1 inch (25 mm) in length parallel to grain that is dried in an oven for 24 hours will achieve practical equilibrium.

Potential Interferences and Interference Management Plan

This procedure is particularly suited to and designed for samples of wood chips of mixed dimensions, rather than specific pieces of wood.

Limitations on specimen size are based primarily on size of immersion tanks available. In very small size specimens, generally less than 1 cm^3 in volume (such as individual wood chips), air bubbles can adhere to the specimen surface and result in substantial errors in the volume measurement and the computed density or specific gravity calculations.

Freshly cut green wood will not absorb appreciable quantities of water during the brief immersion period.

If any drying has taken place, the surface of the specimen needs to be sealed before immersion in water (see Step 3 of the procedures section) to prevent error in the volumetric displacement of the sample by an amount equal to the volume of water absorbed by the wood.

Data to Be Collected with an Explanation of How They will Be Collected (Ex.: "Water Temperature Data for Three Days Using a Continuous Data Logger")

Using the attached data sheet, record at a minimum the following information:

- Name of all experimenters
- Sample type and source-identifying information
- Description of sample, such as all dimensions measured and volume calculations, to verify precision of measurements and all measurements made, all calculations made, and all values determined for density and specific gravity of the sample
- Equipment and methods used to determine moisture content and to dry the sample
- Method used to determine the density and specific gravity

Tools, Equipment, and Supplies Required

- Sample of material to be tested, per theory section
- Vented forced convection drying oven that can be maintained at 103 +/− 2 °C (217 +/− 4 °F)
- Heat-resistant gloves or tongs for removing the sample from the drying oven
- Water tank or graduated cylinder suitable for the selected method of immersion (see Step 3 of the procedures section)

Safety

The drying process utilizes a drying oven at a temperature just above the boiling temperature of water. The oven and the sample in it will be very hot and can cause burns to the skin. Heat-resistant gloves or tongs are needed to handle the sample while placing it in a hot oven or removing it from that oven.

In the event of a burn, immediately run the affected area under cold water for several minutes and contact medical personnel for assistance.

Ass with all laboratory experiments, eye protection, gloves, and a lab coat are recommended for this experiment; additionally, this lab requires the use of heat-resistant gloves.

Planned Experimental Procedures and Steps, in Order

1. Put on appropriate personal protective equipment, then gather samples and tools in a convenient location.
2. Select a representative sample of chips weighing approximately 300–350 g (0.66–0.77 lb).
3. Remove sawdust and undersized chips by shaking the sample on a three-mesh sieve.
4. Obtain the initial mass of the screened chips by weighing them on the balance. Record this value as M_m.
5. Place the screened chips into a graduated beaker of sufficient size, smooth the surface of the sample, and compress the sample with a weight of approximately three times that of the sample and record the volume, using the graduations on the beaker, as the volume of the sample as tested, V_M.
6. Submerge the chips in water at room temperature for at least 1 hour to ensure that they are at their green volume and will not absorb water during the subsequent volume measurement.
7. At the end of the soaking period, remove the chips from the water and allow them to drain in a wire-mesh basket.
8. After the sample has drained, place it in the centrifuge basket.
9. Centrifuge the chips from 800 to 1,200 rpm for 1 to 4 min.
10. Place a container holding enough water to freely submerge the wire mesh basket holding the chips on a balance.
11. Submerge the empty chip holder, except for its wire handle, in the water container. The chip holder must not touch the sides or bottom of the container.
12. Zero the balance.
13. Transfer the chips to the chip holder and slowly lower them into the container of water and remove any entrapped air.
14. After the balance reaches equilibrium, the balance reading represents the volume of water equal to the volume of chip, V_{max}.
15. Oven-dry the chips by drying to practical equilibrium (+/– 0.2 % after several measurements).
16. Weigh the dry chips, M_o.
17. Place the dried chips into a graduated beaker of sufficient size, smooth the surface of the sample, and compress the sample with a weight of approximately three times that of the sample and record the volume, using the graduations on the beaker, as the volume of the oven-dried sample, V_o.

18. Determine and record the moisture content (M) of the specimen in accordance with the data analysis section.
19. Determine and record the density (ρ) and specific gravity (S) of the sample as described in the data analysis section.

Data Collection Format and Data Collection Sheets Tailored to Experiment

See attached data collection sheet.

Data Analysis Plan

The moisture content of the chips is calculated from the following equation:

$$M = ((M_m - M_o)/(M_o)) \times 100$$

where: M = moisture content at the time of testing, %
M_m = initial mass (Step 4)
M_o = oven-dry mass (Step 16)

Density (ρ) is calculated from the following equations:
Density at moisture content M:

$$\rho_M = M_m/V_M$$

Oven-dry density:

$$\rho_o = M_o/V_o$$

Basic density:

$$\rho_b = M_o/V_{max}$$

where: m_o = oven-dry mass of sample (Step 16)
V_M = volume of specimen at moisture content M (Step 5)
V_o = oven-dry volume of specimen (Step 17)
V_{max} = green volume of specimen (Step 14)

The specific gravity (S) of the sample is calculated using the following equations.
Specific gravity at moisture content M:

$$S_M = K_{mo}/V_M$$

Oven-dry specific gravity:

$$S_o = K_{mo}/V_o$$

Basic specific gravity:

$$S_b = K_{mo}/V_{max}$$

where: K = constant determined by the units used to measure mass and volume:
Value of K:

K = 27.680 inches^3/lb when units are mass lb and volume inches^3
= 453.59 cm^3/lb when units are mass lb and volume cm^3
= 453,590 mm^3/lb when units are mass lb and volume mm^3
= 0.061024 inches^3/g when units are mass g and volume inches^3
= 1.000 cm^3/g when units are mass g and volume cm^3
= 1,000 mm^3/g when units are mass g and volume mm^3

The relationships between specific gravity and density values are expressed as follows
Values at moisture content M:

$$S_M = ((\rho_{M/}\rho_w)\ (1 + 0.01\ M))$$

Oven-dry values:

$$S_o = \rho_o/\rho_w$$

Basic values:

$$S_b = \rho_b/\rho_w$$

where: ρ_w = density of water at test temperature; (see Table A.5 in the Appendix)
M = moisture content of specimen, %

Note that when the values of density (ρ) are expressed in g/cm^3, the oven-dried specific gravity (S_o) and oven-dry density (ρ_o) are numerically equal, as are the values of the basic specific gravity (S_b) and basic density (ρ_b). However, the values of specific gravity and density at moisture content M are not equal. These relationships may be useful as quality-control checks for calculations.

Expected Outcomes

The expected outcomes from this experiment will vary based on the actual moisture content of the material tested and the type of wood being tested.

DATA COLLECTION SHEET DENSITY AND SPECIFIC GRAVITY (RELATIVE DENSITY) OF WOOD CHIPS	
Lead Experimenter	
Date of Experiment	
Type of Wood or Material	
Source of Sample	
Initial Mass of Sample – M_m (Step 4)	
Volume of Sample – V_M (Step 5)	
Maximum Volume of Chips – V_{max} (Step 14)	
Weight of Dry Chips – M_o (Step 16)	
Volume of Oven-dried Sample – V_o (Step 17)	
Calculated Moisture Content, M	
Calculated Density	
Calculated Value of S_M	
Calculated Value of S_o	
Calculated Value of S_b	

EXPERIMENT 9.8

DIRECT MOISTURE CONTENT MEASUREMENT OF WOOD BY PRIMARY OVEN-DRYING METHOD

(Consistent with ASTM Method D 4442–20)

Date: ______________________

Principal Investigator:

Name: ___________________________

Email: ___________________________

Phone: ___________________________

Collaborators:

Name: ___________________________

Email: ___________________________

Phone: ___________________________

Name: ___________________________

Email: ___________________________

Phone: ___________________________

Name: ___________________________

Email: ___________________________

Phone: ___________________________

Name: ___________________________

Email: ___________________________

Phone: ___________________________

Objective or Question to Be Addressed

This test method covers the determination of the moisture content (MC) of wood, veneer, and other wood-based materials, including those that contain adhesives and chemical additives.

Theory behind Experiment

Two methods for determining the MC of wood and wood-based products are provided in ASTM Method D 4442–20. They are Method A, designated the primary oven-drying method, and Method B, designated the secondary oven-drying method. The primary oven-drying method is used when the highest accuracy or degree of precision is needed. The secondary oven-drying method is used when the accuracy and precision of the primary procedure is not desired or justified. Test results from the secondary method are generally less precise than those from the primary method. This experiment follows the primary oven-drying method.

For materials that have been chemically treated or impregnated with creosote, petroleum, and their solutions such that the oven-drying procedures could introduce greater bias or dangerous off-gassing during the test, other methods for testing are available and recommended.

The moisture content determined by this experiment is expressed as a percentage of oven-dry mass of the sample (oven-dry basis) or as a percentage of the original mass (wet basis). The method described in this experiment uses the oven-dry basis, the most commonly used basis.

Potential Interferences and Interference Management Plan

Because oven-dry mass is used, moisture content values determined by this experiment may exceed 100 %. That may occur because the term "moisture content", when used with wood or other wood-based materials, can be misleading since those materials frequently contain varying amounts of volatile compounds that are also removed by the heating process, along with the moisture content.

Thermal degradation of the wood fibers is also possible during the heating process, which may cause the final moisture-free mass to be slightly smaller than it would be without that degradation.

The moisture content values determined by this test method generally contain some degree of bias due to varying amounts of volatile compounds or of non-wood chemicals contained in the material and the potential for thermal degradation of the material. That bias is not quantifiable in any meaningful way.

Careful adherence to the stipulated test procedure will minimize the causes of bias in the results.

Data to Be Collected with an Explanation of How They Will Be Collected (Ex.: "Water Temperature Data for Three Days Using a Continuous Data Logger")

Using the attached data sheet, record at a minimum the following information:

- Name of all experimenters
- Sample type and source-identifying information
- Balance model and sensitivity
- Oven model, type, and variance
- Description of specimens and their usual or average oven-dry mass (see Table A in the appendix)
- Moisture content values, including mean value

Tools, Equipment, and Supplies Required

- Forced air-convection oven that can be maintained at a temperature of 103 +/− 2 °C (217 +/− 3 °F). Ovens should be vented to allow the evaporated moisture to escape.
- Balance with a sensitivity of the balance of 1 mg
- Weighing bottles, with stoppers, made of a vapor-tight material that can withstand the drying temperature in the oven. The stoppers should be kept with its assigned bottle in case each stopper has a slightly different weight
- Desiccator: A container filled with moisture-absorbing material (desiccant) shall be used for maintaining moisture-free conditions of weighing bottles and for samples cooling
- Test material: Any conveniently sized piece of wood or wood-based material can be used that will fit into the weighing bottles

Safety

The samples in this experiment are oven-dried at a temperature of 103 +/− 2 °C (217 +/− 3 °F).

Care is required to avoid burns. Use heat-resistant gloves or tongs to put bottles into the oven, to stopper the bottles inside the oven before removing them from the oven, and to remove hot bottles to the desiccator.

In the event of a burn, run the affected area under cold water for several minutes and contact medical personnel for further assistance.

Gloves, eye protection, and a lab coat are required for this experiment.

Planned Experimental Procedures and Steps, in Order

1. Put on the appropriate personal protective equipment and gather all necessary samples and materials in a convenient location.
2. Calibrate and standardize the drying oven in the following manner:
   a. Grind a sample of Douglas fir wood to sawdust and separate the fraction

contained in a 40/60 mesh screen as the calibration sample. The sample origin or drying history is not critical. This is the calibration sample.

b. Tumble or gently shake the sawdust in a closed vapor-tight weighing bottle with a stopper until thoroughly mixed.
c. Divide and transfer the mixed sawdust into eight identical replicate samples and store each in a closed and stoppered storage bottle.
d. Number the sample bottles for control purposes.
e. Weigh each sample in its closed weighing bottle. (Be aware that static electricity affects the mass readings and eliminate by grounding of the bottle if necessary.)
f. Preheat the oven to a temperature of 103 +/− 2 °C (217 +/− 3 °F).
g. Open and place the eight bottles of calibration sample into the oven at third-point positions with respect to height, width, and depth of the oven cavity. This should result in four samples positioned on each of two shelves at one third and two thirds of the cavity height and dry for at least 3 hours.
h. After drying, close the weighing bottles containing the dried specimens before taking them out of the hot oven and store them in a desiccator with fresh desiccant until they have reached room temperature.
i. Weigh the samples in their closed containers.
j. Continue to check the weight of the closed containers until the mass loss in a 3-hour interval is equal to or less than twice the balance sensitivity.
k. Subtract the dry mass of each separate calibration sample from the oven-dry mass of the same sample for all eight aliquots.
l. Determine the sample variability, S_w, by averaging the eight values determined in the previous steps.

3. Determine the combined sample and oven variability, S_o, by repeating Step 2 except that the samples in the set should all be of similar size and shape without grinding.
4. Calculate the combined sample and oven variability using the equations in the data analysis section.
5. Weigh the test sample in its closed weighing bottle. This is A.
6. Preheat the oven to a temperature of 103 +/− 2 °C (217 +/− 3 °F).
7. Place the test sample in its open weighing bottles in the oven within the volume tested for oven variability and dry overnight or for 24 hours.
8. After drying, close the weighing bottle containing the dried sample before taking it out of the hot oven and store it in a desiccator with fresh desiccant until it has reached room temperature.
9. Weigh the sample in its closed container. This is B.
10. Determine the moisture content of the test sample using the equation in the data analysis section.

Data Collection Format and Data Collection Sheets Tailored to Experiment

See attached data collection sheet.

Data Analysis Plan

Calculate the moisture content of the test sample from the following equation:

$$MC, \% = ((A - B)/B) \times 100$$

where: A = original mass, g
B = oven-dry mass, g

Example – A sample of wood weighs 56.70 g. After oven-drying, the mass is 52.30 g.

$$\begin{aligned} MC &= ((56.70–52.30)/52.30) \times 100 \\ &= 8.4\ \% \end{aligned}$$

If the test sample is from wood that has been treated with a nonvolatile chemical, or if it contains a large amount of non-wood chemicals that cannot be neglected, and if the mass of the retained chemical(s) is known, the moisture content may be determined from the following equation.

$$MC, \% = ((A - B)/D) \times 100$$

where: D = B minus the mass of retained chemical in the sample.

Calculate the combined sample and oven variability by the following equation:

$$S_{ow}^2 = S_w^2 + S_o^2$$

Where: S_{ow} = combined specimen and oven variance
S_w = specimen material variance
S_o = oven variance

Expected Outcomes

Table A.8 in the Appendix provides normal moisture content for various wet samples of wood and wood products. Table A.9 provides similar data for previously dried materials. The results from this experiment should be reasonably consistent with the tabulated values for the type of wood tested.

DATA COLLECTION SHEET DIRECT MOISTURE CONTENT MEASUREMENT OF WOOD BY PRIMARY OVEN-DRYING METHOD			
Lead Experimenter			
Date of Experiment			
Mass of Oven Calibration Sample Bottle 1 with Sample (Step 2e)		Mass of Oven Calibration Sample Bottle 1 with Sample (Step 3)	
Mass of Oven Calibration Sample Bottle 2 with Sample (Step 2e)		Mass of Oven Calibration Sample Bottle 2 with Sample (Step 3)	
Mass of Oven Calibration Sample Bottle 3 with Sample (Step 2e)		Mass of Oven Calibration Sample Bottle 3 with Sample (Step 3)	
Mass of Oven Calibration Sample Bottle 4 with Sample (Step 2e)		Mass of Oven Calibration Sample Bottle 4 with Sample (Step 3)	
Mass of Oven Calibration Sample Bottle 5 with Sample (Step 2e)		Mass of Oven Calibration Sample Bottle 5 with Sample (Step 3)	
Mass of Oven Calibration Sample Bottle 6 with Sample (Step 2e)		Mass of Oven Calibration Sample Bottle 6 with Sample (Step 3)	
Mass of Oven Calibration Sample Bottle 7 with Sample (Step 2e)		Mass of Oven Calibration Sample Bottle 7 with Sample (Step 3)	
Mass of Oven Calibration Sample Bottle 8 with Sample (Step 2e)		Mass of Oven Calibration Sample Bottle 8 with Sample (Step 3)	
Mass of Dried Oven Calibration Sample Bottle 1 with Sample (Step 2i)		Mass of Dried Oven Calibration Sample Bottle 1 with Sample (Step 3)	
Mass of Dried Oven Calibration Sample Bottle 2 with Sample (Step 2i)		Mass of Dried Oven Calibration Sample Bottle 2 with Sample (Step 3)	
Mass of Dried Oven Calibration Sample Bottle 3 with Sample (Step 2i)		Mass of Dried Oven Calibration Sample Bottle 3 with Sample (Step 3)	
Mass of Dried Oven Calibration Sample Bottle 4 with Sample (Step 2i)		Mass of Dried Oven Calibration Sample Bottle 4 with Sample (Step 3)	
Mass of Dried Oven Calibration Sample Bottle 5 with Sample (Step 2i)		Mass of Dried Oven Calibration Sample Bottle 5 with Sample (Step 3)	
Mass of Dried Oven Calibration Sample Bottle 6 with Sample (Step 2i)		Mass of Dried Oven Calibration Sample Bottle 6 with Sample (Step 3)	
Mass of Dried Oven Calibration Sample Bottle 7 with Sample (Step 2i)		Mass of Dried Oven Calibration Sample Bottle 7 with Sample (Step 3)	
Mass of Dried Oven Calibration Sample Bottle 8 with Sample (Step 2i)		Mass of Dried Oven Calibration Sample Bottle 8 with Sample (Step 3)	

DATA COLLECTION SHEET DIRECT MOISTURE CONTENT MEASUREMENT OF WOOD BY PRIMARY OVEN-DRYING METHOD			
Bottle 1 Mass from Step 2e minus Mass from Step 2i (Step 2k)		Bottle 1 Mass from Step 2e minus Mass from Step 2i (Step 3)	
Bottle 2 Mass from Step 2e minus Mass from Step 2i (Step 2k)		Bottle 2 Mass from Step 2e minus Mass from Step 2i (Step 3)	
Bottle 3 Mass from Step 2e minus Mass from Step 2i (Step 2k)		Bottle 3 Mass from Step 2e minus Mass from Step 2i (Step 3)	
Bottle 4 Mass from Step 2e minus Mass from Step 2i (Step 2k)		Bottle 4 Mass from Step 2e minus Mass from Step 2i (Step 3)	
Bottle 5 Mass from Step 2e minus Mass from Step 2i (Step 2k)		Bottle 5 Mass from Step 2e minus Mass from Step 2i (Step 3)	
Bottle 6 Mass from Step 2e minus Mass from Step 2i (Step 2k)		Bottle 6 Mass from Step 2e minus Mass from Step 2i (Step 3)	
Bottle 7 Mass from Step 2e minus Mass from Step 2i (Step 2k)		Bottle 7 Mass from Step 2e minus Mass from Step 2i (Step 3)	
Bottle 8 Mass from Step 2e minus Mass from Step 2i (Step 2k)		Bottle 8 Mass from Step 2e minus Mass from Step 2i (Step 3)	
Average Variation across Samples (Step 2l), S_w		Average Variation across Samples (Step 3), S_o	
Combined Oven and Sample Variability from Data Analysis Section (Step 5)		Mass of Test Sample (Step 6), AS	
Mass of Oven-Dried Test Sample (Step 10), B		Calculated Moisture Content from Data Analysis Section	

10 Concrete Testing Experiments

This chapter provides completed designs, for example, experiments that can be used to demonstrate specific engineering phenomena in a university laboratory or field collection site. They may be used as presented or modified by the user to adapt to other desired outcomes. Sections on actual outcomes and data interpretation are necessarily left blank in these examples. They should be completed by the investigators upon collection of the data.

Note that where an experiment indicates that it is consistent with a specified ASTM Method or other standard, this does not mean that the procedure is identical to, nor does it include all the details of, the stated standard method. Outcomes that require strict compliance with the stated standards must use that standard method to conduct the test. Therefore, the experiment procedures outlined in this book are not a substitute for proper compliance with the stated standards, nor are they intended to be.

The experiments in this chapter are:

10.1 Compressive Strength of Concrete Cylinders
10.2 K-slump Test for Freshly Mixed Concrete
10.3 Cone Method Slump Test for Wet Concrete
10.4 Air Content of Freshly Mixed Concrete by Volumetric Method
10.5 Density (Unit Weight), Yield, and Air Content (Gravimetric) of Concrete
10.6 Time of Setting of Concrete Mixtures by Penetration Resistance
10.7 Effect of Water Content on Concrete Strength
10.8 Specific Gravity and Absorbance of Coarse Aggregate
10.9 Specific Gravity and Absorbance of Fine Aggregate

DOI: 10.1201/9781003346685-10

EXPERIMENT 10.1

COMPRESSIVE STRENGTH OF CONCRETE CYLINDERS

(Consistent with ASTM Method C 39)

Date: ______________________________

Principal Investigator:

Name: ______________________________

Email: ______________________________

Phone: ______________________________

Collaborators:

Name: ______________________________

Email: ______________________________

Phone: ______________________________

Name: ______________________________

Email: ______________________________

Phone: ______________________________

Name: ______________________________

Email: ______________________________

Phone: ______________________________

Name: ______________________________

Email: ______________________________

Phone: ______________________________

Objective or Question to Be Addressed

This experiment is designed to outline the procedures for conducting a compression test on a concrete cylinder using a standard, commercial, compression machine. Note that most machines are limited to concrete having density greater than 50 lb/ft^3 (800 kg/m^3).

Theory behind Experiment

The compressive strength of a concrete cylinder is used to provide guidance on the ultimate compressive strength of the concrete structure constructed from the concrete used to make the test samples. The test consists of applying an axial compressive stress to a standard test cylinder and increasing the load incrementally until failure of the cylinder occurs. The compressive strength of the concrete is then defined as the highest strength value recorded divided by the cross-sectional area of the test cylinder. For this test to be reasonably accurate, test cylinders need to be prepared in accordance with ASTM Methods and Practices C 192 and C 617.

Potential Interferences and Interference Management Plan

Cylinders should not be tested if the diameter of the cylinder varies by more than 2 % throughout its vertical surface. In addition, neither end of the cylinder should be tilted at any location by more than 0.5 degrees, and they should be essentially smooth and perpendicular to the vertical axis. Appropriate capping materials are used to ensure a uniform compressive force on the top and bottom of the cylinder.

Data to Be Collected with an Explanation of How They Will Be Collected (Ex.: "Water Temperature Data for Three Days Using a Continuous Data Logger")

Using the attached data sheet, record at a minimum the following information:

- Names of all experimenters
- Date of experiment
- Source and mix design of concrete samples to be tested
- Type and size of cylinder formed (6 inch plastic mold, for example)
- Capping method used
- Cross-sectional area of cast cylinders
- Height of cylinders
- Method of curing the cylinders
- Maximum load exerted on cylinder at failure in lbf (kilonewtons)
- Compressive strength calculated to nearest 10 psi (0.1 MPa)
- Type of failure observed
- Defects found in the sample or the cap
- Age of cylinders when tested

Tools, Equipment, and Supplies Required

- Set of cylindrical cylinder molds with a capacity based on the coarse aggregate, as described in Experiment 10.8

- Commercially available compression machine meeting ASTM Method C 39 standards, including all bearing blocks, spacers, and load indicators

Safety

A commercial compression machine has a lot of safety devices and procedures built in. Do not bypass them.

Compression while adjusting the cylinder in the machine can destroy hands and fingers. Be sure the machine is off when making adjustments.

Older cylinders will fail loudly and explosively, spewing fragments in all directions at high speed. All commercial machines have a cage that surrounds the cylinder during testing. Be sure that cage is closed and latched before starting the compression to avoid serious injury from flying concrete fragments.

Concrete fragments and dust resulting from fracturing of the cylinder can cause hand and eye injuries.

As with all lab experiments, eye protection, gloves, and a lab coat are recommended or required for this experiment; additionally, this experiment requires a dust mask.

Planned Experimental Procedures and Steps, in Order

Note: For best results, compression testing of cylinders should be conducted on several samples over the course of 28 days and comply with the following criteria. Three samples for each test day are suggested, with the values found for all three being averaged to yield the expected strength after that amount of curing.

Samples should be kept moist during the period between removal from moist storage

and testing. They should be tested in the moist condition.

The tolerances for specimen ages are as follows, per ASTM Method C 39:

7 days +/− 6 h
14 days +/− 7 h
21 days +/− 10 h
28 days +/− 20 h

1. Put on appropriate personal protective equipment and gather all required materials, tools, and equipment in a convenient location.
2. Prepare a suitable batch of concrete based on a desired mix design.
3. Prepare a set of cylinders using standard test cylinder molds. Ensure that the tops of the cylinders are smooth and flat.
4. Allow the cylinders to cure in a moist environment for up to 28 days. Sets of four cylinders should be tested. Test one cylinder from each set at 7, 14, 21, and 28 days.
5. Measure and record the size of the cylinders in terms of length and diameter, then place each end into a standard capping mold to create smooth and level ends on the cylinder.
6. Place the lower bearing block, with the hardened face up, on the table or platen of the testing machine.
7. Clean the bearing faces of the upper and lower bearing blocks, any spacers used, and the cylinder ends.

8. Place the cylinder on the *lower* bearing block and align the axis of the cylinder with the center of thrust of the *upper* bearing block.
9. Verify that the load indicator is set to zero and adjust as necessary.
10. *Close and latch the safety cage door around the cylinder.*
11. Begin application of the load and stop at about 5 % of the anticipated maximum load.
12. Verify the alignment of the cylinder after application of the 5 % load. Check to see that the axis of the cylinder does not depart from the vertical by more than 0.5° and that the ends of the cylinder are centered on the bearing blocks. Release the load and adjust as necessary.
    NOTE: An angle of 0.5° is equal to a slope of approximately 1/8 inches in 12 inches (1 mm/100 mm).
13. Reclose and latch the safety cage door.
14. Apply the load continuously and smoothly at a rate of platen movement that applies a stress rate on the cylinder of 35 psi/s +/– 7 psi/s (0.25 MPa/s +/– 0.05 MPa/s).
15. Apply the compressive load until the load indicator shows that the load is decreasing steadily and the cylinder displays a well-defined fracture pattern. Note that older cylinders (older than 14 days, typically) will fracture loudly and explosively, hence the need to be sure that Step 13 is not overlooked.
16. At failure, note the pattern of the failure mode and any unusual conditions inside the cylinder.

Data Collection Format and Data Collection Sheets Tailored to Experiment

Record data for this experiment in any convenient format that will facilitate preparation of the laboratory report.

Data Analysis Plan

Calculate the compressive strength of the specimen as follows:

$$f_{cm} = 4\,P_{\max}/\pi D^2$$

where: f_{cm} = compressive strength, psi (MPa)
P_{max} = maximum load, lbf (kN)
D = average measured cylinder diameter, inch (m)
π = 3.1416 (or more decimal places for greater precision)

The strength of each cylinder will be plotted as compressive strength on the y-axis and time in days on the x-axis to yield a curve of strength vs. time.

Expected Outcomes

It is expected that the compressive strength will increase with curing time on a nonlinear basis. The curve of increasing strength should be relatively steep at the beginning and then slowly flatten to an asymptote near the 28-day mark. That strength is generally accepted for the ASTM purposes as representing approximately 90 % of the ultimate strength of the concrete.

DATA COLLECTION SHEET COMPRESSIVE STRENGTH OF CONCRETE CYLINDERS					
Lead Experimenter					
Date of Experiment					
	7 days	14 days	21 days	28 days	Other
Source of Mix Design Used					
Type and Size of Cylinder Mold					
Capping Method Used					
Diameter of Cylinder, D					
Height of Cylinder					
Cross-sectional Area of Cylinder					
Curing Method Used					
Maximum Load at Failure, P_{max}					
Calculated Compressive Strength, f_{cm}					
Type of Failure Observed (See ASTM Method C 39)					
Defects Found in Cylinder or Cap					

EXPERIMENT 10.2

K-SLUMP OF FRESHLY MIXED CONCRETE

(Consistent with ASTM Test Method C 1890–19)

Date: ______________________

Principal Investigator:

Name: ______________________

Email: ______________________

Phone: ______________________

Collaborators:

Name: ______________________

Email: ______________________

Phone: ______________________

Name: ______________________

Email: ______________________

Phone: ______________________

Name: ______________________

Email: ______________________

Phone: ______________________

Name: ______________________

Email: ______________________

Phone: ______________________

Objective or Question to Be Addressed

A K-slump test is used to quickly determine the approximate concrete consistency and workability. The K-slump apparatus consists of a hollow tube with prescribed perforations and a floating rod with a graduated scale. The tube is inserted into a sample of freshly mixed concrete to a prescribed depth. The mortar fraction of the concrete is allowed to flow into the perforated tube for a period of 60 seconds (some K-slump tester manufacturers prescribe shorter times of about 40 seconds). The floating rod is then lowered onto the surface of the mortar that has penetrated into the tube. The height of the mortar in the tube is read from the scale marked on the portion of the floating rod protruding from the top of the tube, and that is recorded as the slump of the concrete sample.

Theory behind Experiment

The slump of concrete tested by this method is an approximation. The method is not sufficiently accurate to substitute for a standard slump test using a slump cone (see Experiment 10.3), but it is sufficient to demonstrate consistency between batches and for quality control of multiple batch pours. The concept is that the higher the moisture content, and therefore the lower the strength of the concrete, the higher the concrete will rise inside the K-slump tester and the higher the recorded slump value will be.

Potential Interferences and Interference Management Plan

This test relies on the unhindered intrusion of wet concrete into the interior of the tester. The tester should be moistened with clean water before first use, rinsed well with clean water between uses, and maintained free of clogs at the end, or accretions accumulating inside the tester.

Data to Be Collected with an Explanation of How They Will Be Collected (Ex.: "Water Temperature Data for Three Days Using a Continuous Data Logger")

A single reading of the slump from the graduations on the side of the floating rod is collected by eye and recorded. The names of the experimenters, the date of the experiment, and the source of the samples tested will be recorded on the laboratory report.

Tools, Equipment, and Supplies Required

- Commercially available K-slump apparatus meeting the requirements of ASTM Standard C 1890–19
- Small wood float or screed
- Container to hold the fresh concrete sample, such as a wheelbarrow
- Stopwatch

Safety

Wet concrete is highly caustic and can cause serious burns to skin with extended contact and immediate damage to sight if eyes are contacted directly. Eye protection, gloves, and a lab coat are required to be worn for this test.

In the event of skin contact, wash as soon as possible with hot water and soap. Dried concrete and cement dust will rapidly dry the skin, causing chapping and cracking, which may require medical attention to correct. Skin lotion should be applied as soon as the area is washed, unless cracking or chapping of the skin has already occurred.

In the event of contact with the eyes, flush immediately at an eye wash station for up to 15 minutes and contact medical personnel for assistance. *Do not rub the eyes* during this time to avoid grinding cement particles into the eye surface or scratching the lens.

Planned Experimental Procedures and Steps, in Order

1. Put on appropriate personal protective equipment and gather all supplies and apparatus in a convenient location.
2. Verify that the measuring rod indicates a value of 0 when the rod is inserted fully into the empty tube. Adjust the reading scale as needed so that the reading is 0. Wet the apparatus and shake off excess water.
3. Place a sample of the fresh concrete into the container. Use the small wood float or screed to bring the concrete surface to a flat condition. Work the surface as little as possible to avoid formation of a mortar layer.
4. Begin measurement of the K-slump within 1 minute after filling the container. During the K-slump measurement, do not agitate, vibrate, or jar the concrete.
5. Raise the measuring rod and let it rest on the holding pin inside the tube.
6. Insert the apparatus slowly and vertically downward until the collar floater touches the surface of the concrete. Do not rotate the apparatus while inserting it into the concrete.
7. When the collar floater touches the concrete surface, start the stopwatch. At 60 +/− 5 seconds after complete insertion of the tube, lower the measuring rod slowly until it rests on the surface of the mortar that entered into the hollow tube.
8. Read and record the height of the mortar by reading to the nearest ¼ division on the measuring rod. The measured mortar height is the K-slump of the concrete.
9. Wash the apparatus to remove all mortar in the tube.

Data Collection Format and Data Collection Sheets Tailored to Experiment

One value of the K-slump per test is read by eye and recorded.

Data Analysis Plan

The data that are collected are used to empirically judge the quality of the fresh concrete. The readings should relate closely to, but not necessarily exactly correspond to, the mix design slump. Consistency of K-slump readings from continuous operations or from large truckloads of fresh concrete may be used to assess the quality of the fresh concrete over an extended time period or many dumps from the same source. Note that this test is not an acceptable substitute for verification and certification of slump accuracy for overall quality control and certification purposes. A cone slump test is necessary for such purposes (see Experiment 10.3).

Expected Outcomes

The slump measured by this test procedure is approximate and will vary widely depending on the moisture content and water/cement ratio (W/C ratio) of the concrete.

EXPERIMENT 10.3

SLUMP TEST OF HYDRAULIC-CEMENT CONCRETE

(Consistent with ASTM Method C 143/C 143 M – 20)

Date: ______________________________

Principal Investigator:

Name: ______________________________

Email: ______________________________

Phone: ______________________________

Collaborators:

Name: ______________________________

Email: ______________________________

Phone: ______________________________

Name: ______________________________

Email: ______________________________

Phone: ______________________________

Name: ______________________________

Email: ______________________________

Phone: ______________________________

Name: ______________________________

Email: ______________________________

Phone: ______________________________

Objective or Question to Be Addressed

The slump of wet hydraulic concrete is a key element in determining whether the mixture is consistent with the design of that mixture. This test is designed to assess a batch of wet concrete by observing "slump". A sample of freshly mixed, plastic, concrete placed and compacted by rodding in a mold shaped as the frustum of a cone will subside as the mold is raised. This subsidence, called *slump*, is defined and measured as the vertical distance between the original and displaced position of the center of the top surface of the concrete.

Theory behind Experiment

Under laboratory conditions, the slump has generally been found to increase proportionally with the water content of the concrete mixture and inversely related to the ultimate concrete strength. Under field conditions, however, that relationship has not been consistently seen, and results obtained under field conditions should not be related to the strength of the concrete, but rather to its compliance with the design mix characteristics.

Potential Interferences and Interference Management Plan

This test relies upon the cone sliding easily off the compacted concrete without drawing the sides of the concrete mass upward or drawing moisture to the sides of the cone from the mass. Care must be taken to raise the cone as nearly vertically as possible and to do so slowly so as not to draw the surface upwards.

Solid accretions inside the cone from prior tests, or dents in the cone from mishandling will cause the surface of the compacted concrete mass to become deformed during the test. Care must be taken to clean the inside of the cone thoroughly prior to each test and to ensure that the cone is not damaged prior to its use.

Data to Be Collected with an Explanation of How They Will Be Collected (Ex.: "Water Temperature Data for Three Days Using a Continuous Data Logger")

The data to be collected during this experiment consists of one vertical measurement made with a standard ruler or tape measure.

The slump measured will be recorded in a notebook and reported on a laboratory report of the experiment.

Tools, Equipment, and Supplies Required

- Concrete mold
  - In the form of the lateral surface of the frustum of a cone with the base 8 inches (200 mm) in diameter, the top 4 inches (100 mm) in diameter, and the height 12 inches (300 mm). Individual diameters and heights shall be within +/– 1/8 inches (3 mm) of the prescribed dimensions.

- Tamping rod
  - A round, smooth, straight steel rod, rounded at both ends, with a 5/8 inch (16 mm) +/– 1/16 inch (2 mm) diameter. The length of the tamping rod shall be at least 4 inches (100 mm) greater than the depth of the mold, but not greater than 24 inches (600 mm).
- Measuring device
  - Such as a ruler, metal roll-up measuring tape, or similar rigid or semi-rigid length-measuring instrument marked in increments of 1/4 inches [5 mm] or smaller.
- Scoop
  - Large enough so each amount of concrete obtained from the sampling receptacle is representative of the total mix and small enough so that concrete is not spilled during placement in the mold.
- Flat stick or metal slat from which the vertical slump can be measured accurately

Safety

Wet concrete is highly caustic and can cause serious burns to skin with extended contact and immediate damage to sight if eyes are contacted directly. Eye protection, gloves, and a lab coat are required to be worn for this test.

In the event of skin contact, wash as soon as possible with hot water and soap. Drying concrete and cement dust will rapidly dry the skin, causing chapping and cracking, which may require medical attention to correct. Skin lotion should be applied as soon as the area is washed, unless cracking or chapping of the skin has already occurred.

In the event of contact with the eyes, flush immediately at an eye wash station for up to 15 minutes and contact medical personnel for assistance. *Do not rub the eyes* during this time to avoid grinding cement particles into the eye surface or scratching the lens.

Planned Experimental Procedures and Steps, in Order

1. Put on the appropriate personal protective equipment and gather the tools and sample in a convenient location.
2. Dampen the inside of the mold and place it on a rigid, flat, level, moist, nonabsorbent surface, free of vibration, and that is large enough to contain all the slumped concrete.
3. Stand on the two foot pieces attached to the mold or clamp the mold to a base plate.
4. Fill the mold in three layers, each approximately one third the volume of the mold, using the scoop. Place the concrete into the mold carefully to ensure an even distribution of the concrete with minimal segregation.
5. Rod each layer 25 times uniformly over the cross-section with the rounded end of the rod. Rod the bottom layer throughout its depth by inclining the rod slightly and making approximately half of the strokes near the perimeter and then moving with vertical strokes toward the center. Rod through each upper layer and into the layer below approximately 1 inch (25 mm). Fill the

concrete above the top of the mold before rodding is started on the top layer. If the rodding operation results in subsidence of the concrete below the top edge of the mold, add additional concrete to keep an excess of concrete above the top of the mold at all times. After the top layer has been rodded, strike off the surface of the concrete by means of a screeding and rolling motion of the tamping rod.

6. Continue to hold the mold down firmly and remove any loose or spilled concrete from the area surrounding the base of the mold to preclude interference with the movement of the slumping concrete.
7. Remove the mold from the concrete by raising it carefully in a vertical direction. Raise the mold a distance of 12 inches (300 mm) in 5 +/– 2 seconds by a steady upward lift with no lateral or torsional motion. Complete the entire test from the start of the filling through removal of the mold without interruption and complete it within an elapsed time of 2.5 minutes.
8. Immediately place the cone next to the slumped pile of concrete with the base of the cone and the base of the concrete mass at equal elevations. Measure the slump by placing the flat stick or metal slat across the top of the cone in a horizontal position and extending over the top center of the pile of slumped concrete. Measure the vertical distance from the underside of the flat stick to the top of the slumped concrete pile and record that distance as the slump of the sample. Note that if the slumping concrete appears to fall over or shear off from one side or portion of the mass, disregard the test and make a new test on another portion of the sample. If two consecutive tests on a sample of concrete show a falling over or shearing off of a portion of the concrete from the mass of the specimen, it should be reported that the concrete lacks necessary plasticity and cohesiveness for this form of slump test to be applicable.

Data Collection Format and Data Collection Sheets Tailored to Experiment

The distance from the underside of a measuring stick to the top of the slumped concrete mass will be measured using a standard ruler or tape measure.

Data Analysis Plan

The data collected during this experiment consists of one vertical measurement. Analysis consists of rendering an opinion as to whether the slump distance is consistent with the concrete mix design.

Expected Outcomes

In most cases, the concrete will demonstrate a slump reasonably consistent with the concrete mix design. If it does not do so, it may be concluded that an error has been made during the mixing of the concrete. If the slump is less than that designed, the water content of the concrete is most likely insufficient. If the slump is greater than the mix design calls for, the water content most likely exceeds that of the mix design.

EXPERIMENT 10.4

AIR CONTENT OF FRESHLY MIXED CONCRETE BY THE VOLUMETRIC METHOD

(Consistent with ASTM Method C 173–16)

Date: ______________________

Principal Investigator:

Name: ____________________________

Email: ____________________________

Phone: ____________________________

Collaborators:

Name: ____________________________

Email: ____________________________

Phone: ____________________________

Name: ____________________________

Email: ____________________________

Phone: ____________________________

Name: ____________________________

Email: ____________________________

Phone: ____________________________

Name: ____________________________

Email: ____________________________

Phone: ____________________________

Objective or Question to Be Addressed

The determination of the air content is used to characterize fresh concrete, regardless of the type of aggregate used. The objective of this experiment is to determine how much air (what percentage relative to the cement used on a volumetric basis) is entrained in the fresh concrete. Air content impacts the concrete quality and strength, particularly in locations impacted by freezing and thawing.

Theory behind Experiment

This experiment determines the percentage of air contained in the mortar fraction of the concrete, on a volumetric basis, but is not affected by any air that may be present inside porous aggregate particles. This procedure requires the addition of isopropyl alcohol to dispel foam that forms during the experiment procedure when a standard, commercially available meter is initially being filled with water. This is done so that after several "rollings" of the meter to mix the concrete and release the entrained air, little or no foam collects in the neck at the top section of the meter. If more foam is present than that equivalent to 2 % the air above the water level, the test is declared invalid and must be repeated.

Note that the air content of the hardened concrete from which the fresh sample is taken for testing depends upon the methods and amounts of consolidation effort applied to the concrete; the uniformity and stability of the air bubbles in the fresh and hardened concrete; environmental exposure; whether a pump is used to place the concrete in a form; and other factors. Therefore, the air content determined by this procedure is unlikely to be the same as the air content found in the hardened concrete.

Potential Interferences and Interference Management Plan

This experiment may be impacted by the development of foam that otherwise would interfere with an accurate reading of the air meter. Isopropyl alcohol is used to help mitigate the interference of the foam. When the top section of the air meter is initially filled to the zero mark with a mixture of water and isopropyl alcohol, that mixture has a defined volume. When that solution is further mixed with the water present in the concrete, the concentration of alcohol changes and the new solution occupies a volume slightly smaller than it did when the meter was initially filled to the zero mark. For this reason, the meter tends to indicate a higher than actual air content when more than about 2.5 pints (1.2 L) of alcohol is used. When that volume, or greater, of alcohol is required, a correction factor is used to adjust for the volume change by reducing the measured reading by the correction factor. Table 10.4.1, at the end of the procedures section, describes the correction factor to use based on the volume of alcohol used in the experiment.

Data to Be Collected with an Explanation of How They Will Be Collected (Ex.: "Water Temperature Data for Three Days Using a Continuous Data Logger")

Using the attached data collection sheet, record and report, at a minimum, the following information:

- Names of all experimenters
- Date of experiment

- Batch mix design used
- Volume of the measuring bowl from Step 2a
- Volume of isopropyl alcohol added in Step 11
- Initial meter reading from Step 22
- Number of calibrated cups of water added in Step 23, if any
- Final meter reading from Step 25
- Any "new initial reading" from Step 26
- Any new final reading from Step 27
- Correction factor from Table 10.4.1, if used
- Calculated air content to the nearest 0.25 %

Tools, Equipment, and Supplies Required

- Commercially available air meter consisting of a measuring bowl and a top section conforming to ASTM Method C 173–16
- Funnel with a spout of a size permitting it to be inserted through the neck of the top section of the air meter and long enough to extend to a point just above the bottom of the top section
- Standard 5/8 inch (16 mm) × 24 inch (600 mm) long smooth, round, steel tamping rod
- Flat, straight, steel striking bar at least 1/8 bx 3/4 × 12 inches (3 × 20 × 300 mm)
- Metal or plàstic measuring cup with a capacity of 1.00 +/– 0.04 % of the volume of the measuring bowl of the air meter, or being graduated in increments equal to that volume. Note that this cup is used to add water to the air meter only when the concrete air content exceeds 9 % or the calibrated range of the meter
- Graduated cup or beaker for measuring the isopropyl alcohol with a minimum capacity of at least 1 pint (500 mL) with graduations no larger than 4 oz (100 mL)
- Rubber syringe with a capacity of 2 oz (50 mL) or more
- Suitable container of approximately 1 qt (1 L) capacity for dispensing water to the air meter
- Scoop large enough to scoop representative aliquots of the sample concrete from the batch mix to the air meter and small enough to avoid spillage during that transfer process
- Supply of 70 % by volume isopropyl alcohol, commonly sold as "rubbing alcohol", (approximately 65 % by weight)
- Rubber or rawhide mallet of approximately 1.25 +/– 0.5 lb (600 +/– 200 g)

Safety

The mixing of concrete involves the use of cement, which is highly caustic. Cement will dry the skin with relatively short time exposures and cause immediate serious eye damage with exposure. Dust masks, gloves, and eye protection are required when conducting this experiment.

The air meter will become heavy during its use. It is necessary to hold the meter at an angle during the procedure. Care must be taken to avoid dropping the meter. If it were to fall on a toe, foot, or other extremity, serious damage to bone or skin tissue could result.

As with all laboratory experiments, eye protection, gloves, and a lab coat are recommended for use during the conduct of this experiment.

Planned Experimental Procedures and Steps, in Order

1. Put on appropriate personal protective equipment and gather all required tools and supplies in a convenient location.
2. Calibrate the air meter using the following procedure:
   a. Determine the volume of the measuring bowl to an accuracy of at least 0.1 % by measuring the mass of water required to fill the bowl and dividing that mass by the density of the water. See Table A.5 in the Appendix for the density of water at various temperatures.
   b. Determine the accuracy of the graduations on the neck of the top section of the air meter by filling the assembled measuring bowl and top section with water to the mark for the highest air content graduation.
   c. Add water in increments of 1 % of the volume of the measuring bowl to check that the accuracy of the increments throughout the range of air content does not exceed 0.1 % of air.
   d. Verify the volume and gradations of the calibrated cup by adding one or more calibrated cups of water to the assembled air meter and observing the increase in the height of the water column after filling to a given level.
3. Dampen the interior of the measuring bowl of the air meter and remove any standing water from the bottom.
4. Mix a fresh batch of concrete in accordance with a preselected mix design.
5. Using the scoop, fill the measuring bowl of the air meter with freshly mixed concrete in two layers of approximately equal volume. Move the scoop around the perimeter of the measuring bowl opening while placing the concrete in the measuring bowl to ensure an even distribution of the concrete and minimal segregation.
6. Rod each layer 25 times uniformly over the cross-section of the bowl with the rounded end of the tamping rod. Rod the bottom layer throughout its depth, taking care not to damage the bottom of the measuring bowl. For the upper layer, allow the rod to penetrate through the layer being rodded and into the layer below approximately 1 inch (25 mm). After each layer is rodded, tap the sides of the measuring bowl 10–15 times with the mallet to close any voids left by the tamping rod and to release any large bubbles of air that may have been trapped. After tapping the final layer, an excess of 1/8 inch (3 mm) or less of concrete remaining above the rimis acceptable. Add or remove a representative sample of fresh concrete, if necessary, to obtain the required amount of concrete in the bowl.
7. After rodding and tapping of the second layer, strike off the excess concrete with the strike-off bar until the surface is flush with the top of the measuring bowl.
8. Wipe the flange of the measuring bowl clean.

9. Wet the inside of the top section of the meter, including the gasket.
10. Attach the top section of the meter to the measuring bowl and insert the funnel.
11. Add at least 1 pint (0.5 L) of water through the funnel, followed by the selected amount of isopropyl alcohol. Record the amount of isopropyl alcohol added.

    Note that the amount of isopropyl alcohol necessary to obtain a stable reading, while generating a minimum amount of foam immediately above the water column in the meter, depends on a number of factors. Concrete made with less than 500 lb/yd^3 (300 kg/m^3) of cement and having air contents less than 4 % may require less than 0.5 pint (0.25 L) of alcohol. High-cement mixes made with silica fume and containing air contents of 6 % or more may require more than 3 pints (1.5 L) of isopropyl alcohol. It is recommended that 1 pint (0.5 L) be used initially, with adjustments made later, if needed.
12. Continue adding water until it appears in the graduated neck of the top section.
13. Remove the funnel and adjust the liquid level until the bottom of the meniscus is level with the zero mark on the graduated neck. (The rubber syringe is useful for this purpose.)
14. Attach and tighten the watertight cap on the meter.
15. Quickly invert the meter, shake the measuring bowl horizontally, and return the meter to the upright position.

    NOTE: To prevent the aggregate from lodging in the neck of the unit, do not keep it inverted for more than 5 seconds at a time.
16. Repeat the inversion and shaking process for a minimum of 45 seconds and until the concrete has broken free and the aggregate can be heard moving inside the meter as it is inverted.
17. Place one hand on the neck of the meter and the other on the flange. Using the hand on the neck of the meter, tilt the top of the meter at an approximately 45° (0.8 rad) angle from the vertical with the bottom edge of the measuring bowl resting on the floor or on the work surface. *Maintain this position while performing Steps 18, 19, and 20.*
18. Using the hand on the flange to rotate the meter, vigorously roll the meter 1/4 to 1/2 turn forward and back several times, quickly starting and stopping each roll.
19. Turn the measuring bowl about 1/3 turn and repeat Step 18.
20. Repeat Steps 18 and 19 for approximately 1 minute. (The aggregate must be heard sliding in the meter during this process.)

    NOTE: If at any time during the inversion and rolling procedures liquid is found to be leaking from the meter, the test is invalid and a new test must be started.
21. Set the unit upright and slowly loosen the top to allow any pressure to stabilize. Allow the meter to stand while the air rises to the top and the liquid level has stabilized. The liquid level is considered stable when it does not change more than 0.25 % air within a 2- minute period.

NOTE: If it takes more than 6 minutes for the liquid level to stabilize, or if there is more foam than that equivalent to 2 % air content divisions on the meter scale over the liquid level, discard the trial and start a new test. Use a larger addition of isopropyl alcohol than was used in the initial trial in this case.

22. If the liquid level is stable and there is no excessive foam present, read the bottom of the meniscus to the nearest 0.25 % and record this value as the *initial meter reading*.
23. If the air content is greater than the range of the meter, add a sufficient number of calibrated cups of water to bring the liquid level within the graduate range. Read the bottom of the meniscus to the nearest 0.25 % and record the number of calibrated cups of water to be added to the final meter reading.
24. Once an initial meter reading has been obtained, retighten the top and repeat the 1 minute rolling procedure from Steps 18, 19, and 20.
25. When the liquid level is stable again, make a direct reading to the bottom of the meniscus and estimate air content to 0.25 %. If this reading has not changed more than 0.25 % from the initial in Step 22, record it as the *final meter reading* of the sample.
26. If the reading has changed from the *initial meter reading* by more than 0.25 % air, record this reading as a new "*initial reading*" and repeat Steps 18, 19, and 20.
27. Read the new indicated air content. If this reading has not changed by more than 0.25 % air from the "*newest initial reading*", record it as the *final meter reading*.
28. If the reading has changed by more than 0.25 % from the "*newest initial reading*", discard the test and start a new test on a new sample of concrete using more alcohol.
29. Disassemble the air meter by detaching the top section from the measuring bowl. Allow the liquid to discharge from the air meter. Dump the contents of the measuring bowl.
30. Examine the interior of the measuring bowl to be sure that there are no portions of undisturbed, tightly packed concrete present. If portions of undisturbed concrete are found, the test is invalid and will need to be repeated.
31. Note that if more than 2.5 pints (1.25 L) of alcohol are used in the procedure, a correction to the final meter reading is required. Round the volume of alcohol used to the nearest 1 pint (0.5 L) and select the correction factor from Table 10.4.1. The correction factor is subtracted from the final meter

TABLE 10.4.1

Correction for the Effect of Isopropyl Alcohol on Air Meter Reading

Volume of 70 % Isopropyl Alcohol Used		
Pints	**Liters**	**Correction Factor**
2.0 or fewer	32 or fewer	0.0
3.0	48	0.25
4.0	64	0.50
5.0	80	0.75

reading. Note, too, that the correction values given are for air meters that have a measuring bowl volume of 0.075 ft^3 (2.1 L) and a top section that is 1.2 times the volume of the measuring bowl.

Data Collection Format and Data Collection Sheets Tailored to Experiment

See attached data collection sheet.

Data Analysis Plan

Calculate the air content of the concrete in the measuring bowl from the following equation and report the air content to the nearest 0.25 %:

$$A = A_R - C + W$$

where: A = air content, %
A_R = final meter reading, %
C = correction factor from Table 1, %
W = number of calibrated cups of water added to the meter in Step 23

Expected Outcomes

The expected outcomes will depend on the mix design used and the procedures followed here. Most "normal" concrete mixtures should have an air content in the 2–4 % range. However, a value outside this range, determined by this experiment procedures, does not indicate a failed experiment.

DATA COLLECTION SHEET AIR CONTENT OF FRESHLY MIXED CONCRETE BY THE VOLUMETRIC METHOD				
Lead Experimenter				
Date of Experiment				
	Sample 1	Sample 2	Sample 3	Sample 4
Volume of Measuring Bowl (Step 2a)				
Volume of Isopropyl Alcohol Used (Step 11)				
Initial Meter Reading (Step 22)				
Number of Calibrated Cups of Water Added (Step 23)				
Final Meter Reading (Step 25)				
New Initial Meter Reading #1 (Step 26)				
New Final Meter Reading (Step 27)				
Correction Factor from Table 10.4.1, If Used				
Calculated Air Content to Nearest 0.25 %				

EXPERIMENT 10.5

DENSITY (UNIT WEIGHT), YIELD, AND AIR CONTENT OF CONCRETE (GRAVIMETRIC METHOD)

(Consistent with ASTM Method C 138–17 a)

Date: ______________________________

Principal Investigator:

Name: ______________________________

Email: ______________________________

Phone: ______________________________

Collaborators:

Name: ______________________________

Email: ______________________________

Phone: ______________________________

Name: ______________________________

Email: ______________________________

Phone: ______________________________

Name: ______________________________

Email: ______________________________

Phone: ______________________________

Name: ______________________________

Email: ______________________________

Phone: ______________________________

Objective or Question to Be Addressed

This experiment provides methods for characterizing concrete. The methods are designed to answer the questions of how dense (mass per unit volume) the concrete is and how much air is contained in a sample of fresh concrete.

Theory behind Experiment

This experiment includes "*density yield*" and "*air content*". Yield in the context of this experiment is defined as the volume of concrete produced from a mixture of known mass quantities of the component materials, specifically, the mass of coarse aggregate, fine aggregate, cement, water, and any admixtures used. The quantities of the component materials are determined from the mix design. This experiment verifies whether the mixed concrete meets the design specifications.

The term "*density*" is the terminology for what used to be called the "*unit weight*" of the material. Both terms refer to mass per unit volume. The theoretical density is a laboratory calculation. The theoretical density value remains unchanged with every batch of concrete made from the same mix design, but the actual density of the mixed concrete is specific to the batch. The theoretical density and the actual batch density are probably not the same, since the masses measured are subject to imprecise values and the conditions under which the batches are made may alter the masses a bit as well.

Air content is affected by the mixing process used, the time after mixing that the test is conducted (the chemical reactions may impart entrained air) and the admixtures used. Certain admixtures are designed specifically to add or delete entrained air in the fresh concrete.

Potential Interferences and Interference Management Plan

This experiment relies on care when mixing the concrete batch to ensure adherence to the original mix design and attention to details as the experiment is conducted. Care in filling the measuring container with fresh concrete can affect all the calculations, particularly the air content. Take care to ensure adequate cleaning of the exterior of the full measuring container prior to weighing that container.

Data to Be Collected with an Explanation of How They Will Be Collected (Ex.: "Water Temperature Data for Three Days Using a Continuous Data Logger")

Using the attached data sheet, record at a minimum the following information:

- Name of all experimenters
- Date of the experiment
- Design mix details
- Mass and volume of cylindrical molds
- Volume of density to the nearest 0.001 ft^3 (0.01 L)
- Density (unit weight) to the nearest 0.1 lb/ft^3 (1.0 kg/m^3)
- Theoretical density to the nearest 0.1 lb/ft^3 (1.0 kg/m^3)

- Yield to the nearest 0.1 yd^3 (0.1 m^3)
- Relative yield to the nearest 0.01
- Cement content to the nearest 1.0 lb (0.5 kg)
- Air content to the nearest 0.1 %

Tools, Equipment, and Supplies Required

- Scale accurate to 0.1 lb (45 g) from 0 to 160 lb/ft^3 (2,600 kg/m^3), based on the maximum volume to be weighted at one time (typically about 3 ft^3 [0.1 m^3]).
- Standard smooth, straight, round, steel tamping rod, with 5/8 +/– 1/16 inches (16 mm +/– 2 mm) diameter with a rounded tamping end.
- Standard concrete vibrator with a frequency of at least 9,000 vibrations per minute [150 Hz] while operating in the concrete. The diameter of the vibrating element should be between 0.75 and 1.5 inches (19–38 mm).
- Set of cylindrical molds with a capacity based on the coarse aggregate size as indicated by Table 10.5.1. The rim of the measure must be flat and plane. A set of four cylinders per batch is suggested,
- Flat, rectangular, metal plate at least 1.4 inches (6 mm) thick or a glass or acrylic plate at least ½ inch (12 mm) thick with a length and width at least 2 inches (50 mm) greater than the diameter of the measuring container.
- Rubber or rawhide mallet 1.25 +/– 0.5 lb (600 +/– 200 g) if using a measuring container of 0.5 ft^3 (14 L) or smaller, or a similar 2.25 +/– 0.5 lb (1000 +/– 200 g) mallet if using a measure larger than 0.5 ft^3 (14 L).
- Metal scoop large enough to obtain a representative sample of the freshly mixed concrete and small enough to prevent spillage during transfer from the mixer to the cylinders.

Safety

This experiment requires the mixing of a fresh batch of concrete. The aggregates and the cement used are heavy, and the full cylindrical mold of fresh concrete will also be heavy. Care is needed in handling these items to minimize the risk of strains or injury.

Cement and cement dust are highly caustic and will cause drying of the skin from relatively short exposure. Eye contact will result in immediate pain and eye damage. Gloves and eye protection are required during this experiment.

Cement dust and aggregate dust will irritate the throat and lungs and could cause pulmonary distress. A dusk mask is required during this experiment to minimize that risk.

TABLE 10.5.1
Minimum Capacity of Cylindrical Mold Based on Nominal Coarse Aggregate Size

NOMINAL COARSE AGGREGATE SIZE		MINIMUM CYLINDER CAPACITY	
Inches	mm	ft^3	L
1	25	0.2	6
1.5	37.5	0.4	11
2	50	0.5	14
3	75	1.0	28

As with all lab experiments, eye protection, gloves, and a lab coat are recommended or required for this experiment; additionally, this experiment requires a dust mask.

Planned Experimental Procedures and Steps, in Order

1. Put on appropriate personal protective equipment and gather all necessary equipment and supplies in a convenient location.
2. Prepare a batch of fresh concrete in accordance with a desired mix design. Note and record the total mass of the components used in the mix, M.
3. Note and record the total volume of all the components used in the mix, V.
4. Note and record the actual volume of concrete per batch made from the design mix, Y_f.
5. Dampen the interior of the cylinders and remove any standing water from the bottom.
6. Determine the mass of the empty cylinders, M_m.
7. Measure or note and record the volume of the empty cylinders, V_m.
8. Place the cylinder on a flat, level, firm surface.
9. Conduct a slump test on the fresh concrete. See Experiment 10.3 for the procedure for conducting a slump test. Record the slump.
10. Place a representative sample of the concrete in each cylinder using the scoop. Move the scoop around the perimeter of the cylinder opening to ensure an even distribution of the concrete with minimal segregation.
11. Consolidate the sample inside the cylinder.
    NOTE: Selection of the method of consolidation, rodding, or vibration is based on the slump. Rod concrete with a slump greater than 3 inches (75 mm) and vibrate concrete with a slump less than 1 inch (25 mm). Concrete with a slump between 1 inch (25 mm) and 3 inches (75 mm) may be consolidated with either method.
    a. If rodding, place the concrete in the cylinder in three layers of approximately equal volume. Rod each layer with 25 strokes of the tamping rod with 0.5 ft^3 (14 L) or smaller cylinders, 50 strokes when 1 ft^3 (28 L) cylinders are used, and one stroke per 3 $inch^2$ (20 cm^2) of surface for larger cylinders.
    Rod each layer uniformly over the cross-section with the rounded end of the rod. Carefully rod the bottom layer throughout its depth. Allow the rod to penetrate through the layer being rodded and approximately 1 inch (25 mm) into the layer below. After each layer is rodded, tap the sides of the cylinder 10–15 times with the mallet, using sufficient force to close any voids left by the tamping rod and to release any large air bubbles that may have been trapped. Avoid overfilling the cylinders when adding the final layer.
    b. If vibrating, fill and vibrate the cylinders in two approximately equal layers. Place all the concrete for each layer in the cylinder before starting vibration of that layer. Insert the vibrator at three different points for each layer. In compacting the bottom layer, do not allow the vibrator to rest on or touch the bottom or sides of the cylinder. When compacting the final layer, the vibrator should penetrate into the underlying layer approximately 1 inch (25 mm). Ensure that no air pockets are left as the vibrator is withdrawn. Continue vibration only long enough to achieve proper consolidation of

the concrete. Over-vibration may cause segregation and loss of appreciable quantities of intentionally entrained air. Sufficient vibration has usually been applied when the surface of the concrete becomes relatively smooth.

12. Upon completion of the consolidation step, an excess of concrete should protrude approximately 1/8 inch (3 mm) above the top of the cylinder. A small quantity of concrete may be added to correct a deficiency. If there is a great excess of concrete, immediately remove a *representative portion* of the excess concrete with a trowel or scoop.
13. Strike off the top surface of the concrete and finish it smoothly using the flat strike-off plate so that the cylinder is level and full. When completed, the surface should be smooth and level with the top edge of the cylinder.
14. Clean all excess concrete from the exterior of the cylinder and determine the mass of the concrete and cylinder, M_c.

Data Collection Format and Data Collection Sheets Tailored to Experiment

See attached data collection sheet.

Data Analysis Plan

Based on the following definitions, calculate the parameters of this experiment as follows:

A = air content (percentage of voids) in the concrete
C = actual cement content, lb/yd^3 (kg/m^3)
C_b = mass of cement in the batch, lb (kg)
D = density (unit weight) of concrete, lb/ft^3 (kg/m^3)
M = total mass of all materials batched, lb (kg), from Step 2
M_c = mass of the cylinder filled with concrete, lb (kg), from Step 14
M_m = mass of the cylinder, lb (kg), from Step 6
R_y = relative yield
T = theoretical density of the concrete computed on an air-free basis, lb/ft^3 (kg/m^3)
Y = yield, volume of concrete produced per batch, yd^3 (m^3)
Y_d = volume of concrete that the batch was designed to produce, yd^3 (m^3)
Y_f = volume of concrete produced per batch, ft^3 (m^3), from Step 4
V = total absolute volume of the component ingredients in the batch, ft^3 (m^3), from Step 3
V_m = volume of the cylinder, ft^3 (m^3), from Step 7

Calculate the net mass of the concrete by subtracting the mass of the cylinder, M_m, from the mass of the cylinder filled with concrete, M_c. Calculate the density, D, lb/ft^3 (kg/m^3), by dividing the net mass of concrete by the volume of the cylinder, V_m, using the following equation:

$$D = (M_c - M_m)/V_m$$

Calculate the *theoretical* density, T, using the following equation:

$$T = M/V$$

Calculate the absolute volume of each ingredient in cubic meters (m^3) as the mass of the ingredient in kilograms (kg) divided by 1,000 times its *relative* density (specific gravity). Similarly, the absolute volume of each ingredient in cubic feet is equal to the mass of that ingredient divided by the product of its relative density times 62.4 lb/ft^3. For the aggregate components, the relative density and mass are based on the saturated, surface-dry condition. For normal portland cement, use a value of 3.15. The absolute volume of other cements will need to be determined from manufacturer's data or measurement in the lab.

Based on the calculated absolute volumes, calculate the yield, Y, by the following equation:

$$Y(\text{in yd}^3) = M/(D \times 27)$$

Or

$$Y(\text{in m}^3) = M/D$$

Relative yield, R_y, is the ratio of the actual volume of concrete obtained to the volume of concrete the batch was designed to generate, Y_d, calculated by the following equation, with normal units of ft^3/yd^3:

$$R_y = Y/Y_d$$

A value for R_y greater than 1.00 indicates an excess of concrete being produced relative to the design volume, whereas a value less than 1.00 indicates that less volume is produced than predicted by the mix design.

Calculate the actual cement content of the batch, C_b, using the following equation:

$$C = C_{b/}Y$$

Calculate the air content, A, or percentage of voids in the concrete, using the following equations:

$$A = [(T - D)/T] \times 100$$

Or

$$A = [(Y - V)/Y] \times 100 \text{ (if working with SI units)}$$

Or

$$A = [(Y_f - V)/Y_f] \times 100 \text{ (if working with inch–pound units)}$$

Where: Y_f is the actual volume of concrete produced per batch made from the mix design, ft^3 (m^3), see Step 4.

Expected Outcomes

The expected outcome from this experiment will vary significantly based on the original mix design.

DATA COLLECTION SHEET DENSITY (UNIT WEIGHT), YIELD, AND AIR CONTENT OF CONCRETE (GRAVIMETRIC METHOD)					
Lead Experimenter					
Date of Experiment					
	Cylinder 1	Cylinder 2	Cylinder 4	Cylinder 4	Cylinder 5
Total Mass of Design Mix Components, M (Step 2)					
Total Volume of All Components Used in Mix, V (Step 3)					
Actual Volume of Concrete from Design Mix Batch, Y_f (Step 4)					
Mass of Empty Measuring Container, M_m (Step 6)					
Volume of Empty Cylinders, V_m (Step 7)					
Slump of Concrete Sample (Step 9)					
Method of Consolidation of Sample in Mold					
Mass of Concrete Plus Cylinder					
Calculated Density of Concrete, D					
Calculated Theoretical Density of Concrete, T					
Calculated Yield Based on Absolute Volumes, Y					
Calculated Relative Yield, R_y					
Calculated Actual Cement Content, C_b					
Calculated Air Content, A					

EXPERIMENT 10.6

TIME OF SETTING OF CONCRETE MIXTURES BY PENETRATION RESISTANCE

(Consistent with ASTM Method C 403–16)

Date: ______________________________

Principal Investigator:

Name: ______________________________

Email: ______________________________

Phone: ______________________________

Collaborators:

Name: ______________________________

Email: ______________________________

Phone: ______________________________

Name: ______________________________

Email: ______________________________

Phone: ______________________________

Name: ______________________________

Email: ______________________________

Phone: ______________________________

Name: ______________________________

Email: ______________________________

Phone: ______________________________

Objective or Question to Be Addressed

This experiment provides a means for determining the time of setting of concrete. The test involves sieving the mortar from a sample of concrete with a slump greater than zero and using penetration resistance measurements to determine the setting time for the concrete.

Theory behind Experiment

Concrete setting is a function of the mortar that holds the concrete together. By measuring the time it takes for the mortar to set, the time it will take for the concrete to set can be inferred. A sample of the mortar is obtained by sieving a sample of the fresh concrete. The mortar is placed in a suitable container and stored at a specified ambient temperature. At regular time intervals, the resistance of the mortar to penetration by standard needles is measured. A plot of penetration resistance versus elapsed time is developed from which the initial and final setting times are calculated.

Since this test uses the mortar, and not the entire sample of concrete, it will also work for other mortar-like substances, such as non-concrete mortars and grouts, and it is generally applicable to both laboratory analysis and field analysis.

Potential Interferences and Interference Management Plan

The setting of concrete is a gradual process. In this experiment, the times required for the mortar to reach specified values of resistance to penetration are used to define the time for setting. By adjustment of the variables in a concrete mix, the time of setting under each set of conditions can be estimated. The water content, the type of cementitious material used, the relative amounts of cement and admixtures, and so forth all have an effect on the setting time of the concrete. This experiment can be used to examine each of those effects together and independently. Note, however, that changing one factor at a time is essential for the effective evaluation of the impacts of specific variable adjustments.

While this experiment is suitable for prepared mortars, there is evidence to suggest that using a prepared mortar increases the calculated setting time compared to that of mortar sieved from a sample of fresh concrete.

Outliers may occur during the testing because of larger particles in the mortar, the presence of large voids within the penetration zone, interferences from the impressions created by adjacent penetrations, failure to maintain the instrument perpendicular to the test surface during penetration, errors in reading the load, variations in the penetration depths, or variations in rate of loading. The experimenter must exercise reasonable judgment to identify those points that should not be included in the data analysis.

Data to Be Collected with an Explanation of How They Will Be Collected (Ex.: "Water Temperature Data for Three Days Using a Continuous Data Logger")

Using the attached data sheet, record at a minimum the following information:

- Names of all experimenters
- Date of the experiment

- The following information on the concrete mixture:
  - Brand and type of cementitious materials
  - Amounts (by mass) of cementitious materials used, including fine aggregate and coarse aggregate per cubic yard (per cubic meter) of concrete
  - Nominal maximum aggregate size
  - Water-cement ratio
  - The name, type, and amount of any admixture(s) used
  - The air content of the fresh concrete and method used to determine that
  - Slump of concrete
  - Temperature of mortar after sieving
- Record of ambient temperature during the test period
- Report the following information on the time of setting tests:
  - Plot of penetration resistance versus elapsed time for each time of setting test
  - Times of initial and final setting for each test, reported in hours and minutes to the nearest minute
  - Calculated times of initial and final setting reported in hours and minutes to the nearest 5 minute

Tools, Equipment, and Supplies Required

- Rigid, watertight, and non-absorptive containers
  - Use for holding and testing the setting mortar samples. These containers should be oil and grease free and cylindrical or square in area for ease of measurement. Up to ten separate readings of the penetration will be needed, so the surface area needs to be sufficient to allow those to occur without overlap (see Procedure section for spacing of tests) or other disturbance. In general, the lateral dimensions will need to be at least 6 inches (150 mm) and the height at least 6 inches (150 mm).
- Penetration needles
  - Needles need to be compatible with the loading apparatus being used and provide the following bearing areas: 1, 1/2, 1/4, 1/10, 1/20, and 1/40 inches^2 (645, 323, 161, 65, 32, and 16 mm^2). Each needle shank needs to be marked at a distance 1 inch (25 mm) from the bearing area. The length of the 1/40 inches^2 (16 mm^2) needle should be no longer than 3.5 inches (90 mm).
- Loading device to measure the force required to cause penetration of the needles
  - This device, called a *penetrometer*, should be capable of measuring the required penetration force with an accuracy of 62 lbf [10 N] and have a capacity of at least 130 lbf [600 N]. Such devices are commercially available with a variety of measuring mechanisms.
- Standard 5/8 inch (16 mm) tamping rod
  - The rod should be a round, straight, steel rod with at least the tamping end rounded to a hemispherical tip.

- Pipet or other device
  - Suitable for drawing bleed water from the surface of the test sample.
- Thermometer capable of measuring the temperature of the fresh mortar to 60.5 °C (61 °F)

Safety

Concrete mixing involves dusty materials that can cause difficulty breathing and, ultimately, respiratory diseases. Face masks that prevent the inhalation of dust particles are required when preparing and sieving the concrete mixture.

The prepared mortar containers will be heavy, and care is needed when moving or handling them to avoid strains and slipping injuries.

Cement and wet concrete are caustic and will cause drying of the skin and severe eye damage over short exposure times. The use of gloves and eye protection while mixing and handling concrete and mortar is required to avoid this issue.

The penetrometer device is capable of extreme pressures. Care is needed to avoid placing fingers, hands, and or other extremities between the penetrometer needles and the samples.

As with all laboratory experiments, the use of gloves, eye protection, and a lab coat is required for this experiment. A suitable dust mask is also required for this experiment.

Planned Experimental Procedures and Steps, in Order

1. Put on appropriate personal protective equipment and gather the required tools and supplies together in a convenient location.
2. Prepare three samples of concrete for each set of variables being tested. All samples must be from the same batch of concrete.
3. Record the time at which *initial contact* was made between cement and mixing water.
4. Determine and record the slump of the concrete batch. See Experiments 10.2 and 10.3 for slump measurement methods.
5. Determine the air content of the concrete batch. See Experiment 10.4 for a suitable method by which to do that.
6. Wet sieve enough of the concrete not used in the slump and air entrainment tests to fill the test container or containers to a depth of at least 5.5 inches (140 mm) using a 4.75 mm (No. 4) sieve. Experiment 10.4 provides a method for doing that. Place the sieved mortar on a clean, non-absorptive surface.
7. Thoroughly remix the mortar by hand methods on the non-absorptive surface.
8. Measure and record the temperature of the mortar.
9. Place the mortar in the container, or containers, using a single layer.
10. Consolidate the mortar to eliminate air pockets in the sample and level the top surface by rocking the container back and forth, by tapping the sides of the container with the tamping rod, or by rodding the mortar. If rodding the

mortar to level it, use the hemispherical end of the tamping rod and rod the mortar once for each 1 inch^2 (645 mm^2) of top surface area of the container and spacing the strokes uniformly over the surface.

11. Tap the sides of the containers lightly with the tamping rod to close any voids left by the rodding and to finish leveling the surface. Upon completion of this step, the mortar surface should be at least 0.5 inch (10 mm) below the top edge of the container to provide space for the collection and removal of bleed water and to avoid contact between the mortar surface and a protective cover (see Step 14).
12. Store the leveled samples at a temperature between 68 and 77 °F (20–25 °C).
13. Measure and record the ambient air temperature at the start and finish of the test.
14. Cover the samples with a damp burlap or a tight-fitting, water-impermeable cover for the duration of the test, except when bleed water is being removed or penetration tests are being made.
15. Make one penetration test approximately 3–4 hours after the samples are prepared and ready for covering. Cover the sample prior to testing.
    a. Just prior to making a penetration test, remove bleed water from the surface of the samples by means of the pipet or other suitable device. (To facilitate collection of bleed water, the sample may be carefully tilted to an angle of about 10° from the horizontal by placing a block under one side 2 minutes prior to removal of the water.)
    b. Insert a needle of appropriate size, depending upon the degree of setting of the mortar, into the penetrometer and bring the bearing surface of the needle just into contact with the mortar surface.
    c. Gradually and uniformly apply a vertical force downward on the apparatus until the needle penetrates the mortar to a depth of 1 +/– 1/16 inch (25 +/– 2 mm), as indicated by the scribe mark. (The time required to penetrate to the 1 inch (25 mm) depth should be 10 +/– 2 seconds).
    d. Record the force required to produce the 1 inch (25 mm) penetration.
    e. Record the time since initial contact of cement and water.
    f. Calculate the penetration resistance by dividing the recorded force by the bearing area of the needle and record the penetration resistance.

    NOTE: In subsequent penetration tests, take care to avoid areas where the mortar has been disturbed by previous tests. The clear distance between test sites should be at least two diameters of the needle being used, but not less than 1/2 inch (15 mm), and no closer to the edge of the container than 1 inch (25 mm), but no further than 2 inches (50 mm).
16. Repeat Step 15 every ½ to 1 hour until a pattern has emerged from which a suitable curve may be drawn (see data analysis section). A total of about eight tests may reasonably be made in a 6 inch (150 mm) diameter cylinder, one or two more in a square container.
    a. For concrete mixtures containing accelerators, or for samples stored at higher than room temperature, make the initial test after an elapsed time of 1 to 2 hours and subsequent tests at half hour intervals. For concrete mixtures containing retardants, or for samples stored at tem-

peratures below room temperature, the initial test may be delayed for up to 4 to 6 hours. In all cases, time intervals between subsequent tests should be adjusted as necessary to obtain the required number of penetrations.

b. Make at least six penetrations for each time-of-setting test
   Test with time intervals of such duration as to provide a satisfactory curve of penetration resistance versus elapsed time.
c. Continue testing until at least one penetration resistance reading equals or exceeds 4,000 psi (27.6 MPa).
d. A satisfactory curve represents the overall development of penetration resistance and includes points before and after the initial and final setting times. Tests should be done at equally spaced time intervals. Premature penetration testing will result in too many data points earlier than the initial setting time.

Data Collection Format and Data Collection Sheets Tailored to Experiment

See attached data collection sheet.

Data Analysis Plan

Plotting of penetration resistance versus elapsed time is intended to provide information on the rate of setting. It is recommended that the data be plotted as they are being accumulated since they may be used to select the time for subsequent penetration tests and can assist in identifying inaccurate test results.

A graph will be prepared showing penetration resistance as a function of elapsed time since initial contact of water and cement by plotting penetration resistance on the y-axis and elapsed time on the x-axis. Use a scale that allows 500 psi (3.5 MPa) and 1 hour to be plotted at a distance of at least 1/2 inch (15 mm). The limits of penetration resistance should extend from 10 psi (0.1 MPa) to 10,000 psi (100 MPa), and the limits of elapsed time should extend from 10 to 1,000 minutes.

Calculate (or add a trend line to your plot) a line of best fit. Convert the data by taking their logarithms. This will convert the data to a straight, linear, regression line. Disregard obvious outliers.

For each plot, determine the times of initial setting time as the time when the penetration resistance equals 500 psi (3.5 MPa) and the final setting time as the time when the resistance equals 4,000 psi (27.6 MPa), respectively. Record those setting times in hours and minutes to the nearest 5 minute.

Expected Outcomes

The expected outcomes will vary depending upon the variables of the concrete mix from which the mortar is sieved.

DATA COLLECTION SHEET TIME OF SETTING OF CONCRETE MIXTURES BY PENETRATION RESISTANCE						
Lead Experimenter						
Date of Experiment						
Brand and Type of Cementitious Materials						
Amount of Cementitious Material Used						
Amount of Fine Aggregate Used						
Type of Fine Aggregate Used						
Amount of Coarse Aggregate Used						
Amount and Type of Admixtures Used						
Time of Initial Contact between Cement and Water (Step 3)						
Slump of Concrete (Step 4)						
Air Content of Fresh Concrete (Step 5)						
Temperature of Mortar after Sieving (Step 8)						
Ambient Air Temperature at Start of Test (Step 13)						
Ambient Air Temperature at End of Test (Step 13)						
Time Concrete Set in Mold						
	Probe 1	Probe 2	Probe 3	Probe 4	Probe 5	Probe 6
Time of First Penetration Test						
Force Required to Produce 1 inch (25 mm) Penetration						
Calculated Penetration Resistance						
Time of Second Penetration Test						
Force Required to Produce 1 inch (25 mm) Penetration						
Calculated Penetration Resistance						

DATA COLLECTION SHEET TIME OF SETTING OF CONCRETE MIXTURES BY PENETRATION RESISTANCE						
Time of Third Penetration Test						
Force Required to Produce 1 inch (25 mm) Penetration						
Calculated Penetration Resistance						
Time of Fourth Penetration Test						
Force Required to Produce 1 inch (25 mm) Penetration						
Calculated Penetration Resistance						
Time of Fifth Penetration Test						
Force Required to Produce 1 inch (25 mm) Penetration						
Calculated Penetration Resistance						
Time of Sixth Penetration Test						
Force Required to Produce 1 inch (25 mm) Penetration						
Calculated Penetration Resistance						
Time of Seventh Penetration Test						
Force Required to Produce 1 inch (25 mm) Penetration						
Calculated Penetration Resistance						
Time of Eighth Penetration Test						
Force Required to Produce 1 inch (25 mm) Penetration						
Calculated Penetration Resistance						
Time of Ninth Penetration Test						
Force Required to Produce 1 inch (25 mm) Penetration						
Calculated Penetration Resistance						
Time of Tenth Penetration Test						
Force Required to Produce 1 inch (25 mm) Penetration						
Calculated Penetration Resistance						

EXPERIMENT 10.7

EFFECTS OF WATER CONTENT ON CONCRETE STRENGTH AND WORKABILITY

Date: ______________________________

Principal Investigator:

Name: ______________________________

Email: ______________________________

Phone: ______________________________

Collaborators:

Name: ______________________________

Email: ______________________________

Phone: ______________________________

Name: ______________________________

Email: ______________________________

Phone: ______________________________

Name: ______________________________

Email: ______________________________

Phone: ______________________________

Name: ______________________________

Email: ______________________________

Phone: ______________________________

Objective or Question to Be Addressed

This experiment is designed to assess the ultimate strength of concrete by demonstrating the effects of water content and water–cement ratio (W/C).

Theory behind Experiment

Concrete is a mixture of cement, coarse aggregate, fine aggregate, and water. In addition, a variety of admixtures may be included in the design. The most significant factor affecting the ultimate strength of the concrete is the amount of water added to the mix and, in particular, the amount of water in relation to the amount of cement used; the *water–cement ratio*, or W/C ratio. It is the addition of the water that initiates the chemical process that gives concrete strength.

The general, the W/C ratio is going to be in the range of 40–60 %, by weight. Thus, the amount of water to be expected is going to be the selected ratio, say 50 %, for example, times the mass of the cement. If 50 kg (110 lb) of cement is used in a particular mix, then the water required would be 0.5 × 50 kg = 25 kg (55 lb) of water. That conveniently converts to 25 L (6.6 gal) of water. Note that the W/C ratio is generally based on what is known as "*free water*". This is water that is available in the mix for reacting with the cement, as opposed to water bound up in aggregates, for example, that is not available for chemical reactions.

Adding water at a W/C ratio of about 50 % by weight, provides a volume of water at the 23–25 % of the volume of cement necessary to start that chemical reaction. If more water is added beyond that volume, the fresh concrete can become so liquefied that the aggregates do not stay in place as the concrete is poured and the strength will no longer be uniform throughout the concrete mass. Adding only the minimum volume of water, or less than that minimum, will allow the concrete aggregates to remain in place but will also make the concrete so stiff that it may not be possible for the fresh concrete to flow into all the nooks and crannies of the forms or to fill all the voids in the concrete even with proper vibration of the mix. That means that the *workability* of the concrete is decreased, usually to the detriment of the final strength. To overcome these issues, *plasticizers* and *superplasticizers* are used as admixtures. They can provide good workability while maintaining the aggregates in place in the mix by not affecting the W/C ratio and thereby maintaining the ultimate strength of the mix.

Note that a W/C ratio is not a calculated value. It is generally a selected value based on the desired strength of the concrete needed for the ultimate use of the mix, the temperatures at which it is poured, and the exposure conditions to which it will ultimately be subjected. See Table 10.7.1 for examples of typical minimum cement content, maximum W/C ratio, and minimum grade of concrete for various exposure conditions. This table is based on "normal" weight aggregates with a nominal maximum size of about 0.75 inches (20 mm).

It can be seen from Table 10.7.1 that the W/C ratio varies from 0.4 to 0.6 depending on exposure conditions and whether the concrete is reinforced. The amount of cement being added comes from the mix design. The amount of water to add is based on the amount of cement being added, based on the following equations:

$$\text{Required volume of water} = \text{W/C ratio} \times \text{cement volume}$$

Or:

$$\text{Required mass of water} = \text{W/C ratio} \times \text{cement mass}$$

Potential Interferences and Interference Management Plan

The effect of water on the strength of concrete is a function of many factors. The key to the determinations from this experiment is to ensure that the water is clean. Any significant quantities of dissolved or suspended solids can dramatically alter the results of the experiment.

Data to Be Collected with an Explanation of How They Will Be Collected (Ex.: "Water Temperature Data for Three Days Using a Continuous Data Logger")

The data to be collected during this experiment will be the W/C ratio, time from mixing to cylinder testing, and compressive strength of four sets of four cylinders. The W/C ratios will be selected in advance and the water content of the mix adjusted to achieve the selected W/C ratios. The compressive strength will be determined through the use of a standard concrete compression testing machine.

Using the attached data sheet, record at a minimum the following information:

- Name of all experimenters
- Date of the test
- Date the batch was mixed and adjusted
- W/C ratio and strength test results for each cylinder
- Graph of the test data
- Evaluation of the results of the testing and an indication of the expected effects of W/C ratio on the ultimate strength of the concrete

Tools, Equipment, and Supplies Required

- 16 standard 6 inch (152 mm) concrete test cylinders
- Tamping rod for use in filling the cylinders
- Mix design for at least 1 cy (0.75 m^3) of concrete
- Appropriate quantities of cement, coarse aggregate, fine aggregate, and clean water
- 1 yard (0.75 m^3) concrete mixer
- Means for curing the test cylinders under identical environmental conditions for up to 28 days
- Shovels and tools for placing the mix components into the mixer and removing the mixed concrete to the test cylinders

Safety

Concrete cylinders are surprisingly heavy to those who have not yet lifted one. Care is needed when moving the cylinders from the preparation area into the curing environment.

Cement is a caustic substance that will cause serious irritation of eyes and damage to skin from exposure. Eye protection and gloves are required for this experiment.

In the event of eye exposure, flush the eye immediately with copious amounts of water for up to 15 minutes and seek medical evaluation for damage control. In the event of inhalation or ingestion of dust, seek medical assistance for counter measures.

As with all laboratory experiments, gloves, eye protection, a dust mask, and a lab coat are recommended for use during this experiment.

Planned Experimental Procedures and Steps, in Order

1. Prepare a concrete mix design for a convenient ultimate strength. A 2,000 lb (907 kg) mix is suggested. Based on that design, calculate the desired water volume to be added.
2. Put on appropriate personal protective equipment and gather all necessary tools and supplies in a convenient location.
3. Prepare 16 standard 6 inch (152 mm) concrete test cylinders for filling. Mark each cylinder as "–5 %, 7 day", "–5 %, 14 day", –5 %, 21 day, –5 %, 28 day", "design mix, 7 day", design mix 14 day", "design mix 21 day", design mix 28 day", and so forth for +10 % and for +15 %.
4. Prepare the desired mix based on the mix design, *but use 5 % less water than the mix design calls for.*
5. When the mix is properly prepared, fill the first four concrete test cylinders, those marked as "–5 %", with the mix, properly rodding and leveling the cylinders, and place them in a curing environment.
6. Immediately add more water to the mix to bring the water content to the original design mix W/C ratio and remix the batch thoroughly.

   NOTE: Removing enough mix to fill the four cylinders will remove 0.8 ft^3 (0.2 m^3) of material from the original mix volume. The amount of water added back into the mix will need to be adjusted with each addition to maintain the desired W/C ratio accurately.
7. Prepare four more cylinders with this mix, using the "design mix" cylinders, properly rod and level those cylinders and place them in a curing environment.
8. Immediately add more water to the mix to bring the water content to 5 % greater than the original design mix W/C ratio and remix the batch thoroughly.
9. Prepare four more cylinders with this mix, using the "+5 %" cylinders; properly rod and level those cylinders and place them in a curing environment.
10. Immediately add more water to the mix to bring the water content to 10 % greater than the original design mix W/C ratio and remix the batch thoroughly.
11. Prepare four more cylinders with this mix, using the "+10 %" cylinders; properly rod and level those cylinders and place them in a curing environment.
12. At the end of 7 days, 14 days, 21 days, and 28 days after the cylinders are cast, break one cylinder from each set in a standard compression machine,

noting and recording the W/C ratio and compressive strength of each cylinder. See Experiment 10.1 for the procedure for using a compression machine.

13. After all the cylinders have been tested in the compression machine, plot the compressive strength of each set of cylinders on the same chart (four lines) and compare the compressive strength of the mix based on the changing W/C ratio.

Data Collection Format and Data Collection Sheets Tailored to Experiment

See attached data collection sheet.

Data Analysis Plan

The compressive strength of the concrete at 7 days, 14 days, 21 days, and 28 days will be plotted on a graph of compressive strength on the y-axis and time on the x-axis for each set of cylinders. The results will be compared to evaluate the effect of changing the W/C ratio in an otherwise constant mix design.

Expected Outcomes

The outcomes expected from this experiment are that the ultimate strength of the concrete will change with each change in W/C ratio and that the line for the actual design mix will be the strongest.

TABLE 10.7.1
Maximum Cement Content and Maximum Free Water W/C Ratio for Non-Special Purpose Concrete

	Unreinforced Concrete		Reinforced Concrete	
Exposure Conditions	**Maximum Cement Content lb/yd^3 (kg/m^3)**	**Maximum Water/ Cement Ratio**	**Maximum Cement Content lb/yd^3 (kg/m^3)**	**Maximum Water/ Cement Ratio**
Mild	370.8 (220)	0.60	505.7 (300)	0.55
Moderate	404.5 (240)	0.60	505.7 (300)	0.50
Severe	421.4 (250)	0.50	539.4 (320)	0.45
Very Severe	438.2 (260)	0.45	573.1 (340)	0.45
Extreme	472.0 (280)	0.40	606.8 (360)	0.40

Note: Assumes "normal" weight aggregate with maximum size of ¾ inch (20 mm)

DATA COLLECTION SHEET EFFECTS OF WATER CONTENT ON CONCRETE STRENGTH AND WORKABILITY	
Lead Experimenter	
Date of Experiment	
Original Mix Design W/C Ratio, W_o	
W/C Ratio for First Trial (Step 4), W_1	
W/C Ratio for Second Trial (Step 7), W_o	
W/C Ratio for Third Trial (Step 9), W_3	
W/C Ratio for Fourth Trial, W_4	
7-day Compressive Strength, W_o	
7-day Compressive Strength, W_2	
7-day Compressive Strength, W_3	
7-day Compressive Strength, W_4	
14-day Compressive Strength, W_o	
14-day Compressive Strength, W_2	
14-day Compressive Strength, W_3	
14-day Compressive Strength, W_4	
21-day Compressive Strength, W_o	
21-day Compressive Strength, W_2	
21-day Compressive Strength, W_3	
21-day Compressive Strength, W_4	
28-day Compressive Strength, W_o	
28-day Compressive Strength, W_2	
28-day Compressive Strength, W_3	
28-day Compressive Strength, W_4	

EXPERIMENT 10.8

SPECIFIC GRAVITY AND ABSORPTION OF COARSE AGGREGATE

(Consistent with ASTM Method C 127–15)

Date: ______________________________

Principal Investigator:

Name: ______________________________

Email: ______________________________

Phone: ______________________________

Collaborators:

Name: ______________________________

Email: ______________________________

Phone: ______________________________

Name: ______________________________

Email: ______________________________

Phone: ______________________________

Name: ______________________________

Email: ______________________________

Phone: ______________________________

Name: ______________________________

Email: ______________________________

Phone: ______________________________

Objective or Question to Be Addressed

This experiment is designed to determine the specific gravity and absorption of coarse aggregates. Specific gravity (also referred to as relative density) is a dimensionless quantity expressed as oven-dry (OD), saturated-surface-dry (SSD), or as apparent specific gravity. The OD specific gravity is determined after drying the aggregate. The SSD specific gravity and absorption are determined after soaking the aggregate in water for 24 hours.

This experiment is not intended to be used with lightweight aggregates.

Theory behind Experiment

For this experiment, a representative sample of the aggregate is immersed in water for 24 +/– 4 hours to fill the pores in the aggregates. It is then removed from the water and the surfaces of the particles are dried. The mass is determined immediately after surface drying, while the pores are presumed to still be saturated. The volume of the sample is calculated from the volume of water displaced when the sample is submerged in a calibrated container. The sample is oven-dried and the mass determined again. The specific gravity and absorption are calculated from the mass values.

Specific gravity is defined as the ratio of mass of an aggregate to the mass of a volume of water equal to the volume of the aggregate particles (also referred to as the *absolute volume* of the aggregate). It is expressed as the ratio of the density of the aggregate particles to the density of water. Distinction is made between the density of aggregate particles and the bulk density of aggregates (see Experiment 11.10). Bulk density includes the volume of voids between the particles of aggregates; relative density (specific gravity) does not.

Specific gravity is used to calculate the volume occupied by the aggregate in various mixtures containing aggregate, including hydraulic cement concrete, bituminous concrete, and other mixtures that are proportioned or analyzed on an absolute volume basis. SSD specific gravity is used if the aggregate is in a saturated-surface-dry condition, that is, if its absorption has been satisfied. The OD specific gravity is used for computations when the aggregate is dry or assumed to be dry. Apparent specific gravity pertains to the solid material making up the aggregate particles, but not including the pore space within the particles that is accessible to water.

Absorption values are used to calculate the change in the mass of an aggregate due to water absorbed in the pore spaces within the constituent particles, compared to the dry condition, when it is deemed that the aggregate has been in contact with water long enough to satisfy most of the absorption potential.

Potential Interferences and Interference Management Plan

The laboratory standard for absorption is that obtained after submerging dry aggregate for a prescribed period of time, typically 24 +/– 4 hours. Aggregates mined from below the water table commonly have a moisture content greater than the absorption determined by this test method. Aggregates that have not been continuously maintained in a moist condition until used are likely to contain an amount of absorbed moisture less than the 24 hour soaked condition.

TABLE 10.8.1
Minimum and Maximum Mass of Sample to be Used

Nominal Maximum Size, mm (inch)	Minimum Mass of Test Sample, kg (lb)
12.5 (1/2) or less	2 (4.4)
19.0 (3/4)	3 (6.6)
25.0 (1)	4 (8.8)
37.5 (11/2)	5 (11)
50 (2)	8 (18)
63 (21/2)	12 (26)
75 (3)	18 (40)
90 (31/2)	25 (55)

Adapted from ASTM Method C 127–15

For this experiment, the aggregate sample used is a representative composite sample (see Chapter 3) that has been thoroughly mixed and reduced to the approximate quantity needed. In addition, all material passing a 4.75 mm (No. 4) sieve by dry sieving should be rejected and the remainder washed to remove dust or other coatings from the surface. If the coarse aggregate contains a substantial quantity of material finer than the 4.75 mm sieve, the 2.36 mm (No. 8) sieve is used in place of the No. 4 sieve. Alternatively, the material finer than the 4.75 mm sieve may be removed from the sample and tested according to Experiment 10.9.

The minimum and maximum mass of sample to be used is given in Table 10.1. Coarse aggregate may be separated and tested in several size fractions if desired or necessary. Due to the difficulties of handling larger and heavier quantities, it is recommended that sample for this experiment be selected that do not contain fractions larger the 37.5 mm (11/2 inch) sieve. If the sample contains more than 15 % retained on the 37.5 mm (11/2 inch) sieve, the larger fraction should be tested in one or more size fractions separately from the smaller size fractions using the minimum mass of sample for each fraction as determined by the difference between the masses prescribed for the maximum and minimum sizes of that fraction, rather than the mass of the entire sample. For example, a sample containing fractions up to a 75 mm (3 inch) size and separated from the fractions smaller than a 37.5 mm sieve, the larger fraction would use a sample with a minimum mass of 18 kg – 5 kg = 13 kg of sample.

Data to Be Collected with an Explanation of How They Will Be Collected (Ex.: "Water Temperature Data for Three Days Using a Continuous Data Logger")

Using the attached data sheet, record at a minimum the following information:

- Names of all experimenters
- Date of experiment and date of sample collection
- Identification of the type of aggregate (granite, schist, etc.) and the source of the sample (crushed rock, river run, etc.)

- Visual classification of the soil being tested (group name and symbol in accordance with the Uniform Soil Classification system)
- Maximum nominal size of the aggregate
- If any soil or material was excluded from the test specimen, describe the excluded material
- All mass measurements to the nearest 0.05 g by weighing on the balance
- Test temperature to the nearest 0.1 °C, if used
- Report each specific gravity value (OD, SSD, and apparent) from this experiment to the nearest 0.01 and indicate clearly which specific gravity value is reported
- Report absorption values to the nearest 0.1 %

Tools, Equipment, and Supplies Required

- Balance
    - A balance equipped with a suitable apparatus for suspending a sample container in water from the center of the platform or pan of the balance and that is sensitive, readable, and accurate to 0.05 % of the sample mass at any point within the range used for the experiment, or 0.5 g, whichever is greater.
- Wire basket
    - A wire basket of 3.35 mm (No. 6) or finer mesh, or a bucket of approximately equal breadth and height, with a capacity of 4–7 L for 37.5 mm (11/2 inch) nominal maximum size aggregate or smaller
    - An appropriately larger container will be needed for testing larger maximum size aggregate. The container shall be constructed so as to prevent trapping air when the container is submerged
- Watertight tank into which the sample container can be placed while suspended below the balance
- Oven of sufficient size, capable of maintaining a uniform temperature of 110 +/– 5 °C (230 +/– 9 °F)

Safety

Aggregate samples tend to be heavy, and the larger the nominal maximum size, the heavier a given volume of material will be. Care in lifting and handling these materials requires consideration of bending and twisting that can cause muscle strains and back pain.

The volume of water needed to submerge a sample is significant and heavy. Care is required when moving the water container to avoid back strain and to avoid spillage.

The drying oven used in this experiment will operate at a temperature of 110 +/– 5 °C (230 +/– 9 °F). Burns are a significant risk with this equipment. Heat-resistant gloves should be used when putting samples into the oven and when removing them.

In the event of a burn, immediately run the affected area under cold water and seek medical attention for swelling, blisters, or breaking of the skin.

As with all experiments, it is appropriate to wear gloves, eye protection, and a lab coat during this experiment. Most aggregate contains significant quantities of dust that may be released during handling. Eye protection and a dust mask are recommended during the sample preparation procedures.

Planned Experimental Procedures and Steps, in Order

1. Put on appropriate personal protective equipment and gather all needed equipment and supplies in a convenient location.
2. Collect a suitable sample of appropriate size in a container suitable for oven drying.
3. Dry the test sample in the oven to constant mass at a temperature of 110 +/– 5 °C. Three samples collected from the test mass at intervals of 30 minutes that show a change in mass of less than 5 % between readings indicates a suitably dry condition. Record the final mass, A_1.
4. Remove the sample from the oven and cool it in air at room temperature. Cool for 1 to 3 hours for test samples of 37.5 mm (11/2 inch) nominal maximum size or less; longer for larger sizes. The sample is considered cool when the sample may be comfortably handled without heat-resistant gloves (approximately 50 °C [122 °F] or less).
5. Immerse the aggregate in water at room temperature for a period of 24 +/– 4 hours.
6. Remove the test sample from the water and roll it in a large absorbent cloth until all visible films of water are removed from the surfaces of the aggregate pieces. Wipe the larger particles individually. A moving stream of *cool* air is permitted to assist in the drying operation to avoid evaporation of water from aggregate pores during the surface-drying operation.
7. When the surface drying is complete, immediately determine the mass of the test sample in the saturated surface-dry condition.
8. Record this, B, and all subsequent masses to the nearest 0.5 g or 0.05 % of the sample mass, whichever is greater.
9. Immediately after determining the mass in air, place the saturated-surface-dry test sample in the sample container and determine its apparent mass, B_2, in water at 23 +/– 2 °C (73 +/– 3 °F). Shake the container while immersed and before reading the scale to ensure the removal of all air bubbles in the mass.
10. Record the difference between the mass in air and the mass when submerged (B_2 – B) as the mass of water displaced by the sample, C.
11. Remove the sample from the water and dry it in the oven a second time to a constant mass at a temperature of 110 +/– 5 °C (230 +/– 5 °F).
12. Cool in air at room temperature for 1–3 hours, or until the aggregate has cooled to a temperature that is comfortable to handle (approximately 50 °C).
13. Determine and record the mass, A.

Data Collection Format and Data Collection Sheets Tailored to Experiment

See attached data collection sheet.

Data Analysis Plan

Calculate and report the oven-dried specific gravity (OD) by the following equation:

$$\text{Specific gravity (OD)} = A/(B - C)$$

where: A = mass of oven-dry test sample in air, g
B = mass of saturated-surface-dry test sample in air, g
C = apparent mass of saturated test sample in water, g

Calculate and report the saturated-surface-dry specific gravity (SSD) from the following equation:

$$\text{Specific gravity (SSD)} = B/(B - C)$$

Calculate and report the apparent specific gravity using the following equation:

$$\text{Apparent specific gravity} = A/(A - C)$$

Calculate the average specific gravity if the sample is tested in separate size fractions, using the following equation:

$$G = 1/[(P_1/100\ G_1) + (P_2/100\ G_2) + \ldots (P_n\ 100\ G_n)]$$

where: G = average specific gravity
$G_1, G_2 \ldots G_n$ = appropriate average specific gravity values for each size fraction
$P_1, P_2, \ldots P_n$ = mass fraction of each size present in the original sample (not including finer material)

Calculate the percentage of absorption using the following equation:

$$\text{Absorption, \%} = [(B - A)/A] \times 100$$

If the sample is tested in separate size fractions, the average absorption value is the average of the values as computed earlier, weighted in proportion to the mass fraction of each size present in the original sample (not including finer material) as follows:

$$A = (P_1\ A_1/100) + (P_2\ A_2/100) + \ldots (P_n\ A_n/100)$$

where: A = average absorption, %
$A_1, A_2 \ldots A_n$ = absorption percentages for each size fraction
$P_1, P_2, \ldots$ Pn = mass fraction in of each size present in the original sample

Expected Outcomes

The specific gravity and absorbance of aggregate samples is highly variable and dependent on the mix of aggregate types within the sample mass. See Table A.3 in the Appendix for examples of aggregate density and relative density at room temperature.

DATA COLLECTION SHEET SPECIFIC GRAVITY AND ABSORPTION OF COARSE AGGREGATE	
Lead Experimenter	
Date of Experiment	
Source of Aggregate	
Visual USC Classification of Aggregate	
Type of Aggregate	
Final Mass after Oven Drying (Step 3), A_1	
Time at Start of Immersion (Step 5)	
Time at End of Immersion (Step 5)	
Mass of Sample at Surface Dry Condition (Step 7), B	
Apparent Mass in Water (Step 9), B_2	
Mass of Water Displaced by Sample (Step 10), C	
Mass of Oven-Dried Sample (Step 13), A	
Calculated Oven-Dried Specific Gravity (OD)	
Calculated Saturated-Surface-Dry Specific Gravity (SSD)	
Calculated Apparent Specific Gravity	
Calculated Average Specific Gravity (if Tested in Separate Size Fractions)	
Calculated Percentage of Absorption	

EXPERIMENT 10.9

SPECIFIC GRAVITY (RELATIVE DENSITY) AND ABSORPTION OF FINE AGGREGATE (GRAVIMETRIC METHOD)

(Consistent with ASTM C 128–15)

Date: ______________________________

Principal Investigator:

Name: ______________________________

Email: ______________________________

Phone: ______________________________

Collaborators:

Name: ______________________________

Email: ______________________________

Phone: ______________________________

Name: ______________________________

Email: ______________________________

Phone: ______________________________

Name: ______________________________

Email: ______________________________

Phone: ______________________________

Name: ______________________________

Email: ______________________________

Phone: ______________________________

Objective or Question to Be Addressed

This experiment provides a means for determining the specific gravity (relative density) and the absorption of fine aggregates. The specific gravity is a dimensionless quality of the aggregates expressed as oven-dry (OD), saturated-surface-dry (SSD), or as apparent specific gravity. The OD specific gravity is determined after drying the aggregate. The SSD specific gravity and absorption are determined after soaking the aggregate in water for a prescribed duration.

Fine aggregates are those that are those that pass a 4.75 mm (No. 4) sieve and are retained on a 75 μm (No. 200 sieve). This experiment procedure will not work well for lightweight aggregates.

Theory behind Experiment

Specific gravity is defined as the ratio of mass of a volume of aggregate to the mass of an equal volume of water. (This ratio is sometimes referred to as the *absolute volume* of the aggregate.) Specific gravity is also defined as the ratio of the relative density of the aggregate particles to the density of water.

Procedurally, a sample of aggregate is immersed in water for 24 +/– 4 hours to saturate the pore spaces. It is then removed from the water, the surfaces of the particles are dried, and the mass is determined. The sample is then placed in a graduated container and the volume of the sample is determined by the gravimetric or volumetric method. Then the sample is oven-dried and the mass determined again. Using the surface-dried and oven-dried mass values, the specific gravity is calculated from the equations provided in the data analysis section.

Note that there is a difference between the density of aggregate particles and the *bulk density* of aggregates. See Experiment 10.10 for a bulk density determination method. Bulk density includes the volume of the spaces between the soil particles, while relative density does not.

Apparent specific gravity pertains to the solid material making up the constituent particles but not including the pore space within the particles that is accessible to water. This value is not widely used in construction aggregate technology but is easy to calculate in this experiment and relevant for understanding particle characteristics.

Absorption values are used to determine the changes in the mass of an aggregate material related to the amount of water absorbed in the pore spaces compared to the dry condition. The standard for absorption is that obtained after submerging dry aggregate for a prescribed period of time (typically 24 +/– 4 hours). Aggregates mined from below the water table commonly have a moisture content greater than the absorption determined by this procedure unless allowed to dry before use. The general procedures described in this experiment may be used for determining the absorption of aggregates that have had conditioning by means other than a 24 hour soak. Those values and the specific gravity (SSD) values will likely be different from those obtained by the prescribed 24 hour soak.

Potential Interferences and Interference Management Plan

It has been reported that there may be significant differences in the bulk specific gravity and absorption between fine aggregate samples tested with the material finer than 75 µm (No. 200) present vs. not present. Samples that include material finer than 75 µm reportedly show a higher absorption and a lower bulk specific gravity compared with testing the same aggregate from which the material finer than 75 µm has been removed. This appears to be a function of the material finer than 75 µm creating a coating around the coarser fine aggregate particles during the surface drying process. This causes the resultant specific gravity and absorption values to be those of the coated particles and not those of the uncoated parent material.

No other interferences are known or likely to occur with this experiment procedure unless the aggregate is defined as a light aggregate.

Data to Be Collected with an Explanation of How They Will Be Collected (Ex.: "Water Temperature Data for Three Days Using a Continuous Data Logger")

Using the attached data sheet, record at a minimum the following information:

- Names of all experimenters
- Date of experiment and date of sample collection
- Identification of the type of aggregate (granite, schist, etc.) and the source of the sample (crushed rock, river run, etc.)
- The visual classification of the soil being tested (group name and symbol in accordance with the Uniform Soil Classification system)
- Maximum nominal size of the aggregate
- If any soil or material was excluded from the test specimen, describe the excluded material
- All mass measurements to the nearest 0.05 g by weighing on the balance
- Test temperature to the nearest 0.1 °C, if used
- Report each specific gravity value (OD, SSD, and apparent) from this experiment to the nearest 0.01 and indicate clearly which specific gravity value is reported
- Report absorption values to the nearest 0.1 %

Tools, Equipment, and Supplies Required

- Balance
  - A balance or scale having a capacity of 1 kg or more, sensitive to 0.1 g or less, and accurate within 0.1 % of the test load at any point within the range of use
- Pycnometer (gravimetric procedure)
  - A flask or other suitable container into which the test sample can be readily introduced and in which the volume content can be reproduced within +/– 0.1 cm^3 (0.006 inches^3). The volume of the container should be at least 50 % greater than that required to accommodate the test sample. (A volumetric flask of 500 cm^3 capacity or a fruit jar fitted with a pycnometer top is satisfactory for a 500 g sample of most fine aggregates.)

- Metal mold and tamper
  - A commercially available standard surface moisture test mold in the form of a frustrum of a cone is recommended
- Oven
  - An oven of sufficient size, capable of maintaining a uniform temperature of 110 +/– 5 °C (230 +/– 9 °F)

Safety

An aggregate sample from which the test sample is to be selected can be heavy. Care in lifting and handling these materials requires consideration of bending and twisting that can cause muscle strains and back pain.

The volume of water needed to submerge a sample can be significant and heavy. Care is required when moving the water containers to avoid back strain and to avoid spillage.

The drying oven used in this experiment will operate at a temperature of 110 +/– 5 °C (230 +/– 9 °F). Burns are a significant risk with this equipment. Heat-resistant gloves should be used when putting samples into the oven and when removing them.

In the event of a burn, immediately run the affected area under cold water and seek medical attention for swelling, blisters, or breaking of the skin.

As with all experiments, it is appropriate to wear gloves, eye protection, and a lab coat during this experiment. Most fine aggregate contains significant quantities of dust that may be released during handling. Eye protection and a dust mask are recommended during the sample preparation procedures.

Planned Experimental Procedures and Steps, in Order

1. Put on appropriate personal protective equipment and gather all materials and supplies in an appropriate location.
2. Collect a representative sample of aggregate and thoroughly mix and reduce it to yield a test sample of approximately 1 kg.
3. Place the sample in a suitable pan or vessel and dry it in the oven to constant mass at a temperature of 110 +/– 5 °C (230 +/– 9 °F). A constant mass is achieved when three consecutive weights of the sample are all within 0.1 % of each other. This is A.
4. Allow the sample to cool to comfortable handling temperature (approximately 50 °C).
5. Cover with water by immersion and permit to stand for 24 +/– 4 hours.
6. Decant excess water with care to avoid loss of fines, spread the sample on a flat nonabsorbent surface exposed to a gently moving current of warm air, and stir frequently to secure homogeneous surface drying.
7. Continue the air-drying step until the test specimen approaches a free-flowing condition indicating that the sample has reached a saturated, surface-dry condition.
8. Use the following procedure to determine when a free-flowing condition has been achieved.
   a. Hold a surface moisture test mold firmly on a smooth nonabsorbent surface with the large diameter down.

b. Place a portion of the partially dried fine aggregate loosely in the mold by filling it to overflowing and piling additional material above the top of the mold by holding it with the cupped fingers of the hand holding the mold or with a small funnel with an opening of the same size as the top of the mold.
c. Lightly tamp the fine aggregate into the mold with 25 light drops of the tamper. Start each drop approximately 5 mm (0.2 inches) above the top surface of the aggregate. Let the tamper fall freely under its own weight on each drop. Adjust the starting height to the new surface elevation after each drop and distribute the drops over the surface.
d. Remove loose sand from around the base of the mold and lift the mold vertically.
e. If surface moisture is still present, the fine aggregate will retain the molded shape. Slight slumping of the molded fine aggregate indicates that it has reached a surface-dry, or free-flowing condition.
f. Determine the mass of the saturated, surface-dry sample, S.
g. Some fine aggregates do not slump in the cone test upon reaching the surface-dry condition. Test by dropping a handful of the fine aggregate from the cone test onto a surface from a height of 100–150 mm (4–6 inches) and watch for fines becoming airborne. The presence of airborne fines indicates the presence of this non-slumping problem. For these materials, consider that the saturated surface-dry condition has been achieved when one side of the fine aggregate slumps slightly upon removing the mold.

9. Make the first trial for surface moisture when there is still some surface water in the sample. Continue drying with constant stirring and test at frequent intervals until the test indicates that the sample has reached a surface-dry condition.

   NOTE: If the first trial of the surface moisture test indicates that moisture is not present on the surface, it has been dried past the saturated surface-dry condition. In this case, thoroughly mix a few milliliters of water with the fine aggregate and permit the specimen to stand in a covered container for 30 minutes. Then resume the process of drying and testing at frequent intervals for the onset of the surface-dry condition.
10. Partially fill the pycnometer with water.
11. Introduce 500 +/− 10 g of saturated surface-dry sample into the pycnometer and fill with additional water to approximately 90 % of the pycnometer capacity.
12. Agitate the pycnometer by gently rolling, inverting, or agitating the pycnometer (or use a combination of these actions) to eliminate visible air bubbles.

    NOTE: About 15–20 minutes are normally required to eliminate the air bubbles. Dipping the tip of a paper towel into the pycnometer has been found to be useful in dispersing the foam that sometimes builds up when eliminating the air bubbles.
13. After eliminating all air bubbles, adjust the temperature of the pycnometer and its contents to 23 +/− 2 °C (73 +/− 3 °F), if necessary, by partial immersion in a circulating water bath and bring the water level in the pycnometer to its calibrated capacity.

14. Determine the total mass, C, of the pycnometer, specimen, and water.
15. Remove the aggregate from the pycnometer, dry the aggregate in the oven to constant mass at a temperature of 110 +/– 5 °C (230 +/– 9 °F).
16. Cool in air at room temperature for 1 +/– 1/2 hour and determine the mass.
17. Determine the mass of the pycnometer, B, filled to its calibrated capacity with water at 23 +/– 2 °C.

Data Collection Format and Data Collection Sheets Tailored to Experiment

See attached data collection sheet.

Data Analysis Plan

The following symbols are used for all of the subsequent calculations:

A = mass of oven dry specimen, g
B = mass of pycnometer filled with water, to calibration mark, g
C = mass of pycnometer filled with specimen and water to calibration mark, g
S = mass of saturated surface-dry specimen, g

Calculate the specific gravity on the basis of oven-dry aggregate as follows:

$$\text{Specific gravity (OD)} = A/(B + S - C)$$

Calculate the specific gravity on the basis of saturated surface-dry aggregate as follows:

$$\text{Specific gravity (SSD} = S/(B + S - C)$$

Calculate the apparent specific gravity as follows:

$$\text{Apparent specific gravity} = A/(B + A - C)$$

Calculate the percentage of absorption as follows:

$$\text{Absorption, \%} = 100\ [(S - A)\ A]$$

Expected Outcomes

The expected outcomes from this experiment are variable because all aggregates are not the same. See Table A.3 in the Appendix for the range of dry unit weight values generally associated with various soil types. The dry unit weight divided by the unit weight of water will yield the specific gravity of those soils. The unit weight of water at various temperatures is provided in Table A.5 in the Appendix.

DATA COLLECTION SHEET SPECIFIC GRAVITY (RELATIVE DENSITY) AND ABSORPTION OF FINE AGGREGATE (GRAVIMETRIC METHOD)	
Lead Experimenter	
Date of Experiment	
Date of Sample Collection	
Source of Sample	
Visual USC Classification of Sample	
Identification of Aggregate Type	
Oven-dried Mass of Sample (Step 3), A	
Time Immersion Started (Step 5)	
Time Immersion Ended (Step 5)	
Mass of Saturated, Surface-dry Sample (Step 8 f), S	
Mass of Pycnometer and Sample (Step 14), C	
Mass of Oven-Dried Sample Removed from Pycnometer (Step 16) This Mass Should Be Essentially Equal to S	
Mass of Pycnometer Filled to Capacity with Water (Step 17), B	
Calculated Specific Gravity of Aggregate on the Basis of Oven-dry Aggregate	
Calculated Specific Gravity of Aggregate on the Basis of Saturated Surface-dry Aggregate	
Calculated Apparent Specific Gravity	
Calculated Percentage of Absorption	

11 Soil Testing Experiments

This chapter provides completed designs for example experiments that can be used to demonstrate specific engineering phenomena in a university laboratory or field collection site. They may be used as presented or modified by the user to adapt to other desired outcomes. Sections on actual outcomes and data interpretation are necessarily left blank in these examples. They should be completed by the investigators upon collection of the data.

Note that where an experiment indicates that it is consistent with a specified ASTM method or other standard, this does not mean that the procedure is identical to, nor does it include all the details of, the stated standard method. Outcomes that require strict compliance with the stated standards must use that standard method to conduct the test. Therefore, the experiment procedures outlined in this book are not a substitute for proper compliance with the stated standards, nor are they intended to be.

The experiments in this chapter are:

11.1 Water Content of Soil and Rock by Mass
11.2 Water Content of Soil by Direct Heating
11.3 Water Content of Soil and Rock by Microwave Heating
11.4 Density and Unit Weight of Soil by Water Displacement
11.5 Density and Unit Weight of Soil by Direct Measurement
11.6 Density and Unit Weight of Soil In-Situ by Sand Cone Method
11.7 Liquid Limit, Plastic Limit, and Plasticity Index of Soil
11.8 Specific Gravity of Soil by Water Pycnometer
11.9 Determination of Porosity and Void Ratio in Soil
11.10 Bulk Density and Voids in Aggregate

DOI: 10.1201/9781003346685-11

EXPERIMENT 11.1

DETERMINATION OF WATER CONTENT OF SOIL AND ROCK BY MASS

(Consistent with ASTM Method D 2216–19)

Date: ______________________________

Principal Investigator:

Name: ______________________________

Email: ______________________________

Phone: ______________________________

Collaborators:

Name: ______________________________

Email: ______________________________

Phone: ______________________________

Name: ______________________________

Email: ______________________________

Phone: ______________________________

Name: ______________________________

Email: ______________________________

Phone: ______________________________

Name: ______________________________

Email: ______________________________

Phone: ______________________________

Objective or Question to Be Addressed

This experiment demonstrates a procedure for determining the water content of soil and rock samples by mass analysis. In this experiment, the terms "sample" and "soil sample" are used interchangeably to indicate soil, rock fragments, and other soil-based materials.

Theory behind Experiment

The water content of soil and rock samples is defined as the ratio of the mass of water contained in the pore spaces of the sample, to the solid mass of the soil particles in the sample, expressed as a percentage. Therefore, the water content of samples containing non-soil components may require different handling procedures, not defined in this experiment, or a modified definition of the water content. In addition, some organic materials may be decomposed by oven drying at the standard drying temperature for this method (110 +/– 5 °C) (230 +/– 5 °F). This procedure is applicable for most soil types. However, for soils containing significant amounts of hydrated materials, highly organic soils, or soils in which the pore water contains significant amounts of dissolved solids, this test method may not yield reliable values due to the potential for heating above 110 °C (230 °F), yielding deposition of suspended solids as accretions within the soil, and the lack of a reliable way to account for the presence of those previously dissolved solids.

In order to reduce the degree of dehydration of minerals such as gypsum, if known or suspected to be present, or to reduce decomposition of highly fibrous organic soils, drying the sample at 60 °C (140 °F) or in a desiccator at room temperature is recommended. When a drying temperature is used that is different from the standard, the resulting water content may be different from the standard water content determined at the standard drying temperature.

Potential Interferences and Interference Management Plan

There are no known interferences with this procedure. However, sample moisture may be affected by inadvertent drying during sampling, handling, and storage of the sample. Storage immediately after sampling in a container with a tightly fitting or screw-on lid will minimize this potential.

Similarly, a dried sample may regain moisture between drying and final weighing unless care is taken to maintain storage in a container with a tightly fitting or screw-on lid.

There is a potential for some soils to disaggregate, due to explosion of pore water or ignition of organic components during drying. Careful observation of the sample during and after drying will minimize the unnoticed occurrence of this phenomenon. A *loose* cover over or around the container in which the sample is dried will minimize the loss of small particles caused by rapid disintegration of a larger particle. Note that a tight cover during drying will allow excess pressure to build up inside the container and lead to an explosion and loss of a lot of sample, along with potential damage to the oven.

Data to Be Collected with an Explanation of How They Will Be Collected (Ex.: "Water Temperature Data for Three Days Using a Continuous Data Logger")

At a minimum, the following general information will be recorded as measured by the instruments and calculated by the equations provided in the data analysis section.

Using the attached data sheet, record at a minimum the following information:

- Name of all experimenters
- Date of the test
- Identification of the source of the sample being tested
- Water content of the specimen to the nearest 0.1 %
- Mass of the original, undried, sample
- Drying temperature if different from that specified herein
- Indication of whether any material (size and amount) was excluded from the test sample, and why

Tools, Equipment, and Supplies Required

- Vented, thermostatically controlled drying oven, preferably of the forced-draft type and capable of maintaining a uniform temperature of (110 +/− 5 °C) (230 +/− 5 °F) throughout the drying chamber
- Balances with a readability to 0.01 g and a capacity of more than 500 g
- Sample containers with tightly fitting lids or caps, uniquely identified for each water content determination
- Desiccator (optional) containing silica gel or anhydrous calcium sulfate
- Container-handling tools, such as heat-resistant gloves and tongs for moving and handling hot containers after drying
- Miscellaneous, knives, spatulas, scoops, quartering cloth, wire saws, etc., as needed for sample preparation
- Soil samples of 400–600 g (as collected), stored in tight-fitting containers to minimize the loss or gain of moisture content between sampling and testing

Safety

Containers will be hot when removed from the drying oven. Handle hot containers with heat-resistant gloves or tongs. Some soil types can retain considerable heat, and serious burns could result from improper handling.

In the event of a burn, run the affected area under cold water immediately and seek medical assistance for pain and/or broken skin or blisters.

Suitable eye protection is recommended due to the possibility of particle shattering during the heating, mixing, or mass determinations.

Highly organic soils and soils containing oil or other contaminates may ignite into flames during drying. Means for smothering flames to prevent operator injury or oven damage should be available during testing.

Fumes given off from contaminated soils or wastes may be toxic, and the oven should be vented accordingly.

Due to the possibility of steam explosions, or thermal stress shattering porous or brittle aggregates, a *loose* covering over the sample container during drying may be appropriate to prevent operator injury or oven damage. A cover also prevents scattering of the test sample in the oven during the drying cycle. Paper or cardboard covers may ignite if overheated and should be avoided.

Planned Experimental Procedures and Steps, in Order

1. Put on appropriate personal protective equipment and gather all materials and supplies needed in a convenient location.

2. Weigh and record the mass, M_c, of the clean and dry sample container and its lid, along with its identification number.
3. Place the moist test sample in the container and set the lid securely in place. Weigh the container and record the mass of the sample, M_1. Record this value to four decimal places (10^{-4}). Samples of large particles or rocks should be placed in pans and the material broken up into smaller pieces for more uniform and complete drying.
4. Remove the lid (if used) and place the container with the moist specimen in the drying oven. Dry the specimen to a constant mass. Maintain the drying oven at 110 +/− 5 °C (230 +/− 5 °F) during the drying process. The type of material tested, the size of the sample, the oven type and capacity, and other factors will influence the time required to obtain constant mass. In most cases, drying a test sample overnight (about 12–16 hours) will normally be sufficient, when a forced draft oven is used. Particularly moist samples and samples larger than 500 g may require longer drying times.
5. Test for sample dryness by placing a small strip of torn paper on top of the material immediately upon removal from the oven. If the paper strip curls, the material is not dry.
6. After the specimen has dried to a constant mass, remove the container from the oven and immediately replace the lid or place the sample in a desiccator.
7. Allow the sample and container to cool to room temperature.
8. Weigh the container and oven-dried specimen using the same balance used in Step 2 and record this value, M_2.
9. Calculate the moisture content of the original sample using the equations in the data analysis section.

Data Collection Format and Data Collection Sheets Tailored to Experiment

See attached data collection sheet.

Data Analysis Plan

The water content of the sample is calculated as follows. All three equations say the same thing but are provided to illustrate the evolution of the last, most simplistic rendering of the equations.

$$w = ((\text{mass of water})/(\text{mass of oven-dried sample})) \times 100$$
$$w = ((M_1 - M_2)/(M_2 - M_c) \times 100$$
$$w = (M_w/M_s) \times 100$$

where: w = water content, %
M_1 = mass of container and moist sample, g
M_2 = mass of container and oven-dried sample, g
M_c = mass of container, g
M_w = mass of water, g
M_s = mass of oven-dried soil, g

Expected Outcomes

Outcomes will depend on the water content and minerology of the original sample. Multiple tests of samples from the same source, collected at the same time, should yield reasonably consistent results.

DATA COLLECTION SHEET DETERMINATION OF WATER CONTENT OF SOIL AND ROCK BY MASS	
Lead Experimenter	
Date of Experiment	
Mass of Clean, Dry Soil Container (Step 2), M_c	
Mass of Container and Wet Sample (Step 3), M_1	
Mass of Oven-dried Sample and Container (Step 8), M_2	
Calculated Moisture Content (Step 9)	

EXPERIMENT 11.2

DETERMINATION OF WATER CONTENT OF SOIL BY DIRECT HEATING

(Consistent with ASTM Method D 4959–16)

Date: ______________________________

Principal Investigator:

Name: ______________________________

Email: ______________________________

Phone: ______________________________

Collaborators:

Name: ______________________________

Email: ______________________________

Phone: ______________________________

Name: ______________________________

Email: ______________________________

Phone: ______________________________

Name: ______________________________

Email: ______________________________

Phone: ______________________________

Name: ______________________________

Email: ______________________________

Phone: ______________________________

Objective or Question to Be Addressed

This experiment determines the water content of soils by drying with direct heat, such as using a hotplate, stove, blowtorch, or similar device. The method is applicable for most soil types. With soils containing significant amounts of hydrated materials, highly organic soils or soils that contain dissolved solids, this experiment may yield unreliable results due to the potential for precipitation of solids that were previously dissolved.

Theory behind Experiment

This experiment provides a means for determining the water content of a soil sample by directly heating the sample. When a sample of moist soil is subjected to drying by the application of direct heat, the difference between the mass of the moist sample and the mass of the dried sample is assumed to be equal to the mass of water contained in the specimen.

Potential Interferences and Interference Management Plan

The most likely interference with this method is the possibility of overheating the soil, thereby yielding a calculated water content higher than would be determined by Test Methods D 2216, the most accurate of the various methods provided by ASTM Standards. See Experiment 11.1 for a procedure consistent with D 2216. Incremental drying is used in this experiment to minimize that risk.

When testing sand and gravel size particles, additional care needs to be taken to avoid the possibility of particle shattering. Also, localized high temperatures in the soil during heating may alter other physical characteristics of the soil, such as the breakdown of individual particles, vaporization or oxidation of organics, and chemical changes, yielding poor results.

Data to Be Collected with an Explanation of How They Will Be Collected (Ex.: "Water Temperature Data for Three Days Using a Continuous Data Logger")

Using the attached data sheet, record at a minimum the following information:

- Name of all experimenters
- Date of the test
- Identification of the sample source and USC classification
- Mass of the sample prior to drying
- Mass after each incremental drying period
- Description of the type of direct heat source used
- Drying settings used
- Drying times for each cycle
- Number of drying cycles used
- Calculated water content of the sample to the nearest 0.1 %

Tools, Equipment, and Supplies Required

- Source of heat – directed to the soil sample to raise the temperature of the sample to 110 °C (230 °F)

    - Normally, a stove, hotplate, blowtorch, heat lamp, or hair dryer is used for this purpose. Do not use heat sources that directly apply open flame to the sample, as this may cause extreme degradation of the specimen along with oxidation of and depositing of soot in the sample.
- Balance – readability is generally required to the nearest 0.1 g and a capacity greater than the mass of the moist sample
- Suitable container for each water content determination
- Gloves, tongs, or other suitable device or mechanism for moving hot containers after drying
- Desiccator cabinet or jar of suitable size containing silica gel, anhydrous calcium phosphate, or equivalent
- Miscellaneous mixing tools, dry light-weight paper or tissue, and small knives
- Test sample of adequate size, based on the following table

RECOMMENDED MINIMUM SAMPLE SIZE FOR THIS EXPERIMENT

Sieve Size Retaining >10 % of Sample – mm	Minimum Mass of Sample to Use
2.0 (No. 10 sieve)	250 +/– 50 g
4.75 (No. 4 sieve)	400 +/– 100 g
19.0 (No. ¾ sieve)	750 +/– 250 g

Note: Larger-size samples are recommended to minimize testing inaccuracies.

Safety

Heating devices tend to cause burns that can be serious. Heat-resistant gloves, tongs, or other devices for handling hot samples and containers are required. Note that both the containers and the soil sample will be hot enough to cause burns.

In the event of a burn, run the affected area under cold water for several minutes and seek medical assistance for blistering, continued pain, or broken skin caused by a burn.

Some soil particles may explode from the rapid heating of contained moisture. Eye protection is required during the heating and cooling portions of the experiment, at a minimum.

Soils with a high organic content, or those containing flammable contaminants, may ignite during drying. A means for smothering flames should be available during testing. Fumes from such contaminants may be toxic, and the experiment should be conducted inside a fume hood if contamination is known or expected to be present.

Scattering of the test specimen and shattering of the aggregates may occur during the drying cycle. A vented covering over the sample to contain wayward particles is strongly recommended. This should also help with uniform heating of the sample.

As with all experiments, the use of gloves, eye protection, and a lab coat is recommended during the conduct of this experiment.

Planned Experimental Procedures and Steps, in Order

1. Put on appropriate personal protective equipment and gather all necessary tools, equipment, and supplies in a convenient location.

2. Determine and record the mass of a clean, dry sample container of suitable size, M_c.
3. Place the soil sample in the container and immediately record the mass of the soil and container, M_1.
4. Apply heat to the soil specimen and container, taking care to avoid localized overheating.
5. Continue heating while stirring the specimen to obtain even heat distribution until the specimen first appears dry. This condition should be evidenced by a comparatively uniform color of the sample during stirring. There should be no localized burnt or darkened appearance of any part of the soil observable during the mixing and stirring.
   NOTE: A piece of dry, light-weight paper or tissue, such as cigarette paper, placed on the surface of the apparently dry soil will quickly curl or ripple if the soil still contains significant water.
6. After the soil appears dry, remove the container and soil from the heat source and cool to room temperature in a desiccator.
7. Determine and record the mass of the soil and container, M_{test}.
8. Return the container and soil to the heat source for an additional application of heat.
9. Repeat Steps 5–7 until the change between two consecutive mass determinations is 0.1 % or less of the dry mass of the soil for the last two determinations.
10. Use the final dry mass of the soil and container in calculating the water content, M_2.
11. Calculate the moisture content per the data analysis plan.

Data Collection Format and Data Collection Sheets Tailored to Experiment

See attached data collection sheet.

Data Analysis Plan

Calculate the water content of the soil as follows:

$$w = [(M_1 - M_2)/(M_2 - M_c)] \times 100$$
$$\text{or: } w = (M_{w/}M_s) \times 100$$

where: w = water content, %
M_1 = mass of container and moist specimen, g
M_2 = mass of container and dried specimen, g
M_c = mass of container, g
M_w = mass of water, g ($M_w = M_1 - M_2$)
M_s = mass of solid particles, g. ($M_s = M_2 - M_c$)

Expected Outcomes

The outcomes from this experiment will vary depending on the original moisture content of the soil sample. Doing the experiment on more than one sample of the same soil at the same time will serve to verify the experimental procedures and outcomes.

DATA COLLECTION SHEET DETERMINATION OF WATER CONTENT OF SOIL BY DIRECT HEATING	
Lead Experimenter	
Date of Experiment	
Type of Sample	
USC Classification of Sample	
Source of Sample	
Mass of Dry Sample Container (Step 2), M_c	
Mass of Moist Sample and Sample Container, M_1	
Mass after First Drying Attempt, M_{int1}	
Mass after Second Drying Attempt, M_{int2}	
Mass after Third Drying Attempt, M_{int3}	
Mass after Fourth Drying Attempt, M_{int4}	
Mass after Fifth Drying Attempt (Step 9), M_{int5}	
Mass after Final Drying Attempt, M_2	
Calculated Water Content per Data Analysis Plan (Step 11), w	

EXPERIMENT 11.3

DETERMINATION OF WATER CONTENT OF SOIL AND ROCK BY MICROWAVE OVEN HEATING

(Consistent with ASTM Method D 4643–17)

Date: ______________________________

Principal Investigator:

Name: ______________________________

Email: ______________________________

Phone: ______________________________

Collaborators:

Name: ______________________________

Email: ______________________________

Phone: ______________________________

Name: ______________________________

Email: ______________________________

Phone: ______________________________

Name: ______________________________

Email: ______________________________

Phone: ______________________________

Name: ______________________________

Email: ______________________________

Phone: ______________________________

Objective or Question to Be Addressed

This experiment demonstrates a procedure for determining the water content of soils by incrementally drying a sample in a microwave oven. This procedure may be used as a substitute for Test Method D 2216 when more rapid results are desired, although this method produces slightly less accurate results.

This procedure is applicable for most soil types. However, for soils containing significant amounts of hydrated materials, highly organic soils, or soils in which the pore water contains significant amounts of dissolved solids, this test method may not yield reliable values due to the potential for heating above 110 °C (230 °F), yielding deposition of suspended solids as accretions within the soil, and the lack of a reliable way to account for the presence of those previously dissolved solids.

Theory behind Experiment

A soil sample is placed in a suitable (non-metallic) container and weighed. It is then placed in a microwave oven and dried. After a defined time of drying, the sample is removed from the oven and weighed again. This procedure is repeated until the weight of the sample becomes nearly constant. The difference between the weight (mass) of the original sample and the dried sample is used as the mass of water originally contained in the sample. The water content is determined by dividing the mass of water removed during drying by the mass of the dried soil, multiplied by 100.

The principal problem with the use of this procedure for water-content determination in soil, as opposed to the historic oven-drying procedure described in ASTM Standard D 2216 (see Experiment 11.1), has been the possibility of overheating the soil, thereby yielding a calculated water content higher than would be determined by Test Method D 2216 because of the potential for the volatilization of soil components along with the pore water. The incremental drying procedure described in this experiment is intended to minimize that effect.

Microwave ovens can heat materials much faster than other types of ovens and could, hypothetically, dry the samples being tested for water content much more quickly than a conventional drying oven. There is limited evidence that drying the samples in this fashion either does or does not destroy any portion of the organic fractions during the drying period. It is also unclear what intensity setting on the microwave would yield the best results and over what time period the heating should be done for optimal efficiency and efficacy.

The frequency of the microwave used needs to be considered. Most consumer microwave ovens operate with a wavelength at or around a nominal 2.45 gigahertz (GHz), or about 12.2 cm (4.80 inches) in the 2.4 GHz to 2.5 GHz ISM band. The actual wavelength may be anywhere within that range of 2.4 to 2.5 GHz. Each oven usually includes a statement somewhere on a data panel indicating the actual wavelength for that oven. Commercial and industrial ovens often use 915 megahertz (MHz), or 32.8 cm (12.9 inches).

The reason for the big difference is that the frequency at which consumer microwave ovens operate is determined by the need for those microwaves to cook food. The approximate resonant frequency of the free water molecule is 2.45 GHz. The higher the frequency, the less effectively the energy would penetrate, and at high frequencies the food would tend to cook less evenly. Lower frequencies penetrate better, but they are generally absorbed so weakly by food that they would not cook well. Cooking food is the defined purpose for these ovens, so the 2.4–2.5 GHz range is a good compromise that allows food to cook in a short time, but also cook through evenly. Commercial microwave ovens are often used primarily to heat pre-cooked food, but not to cook or re-cook food, so they can use a lower frequency (and therefore operate at a lower cost) without loss of effectiveness. Vending machines that sell hot food, for example, will heat the selected package for a specified length of time.

Most ovens also include power options. The power options indicate the approximate percentage of the time that the oven is producing microwaves while it is operating. When the oven is set to "Low" it normally produces microwaves for about 30 % of the time it is on and shuts off the microwave production for 70 % of that time in equal timed cycles. For example, an oven set to "Low" for 10 minutes would be expected to produce microwaves for 18 seconds, shut off the microwaves for 42 seconds, turn the microwaves on again for 18 seconds, and so forth for the full 10 minutes. During the periods when the microwaves are shut off, the materials inside the oven continue to heat through by convection within the materials, but not overheat. For these reasons, consumer microwave ovens are generally better for water-content determination in soils than commercial ovens.

At a power setting of "Medium", a microwave oven will normally produce microwaves for 50 % of the time it is on, shutting off the microwaves for the remaining 50 % in equal timed cycles. At a setting of "High", microwaves are generated for 100 % of the time that the oven is operating. Some ovens also include intermediary settings between these three, and researchers are encouraged to do additional testing at those power levels for better understanding of the efficacy of this procedure.

Potential Interferences and Interference Management Plan

The effects of microwave energy on soil depends on the mineralogy of the soil. As a result, no microwave energy pattern or procedure is applicable for all types of soil. The procedure described in this experiment is intended for minus 4.75 mm (No. 4) sieve sized material. Larger-size particles can be tested; however, care must be taken because of the increased chance of particle shattering. That will not affect the water-content calculation if the shattering is contained and all soil particles are accounted for.

Due to the localized high temperatures that the sample is exposed to in microwave heating, the physical characteristics of the soil may be altered. Degradation of individual particles may occur, along with vaporization or chemical transition. It is therefore recommended that samples used in this procedure not be used for other tests subsequent to drying.

Samples should be tested as quickly as possible after collection to minimize unrecorded moisture changes that will result in erroneous water-content determinations. All samples should be stored in sealed containers at the time of collection to prevent the addition or loss of moisture.

The samples should be cut or broken into small-size aggregations after the initial weighing to aid in obtaining more uniform drying. Care needs to be taken to avoid loss of broken particles during this procedure.

If experience with a particular soil type and sample size indicates that shorter or longer initial drying times from those provided in this experiment procedures section can be used without overheating, the initial and subsequent drying times indicted by this procedure may be adjusted accordingly. In addition, most ovens have a variable power setting. For the majority of soils tested, a setting of "high" should be satisfactory and accomplish the drying most quickly; however, for some soils such a setting may be too severe, causing shattering of the sample particles or burning of the soil contents. The proper setting can be determined only through the use of, and experience with, a particular oven for various soil types and sample sizes. The energy output of microwave ovens may decrease with age and usage; therefore, power settings and drying times should be established and regularly verified for each oven.

Data to Be Collected with an Explanation of How They Will Be Collected (Ex.: "Water Temperature Data for Three Days Using a Continuous Data Logger")

Using the attached data sheet, record at a minimum the following information:

- Name of individual(s) performing test
- Date of the test
- Means of identifying the sample (material) being tested
- At least the following test sample data:
    - Water content of the sample to the nearest 1 % or 0.1 %
    - An indication of whether the sample has a mass less than 100 g
    - An indication of whether the sample contains more than one soil type (layered, and the like)
    - An indication of any material (size, amount, and layer or layer sequences) excluded from the test sample
- Time and setting of the initial drying period and all subsequent incremental drying periods
- Initial mass of the test sample prior to drying and the mass after the final incremental drying periods
- Describe if a desiccator was used and for how long
- Identification of the microwave oven and the drying settings and cycles used when standardized drying is utilized

Tools, Equipment, and Supplies Required

- Microwave oven, preferably with a vented chamber
    - Ovens with variable power controls and input power ratings of about 700 W have been found to be adequate for this use. Variable power controls are important and reduce the potential for overheating of the test sample. Note that commercial microwave ovens equipped with built-in scales and computer controls have recently been developed specifically for use in drying soils and may be used for this procedure with good results.
- Balance with 0.01 g readability
- Sample containers: microwave safe, nonmetallic containers of suitable size for the samples being tested
- Container-handling equipment, such as heat-resistant gloves or a clamp-style holder, suitable for removing hot containers from the oven

- Desiccator (optional) cabinet or jar of suitable size containing silica gel, anhydrous calcium phosphate, or equivalent
- Heat sink
    - A material or liquid that can be placed in the microwave to absorb energy after the moisture has been driven from the test sample. The heat sink reduces the possibility of overheating the sample, causing damage to the sample or the oven. Glass beakers filled with water and materials that have a boiling point above water, such as nonflammable oils, have reportedly been used successfully. Note that pure water would add to the humidity in the oven and reduce the ability of the microwave energy to dry the sample.
- Stirring tools: spatulas, putty knives, and glass rods for cutting, breaking up, and stirring the test sample before and during the test

Safety

Containers will be hot after microwaving. Handle hot containers with heat-resistant gloves. Some soil types can retain considerable heat, and serious burns could result from improper handling.

In the event of a burn, run the affected area under cold water immediately and seek medical assistance for pain and/or broken skin or blisters.

Suitable eye protection is recommended due to the possibility of particle shattering during the heating, mixing, or mass determinations.

Microwave drying of soils containing metallic materials may cause arcing in the oven.

Continued operation of the oven after the soil has reached constant weight may cause damage or premature failure of the microwave oven.

Highly organic soils and soils containing oil or other contaminates may ignite into flames during microwave drying. Means for smothering flames to prevent operator injury or oven damage should be available during testing.

Fumes given off from contaminated soils or wastes may be toxic, and the oven should be vented accordingly.

Due to the possibility of steam explosions, or thermal stress shattering of porous or brittle aggregates, a loose or vented covering over the sample container may be appropriate to prevent operator injury or oven damage. A cover also prevents scattering of the test sample in the oven during the drying cycle. Commercial plastic covers are available for use in microwave ovens that will do well for this purpose. Do not use metallic containers in a microwave oven because arcing and oven damage can result. Paper or cardboard covers may ignite if overheated and should be avoided.

Placing the test sample directly on the glass tray provided with some ovens should be avoided because the concentrated heating of the sample may result in the glass tray shattering, possibly causing injury to the operator. The tray is designed to rotate, in most cases, and that is desirable for even heating of the sample. Therefore, a similarly sized, non-metallic, shatterproof plastic dish, with a high melting temperature, should be substituted for the glass tray that comes with the microwave unit.

Planned Experimental Procedures and Steps, in Order

1. Put on appropriate personal protective equipment and gather all tools and samples in a convenient location.

2. Determine and record the mass of a clean, dry sample container, M_1.
3. Place a soil sample of at least 100 grams in the sample container, and immediately determine and record the mass.
4. Place the sample container with the sample in a microwave oven with the heat sink and turn the oven on for 3 minutes.
5. After the set time has elapsed, remove the sample container with the sample from the oven, either weigh the sample container with sample immediately, or cool (preferably in a desiccator) to allow handling and to prevent damage to the balance.
6. Determine and record the mass.
7. With a small spatula or knife or short length of glass rod, carefully mix the soil, being careful not to lose any soil.
8. Return the sample container with the sample to the oven and reheat in the oven for 1 minute.
9. Repeat Steps 5–7 until the change between two consecutive mass determinations is achieved of 0.1 % or less of the initial wet mass of the soil.
10. Use the final mass determination in calculating the water content. Obtain this value immediately after the heating cycle, or, if the mass determination is to be delayed, after cooling in a desiccator.

Data Collection Format and Data Collection Sheets Tailored to Experiment

See attached data collection sheet.

Data Analysis Plan

The water content of the sample is calculated as follows. All three equations say the same thing, but are provided to illustrate the evolution of the last, most simplistic rendering of the equations.

$$W = ((\text{mass of water})/(\text{mass of oven-dried sample})) \times 100$$

$$w\ ((M_1 - M_2)/(M_2 - M_c) \times 100$$

$$w = (M_{w/}M_s) \times 100$$

where: w = water content, %
M_1 = mass of container and moist sample, g
M_2 = mass of container and oven-dried sample, g
M_c = mass of container, g
M_w = mass of water, g
M_s = mass of oven-dried soil, g

Expected Outcomes

Outcomes will depend on the water content and minerology of the original sample. Multiple tests of samples from the same source, collected at the same time, should yield reasonably consistent results.

DATA COLLECTION SHEET DETERMINATION OF WATER CONTENT OF SOIL AND ROCK BY MICROWAVE OVEN HEATING	
Lead Experimenter	
Date of Experiment	
Mass of Clean, Dry Soil Container (Step 2), M_c	
Mass of Soil Container and Wet Sample (Step 3), M_1	
Oven-dried Mass of Soil and Container, First Try	
Oven-dried Mass of Soil and Container, Second Try	
Oven-dried Mass of Soil and Container, Third Try	
Oven-dried Mass of Soil and Container, Fourth Try	
Oven-dried Mass of Soil and Container, Fifth Try	
Final Mass of Oven-Dried Soil and Container (Step 9), M_2	
Calculated Water Content Using Data Analysis Plan (Step 10), w	

EXPERIMENT 11.4

DENSITY AND UNIT WEIGHT OF SOIL SAMPLES BY WATER DISPLACEMENT

(Consistent with ASTM Method D 7263–21)

Date: ______________________

Principal Investigator:

Name: ______________________

Email: ______________________

Phone: ______________________

Collaborators:

Name: ______________________

Email: ______________________

Phone: ______________________

Name: ______________________

Email: ______________________

Phone: ______________________

Name: ______________________

Email: ______________________

Phone: ______________________

Name: ______________________

Email: ______________________

Phone: ______________________

Name: ______________________

Email: ______________________

Phone: ______________________

Objective or Question to Be Addressed

This test method uses a water displacement methodology to determine characteristics of intact, disturbed, remolded, or compacted soil, including total/moist/bulk density, dry density, and dry unit weight. With this method, a sample that is not susceptible to wax intrusion is coated in wax and then placed in water to measure the volume by determining the quantity of water displaced. The density and unit weight are then calculated based on the mass and volume measurements.

Theory behind Experiment

A test sample is collected and its mass in air is measured. It is then coated in wax and its mass is measured again. The wax-coated sample is then placed in a wire basket that is attached to a balance and fully submerged in a container of water in which the volume is measurable before and after submerging the sample. Its mass in the water is measured. The volume of water displaced (measured by graduations on the side of the water tank) defines the volume of the sample. The density and unit weight are then calculated from the mass and volume measurements.

Unit values are stated in SI units with inch-pound units in parentheses. The terms "density" and "unit weight" are often used interchangeably. However, *density* refers to the mass per unit volume, whereas *unit weight* refers to force per unit volume. In this experiment, density is determined in SI units. After the density has been determined, the unit weight is calculated in SI or inch-pound units, or both.

Density plays an important role in the relationships of mass and volume of soil and rock. When particle density, that is, specific gravity (see Experiment 11.8) is also known, the dry density can be used to calculate porosity and void ratio. Dry density measurements are also useful for determining the degree of soil compaction. The water content of soils is highly variable. Moreover, since the volume of swelling soils shrinks with drying, the total density of soils will vary with the water content. Therefore, the water content of the soil should be determined at the time of sampling.

The density and unit weight of remolded/reconstituted samples are used to evaluate the degree of compaction of earthen fills, embankments, levees, and other earthen structures. Dry density values are used to calculate the dry unit weight values needed to create a compaction curve.

Potential Interferences and Interference Management Plan

This experiment assumes a test sample that is not subject to wax intrusion. Any wax that does intrude into the sample will skew the data and the sample should be tested by a different method.

Data to Be Collected with an Explanation of How They Will Be Collected (Ex.: "Water Temperature Data for Three Days Using a Continuous Data Logger")

Using the attached data sheet, record at a minimum the following information:

- The type of sample: intact, reconstituted (compacted), remolded, or clod

- The shape of the sample: cylindrical, cuboidal, or irregular
- The water content of the trimmings/excess soil and final water content determinations, including the masses used to make the determinations
- The mass of the moist sample, mass of the wax-coated sample, mass of the wax-coated submerged sample, density of the wax, temperature of the water when zeroing the balance and with sample in the wire basket, and the density of the water at test temperature (see Table A.5 in the Appendix)
- Volume of the soil sample
- Density: moist (total) and dry
- Unit weight: moist (total) and dry

Tools, Equipment, and Supplies Required

- Balance capable of measuring the mass of the sample to 0.01 g
  - The capacity of this balance will need to exceed the mass of the sample suspended in water. A balance having a below-balance port using a weighing hook or a yoke assemblage for top-loading balances is typically used to make this measurement. In general, a balance with a minimum capacity of 1,000 g is sufficient. A higher capacity balance may be needed when determining the mass of an un-extruded sample.
- Vented, thermostatically controlled drying oven, preferably of the forced-draft type, capable of maintaining a uniform temperature of 110 +/− 5 °C (230 +/− 9 °F) throughout the drying chamber
- Non-shrinking paraffin, microcrystalline, or other suitable wax mixture that does not become brittle when dry and does not shrink during solidification. The density of the wax must be known to three significant figures and have a relatively constant density, ρx. The density must not appreciably change after repeated melting and solidification cycles
  - A 50/50 mixture of paraffin wax and petroleum jelly by mass has been shown to provide an adequate alternative. Paraffin wax is commercially available and has a typical density of 900 kg/m^3 (0.900 g/cm^3).
- Wax-melting container or device capable of melting the wax without overheating it. A heater that uses hot water and a container/device that is thermostatically controlled is preferred
- Wire basket of 3.35 mm (0.132 inches) or finer mesh of approximately equal width and height of sufficient size to contain the sample to be tested. A hairnet may also be used in lieu of the basket for smaller soil samples
- Water tank of sufficient size to contain the submerged basket and soil sample
- Thermometric device capable of measuring the temperature range within which the experiment is being performed readable to 0.1 °C or better and having an accuracy of at least +/− 0.5 °C
- Miscellaneous items such as a paintbrush, trimming and carving tools (such as a wire saw, steel straightedge, miter box, and vertical trimming lathe), apparatus for preparing remolded or reconstituted samples, sample extruder, sample containers for water contents, plastic wrap, aluminum foil, plastic bags, gloves, and tongs, may be necessary or useful
- Soil sample meeting the following requirements:

- Cylindrical samples must have a minimum diameter of 33 mm (1.3 inches) and be sufficiently cohesive and able to maintain shape during the measuring procedure.
- Average height to average diameter ratio should be, but is not required to be, between 2 and 2.5.
- Largest particle size must be smaller than 1/6 the sample diameter.
- Cubical/cuboidal samples must have minimum dimensions (height, width, and length) of 33 mm (1.3 inches), and the largest particle size must be smaller than 1/10 of the sample's smallest dimension. If, after completion of a test on an intact sample, it is found based on visual observation that oversize particles are present, indicate this information in the remarks section of the data sheet.
- Irregularly shaped (clods) samples must be of sufficient size to adequately represent the soil under evaluation. Avoid selecting a sample that is too small since it would not be representative.

Safety

Wax-melting equipment and hot wax may burn unprotected skin. Overheated wax may burst into flames. Extreme care should be exercised when working with hot wax.

Vapors given off by molten wax will ignite spontaneously at temperatures above 205 °C (400 °F), and wax can ignite if allowed to come in contact with a heating element or open flame. *Do not use an open-flame device to heat the wax.*

Eye protection, heat-resistant gloves, and a lab coat are required personal protection equipment.

In the event of skin contact with hot wax, immediately run the affected area under cold water until the wax cools and sets. If the wax peels of easily, peel it off and seek medical attention for the underlying burn. Otherwise leave the cooled wax on the skin and seek immediate medical assistance to avoid tearing the skin and causing an infection or excessive bleeding.

In the event the wax ignites, cool with water or cover with a nonflammable cover to exclude oxygen and notify lab safety personnel immediately for a safety inspection.

Planned Experimental Procedures and Steps, in Order

1. Put on appropriate personal protective equipment and gather all materials and supplies in a convenient location.
2. Determine and record the density of the wax being used to coat the sample to three significant figures. This is provided by manufacturer or measured in the lab and weighed in the lab.
3. Obtain the test sample. Cylindrical and cuboidal samples must have a fairly regular shape with smoothed sides. Re-entrant angles should be avoided for irregularly shaped samples (clods).
4. Measure and record the mass of the sample to four significant figures in grams.
   a. If the sample has not been extruded from a sampling tube, measure and record the mass of the mold plus sample, M_1, to four significant figures in grams.

b. Cylindrical samples: Using the measuring devices, take and record a minimum of three height measurements 120° apart and at least three diameter measurements at the quarter points of the height to four significant figures in millimeters. For samples that are not extruded from a sampling tube but are trimmed, measure and record the height and inside diameter of the mold to four significant figures in millimeters, as they are considered to be the sample's physical dimensions. Then, extrude the sample and clean off/wipe out the mold. Calculate the volume of the sample, V_s.
c. Measure and record the mass of the mold, M_2, to four significant figures in grams.
d. Calculate the mass of a cylindrical sample, as $M_1 – M_2 = M_s$.
e. Cuboidal samples: Using the measuring devices, take and record a minimum of three measurements for each height, width, and length of the sample to four significant figures in millimeters. Calculate the volume of the sample, V_s.
f. Calculate and record the mass of the cuboidal sample to four significant figures as $M_1 – M_2 = M_s$.

NOTE: It is common for the density and unit weight of the sample to be less than the value based on the volume of the mold after removal from a mold. This situation occurs as a result of the sample swelling after removal of the lateral confinement due to the mold.

5. Melt the wax to only slightly above its melting point to avoid flashing of the wax vapors and to permit quick forming of a uniform surface coating of the wax.
6. Cover the sample with a thin coat of melted wax, either with a paintbrush or by dipping the sample in a container of melted wax. Apply a second coat of wax after the first coat has hardened. The wax should be sufficiently warm to flow when brushed on the sample, but not so hot that it dries the soil. If overheated wax comes in contact with the soil sample, it may cause the moisture to vaporize and form air bubbles under the wax. If that happens, bubbles may be trimmed out and filled with wax.
7. Measure and record the mass of the wax-coated sample in air, (M_c), to four significant figures in grams.
8. Attach the wire basket to the below-balance port or to the yoke assemblage of the balance and submerge it in the water tank. Then zero the balance. The basket must be fully submerged and not touching the sides or the bottom of the water tank while submerged. The wire basket must also be at the same depth for both zeroing the balance and taking the measurement with the sample.
9. Before removing the wire basket, measure and record the temperature of the water to the nearest 0.1 °C (0.2 °F). The temperature of the water must remain within +/– 3 °C (5 °F) of this temperature between zeroing the balance and submerging the sample in the basket.
10. Lift the wire basket out of the water and place the wax-coated sample in the wire basket and submerge it in the water tank. Make sure the basket plus sample is fully submerged and not touching the sides or the bottom of the water tank while submerged.

11. Measure and record the mass of the wax-coated sample submerged in water, (M_{sub}), to four significant figures in grams.
12. Measure and record the temperature of the water to the nearest 0.1 °C (0.2 °F) to confirm that the temperature has not changed more than +/− 3 °C since the balance was zeroed.
13. Remove the sample from the wire basket and remove the wax from the sample. It can be peeled off after a break is made in the wax surface. Use the sample without wax to measure the water content. If, however, the wax becomes difficult to remove from the sample, the water content can be measured from an adjacent piece of soil or from trimmings providing the adjacent soil or trimmings have been kept sealed and are representative of the water content of the soil. Indicate on the data sheet if the water content is not from the sample itself.
14. After the wax is removed, calculate the wet and dry density and the wet and dry unit weight of the sample in accordance with the data analysis section of this experiment.

Data Collection Format and Data Collection Sheets Tailored to Experiment

See data collection sheet attached.

Data Analysis Plan

1. Calculate the volume of the test sample using the following equation:

$$V = ((M_c - M_{sub})/\rho_w) - ((M_c - M_t)/\rho_x)$$

where: V = volume of test sample, 4 significant figures, cm^3
M_c = mass of wax-coated sample, 4 significant figures, g
M_{sub} = mass of wax-coated sample submerged in water, 4 significant figures, g
ρ_w = density of water at test temperature (see Table A.5 in the Appendix), g/cm^3 (equivalent to g/mL)
M_t = mass of moist (total) sample, 4 significant figures, g
ρ_x = density of wax, 3 significant figures, g/cm^3

2. Calculate the moist (total) density, ρ_t, using the following equation:

$$\rho_t = M_t/V$$

where: ρ_t = moist (total) density, 4 significant figures, g/cm^3

3. Calculate the dry density using the following equation:

$$\rho_d = \rho_t/(1+(w/100))$$

where: ρd = dry density of soil sample, 4 significant figures, g/cm^3
w = water content of soil sample, to nearest 0.1 %

4. Calculate the moist (total) unit weight, γ_t, to 4 significant figures using the following equation:

$$\gamma_t = \rho_t\,(9.8066) \text{ in kN/m}^3$$
$$\text{or } \gamma_t = \rho_t\,(62.428) \text{ in lbf/ft}^3$$

5. Calculate the dry unit weight, γ_d, to 4 significant figures using the following equation:

$$\gamma_d = \rho_d\,(9.8066) \text{ in kN/m}^3$$
$$\text{or } \gamma_d = \rho_d\,(62.428) \text{ in lbf/ft}^3$$

6. Record the volume of the soil sample, the density of the sample – moist (total) and dry – and the unit weight of the soil – moist (total) and dry.

Expected Outcomes

The unit outcomes from this experiment should be reasonably consistent with the values shown in Table A.3 in the Appendix.

DATA COLLECTION SHEET DENSITY AND UNIT WEIGHT OF SOIL SAMPLES BY WATER DISPLACEMENT	
Lead Experimenter	
Date of Experiment	
Type of Sample	
Shape of Sample	
Source of Sample	
Density of Wax (from Manufacturer)	
Moist Mass of Sample, M_t	
Temperature of Water in Tank at Time of Experiment	
Mass of Submerged Sample	
Temperature of Water in Tank (for Check)	
Mass of Sample (Step 4)	
Mass of Cylindrical Mold Plus Sample (Step 4a)	
Height of Cylinder (First) (Step 4b)	
Height of Cylinder (Second) (Step 4b)	
Height of Cylinder (Third) (Step 4b)	
Average Height of Cylinder (Step 4b)	
Diameter of Cylindrical Sample (First) (Step 4b)	
Diameter of Cylindrical Sample (Second) (Step 4b)	
Diameter of Cylindrical Sample (Third) (Step 4b)	
Average Diameter of Cylindrical Sample (Step 4b)	
Calculated Volume of Cylindrical Sample (Step 4b)	
Mass of Cylinder (Step 4c)	
Calculated Mass of Cylindrical Sample (Step 4d)	

Mass of Cuboidal Sample plus Mold (Step 4a)	
Height of Cuboidal Sample (First) (Step 4e)	
Width of Cuboidal Sample First) (Step 4e)	
Length of Cuboidal Sample First) (Step 4e)	
Height of Cuboidal Sample (Second) (Step 4e)	
Width of Cuboidal Sample (Second) (Step 4e)	
Length of Cuboidal Sample (Second) (Step 4e)	
Height of Cuboidal Sample (Third) (Step 4e)	
Width of Cuboidal Sample (Third) (Step 4e)	
Length of Cuboidal Sample (Third) (Step 4e)	
Average Height of Cuboidal Sample (Step 4e)	
Average Width of Cuboidal Sample (Step 4e)	
Average Length of Cuboidal Sample (Step 4e)	
Calculated Volume of Cuboidal Sample (Step 4e)	
Mass of Cuboidal Mold (Step 4c)	
Mass of Cuboidal Sample (Step 4f)	
Mass of Wax-Coated Sample (Step 7), M_c	
Temperature of Water in Tank at Time of Experiment (Step 9)	
Mass of Submerged Sample (Step 11), M_{sub}	
Temperature of Water in Tank (Step 12)	
Calculated Volume of the Sample from Data Analysis Step 1, V	
Calculated Moist Density from Data Analysis Section Step 2, ρ_t	
Calculated Dry Density from Data Analysis Section Step 3, ρ_d	
Calculated Moist Unit Weight from Data Analysis Section Step 4, γ_t	
Calculated Dry Unit Weight from Data Analysis Section Step 5, γ_d	

EXPERIMENT 11.5

DENSITY AND UNIT WEIGHT OF SOIL SAMPLES BY DIRECT MEASUREMENT

(Consistent with ASTM Method D 7263–21)

Date: ______________________________

Principal Investigator:

Name: ______________________________

Email: ______________________________

Phone: ______________________________

Collaborators:

Name: ______________________________

Email: ______________________________

Phone: ______________________________

Name: ______________________________

Email: ______________________________

Phone: ______________________________

Name: ______________________________

Email: ______________________________

Phone: ______________________________

Name: ______________________________

Email: ______________________________

Phone: ______________________________

Objective or Question to Be Addressed

This test method uses a water displacement methodology to determine characteristics of intact, disturbed, remolded, or compacted soil, including total/moist/bulk density, dry density, and dry unit weight. This method measures the dimensions and mass of a soil sample. The density and unit weight are then calculated using these direct measurements. The density and unit weight are then calculated based on the mass and volume measurements.

Theory behind Experiment

A test sample is obtained from a selected source. Usually, the sample has a cylindrical or cuboidal shape. Intact and reconstituted/remolded samples may be tested by this method in conjunction with strength, permeability/hydraulic conductivity (air/water) and compressibility determinations.

The test sample shape impacts the volume calculations. If the test sample is cylindrical in shape, its mass, height, and diameter are measured. If it is cuboidal in shape, its mass, height, width, and length are measured. The density and unit weight are then calculated based on the physical dimensions and mass of the sample.

Unit values are stated in SI units with inch-pound units in parentheses. The terms "density" and "unit weight" are often used interchangeably. However, *density* refers to the mass per unit volume, whereas *unit weight* refers to force per unit volume. In this experiment, density is determined in SI units. After the density has been determined, the unit weight is calculated in SI or inch-pound units, or both.

Density plays an important role in the relationships of mass and volume soil. When particle density, that is, specific gravity (see Experiment 11.8) is also known, the dry density can be used to calculate porosity and void ratio. Dry density measurements are also useful for determining the degree of soil compaction. The water content of soils is highly variable. Moreover, since the volume of swelling soils shrinks with drying, the total density of soils will vary with the water content. Therefore, the water content of the soil should be determined at the time of sampling.

The density and unit weight of remolded/reconstituted samples are used to evaluate the degree of compaction of earthen fills, embankments, levees, and other earthen structures. Dry density values are used to calculate the dry unit weight values needed to create a compaction curve.

Potential Interferences and Interference Management Plan

There are no known interferences that are likely to occur with this experiment.

Data to Be Collected with an Explanation of How They Will Be Collected (Ex.: "Water Temperature Data for Three Days Using a Continuous Data Logger")

Using the attached data sheet, record at a minimum the following information:

- The type of sample: intact, reconstituted (compacted), remolded, or clod
- The shape of the sample: cylindrical, cuboidal, or irregular

- The water content of the trimmings/excess soil and final water content determinations, including the masses used to make the determinations
- The mass of the sample, the height, diameter (cylindrical samples), width, and length (cuboidal samples). Include the averaged values of height, diameter, width, and length as applicable
- Volume of the soil sample
- Density: moist (total) and dry
- Unit weight: moist (total) and dry

Tools, Equipment, and Supplies Required

- Balance capable of measuring the mass of the sample to 0.01 g
  - In general, a balance with a minimum capacity of 1,000 g is sufficient. A higher-capacity balance may be needed when determining the mass of an un-extruded sample.
- Vented, thermostatically controlled, drying oven, preferably of the forced-draft type, capable of maintaining a uniform temperature of 110 +/– 5 °C (230 +/– 9 °F) throughout the drying chamber
- Measuring devices with which to measure the physical dimensions, such as height, width, length, and diameter, of the sample to be tested to four significant figures. The devices shall be constructed such that their use will not disturb/deform, indent, or penetrate the sample
- Miscellaneous items, such as a paintbrush, trimming and carving tools (such as a wire saw, steel straightedge, miter box, and vertical trimming lathe), apparatus for preparing remolded or reconstituted samples, sample extruder, sample containers for water contents, plastic wrap, aluminum foil, plastic bags, gloves, and tongs, may be necessary or useful
- Soil sample meeting the following requirements:
  - Cylindrical samples must have a minimum diameter of 33 mm (1.3 inches) and be sufficiently cohesive and able to maintain shape during the measuring procedure.
  - Average height to average diameter ratio should, but is not required to, be between 2 and 2.5. Largest particle size must be smaller than 1/6 the sample diameter.
  - Cubical/cuboidal samples must have minimum dimensions (height, width, and length) of 33 mm (1.3 inches), and the largest particle size must be smaller than 1/10 of the sample's smallest dimension.
  - If after completion of a test on an intact sample, it is found based on visual observation that oversize particles are present, indicate this information in the remarks section of the data sheet.
  - Irregularly shaped (clods) samples must be of sufficient size to adequately represent the soil under evaluation. Avoid selecting a sample that is too small since it would not be representative.

Safety

Eye protection, protective gloves, and a lab coat are required personal protection equipment for this and all laboratory experiments.

Planned Experimental Procedures and Steps, in Order

1. Put on appropriate personal protective equipment and gather all materials and supplies in a convenient location.
2. Obtain the prepared cylindrical or cuboidal test sample.
3. Place the sample on a zeroed/tared balance.
4. Measure and record the mass of the sample to four significant figures in grams (g).
   a. If the sample has not been extruded from a sampling tube, measure and record the mass of the mold plus sample to four significant figures in grams, as M_1.
   b. Cylindrical samples: Using the measuring devices, take and record a minimum of three height measurements 120° apart and at least three diameter measurements at the quarter points of the height to four significant figures in millimeters. For samples that are not extruded from a sampling tube but are trimmed, measure and record the height and inside diameter of the mold to four significant figures, in millimeters, as they are the sample's physical dimensions. Then, extrude the sample and clean off/wipe out the mold. Calculate the volume of the sample, V_s.
   c. Measure and record the mass of the mold, M_2, to four significant figures in grams.
   d. Calculate the mass of a cylindrical sample, as $M_1-M_2 = M_s$.
   e. Cuboidal Samples: Using the measuring devices, take and record a minimum of three measurements for each height, width, and length of the sample to four significant figures in millimeters. Calculate the volume of the sample, V_s.
   f. Measure and record the mass of the cuboidal sample to four significant figures, as $M_1-M_2 = M_s$.

   NOTE: It is common for the density and unit weight of the sample to be less than the value based on the volume of the mold after removal from a mold. This situation occurs as a result of the sample swelling after removal of the lateral confinement due to the mold.
5. After the physical dimensions are measured and recorded, determine the final water content of the sample to the nearest 0.1 % in accordance with the data analysis section of this design. See Experiment 11.2 for a detailed description of this procedure. If the sample is to be used for other testing, such as strength or permeability/hydraulic conductivity, refer to the applicable method for when to take the final water content determination. When no other testing is needed, use the whole sample to make the final water content determination. If other testing, such as classification testing (Atterberg limits, particle-size analysis, and so on), is needed, it is acceptable to determine the water content of the sample by taking a representative slice of the sample. This is w.

Data Collection Format and Data Collection Sheets Tailored to Experiment

See data collection sheet attached.

Data Analysis Plan

1. Water content, w: Calculate the water content taken during the testing in accordance with ASTM Test Method D 2216 to the nearest 0.1 % (see Experiment 11.2).

2. Calculate the average height and average diameter for cylindrical samples to four significant figures in millimeters and the average height, average width, and average length for cuboidal samples to four significant figures in millimeters.
3. Calculate the volume of cylindrical test samples using the following equation:

$$V = (\pi\ (d^2)\ (h))$$

where: V = volume of moist (total) test sample, 4 significant figures, cm^3
d = average sample diameter, 4 significant figures, cm
h = average sample height, 4 significant figures, cm

4. Calculate the volume of cuboidal test samples using the following equation:

$$V = (l \times w \times h)$$

where: V = volume of moist (total) test sample, 4 significant figures, cm^3
l = average length, 4 significant figures, cm
w = average width, 4 significant figures, cm
h = average height, 4 significant figures, cm
5. Calculate the moist (total) density, ρ_t, using the following equation:

$$\rho_t = M_{t/}V$$

where: ρ_t = moist (total) density, 4 significant figures, g/cm^3
6. Calculate the dry density using the following equation:

$$\rho_d = \rho_t/(1+(w/100))$$

where: ρd = dry density of soil sample, 4 significant figures, g/cm^3
w = water content of soil sample, to nearest 0.1 %
7. Calculate the moist (total) unit weight, γ_t, to 4 significant figures using the following equation:

$$\gamma_t = \rho_t\ (9.8066) \text{ in kN/m}^3$$
$$\text{or } \gamma_t = \rho_t\ (62.428) \text{ in lbf/ft}^3$$

8. Calculate the dry unit weight, γ_d, to 4 significant figures using the following equation:

$$\gamma_d = \rho_d\ (9.8066) \text{ in kN/m}^3$$
$$\text{or } \gamma_d = \rho_d\ (62.428) \text{ in lbf/ft}^3$$

Expected Outcomes

The unit outcomes from this experiment should be reasonably consistent with the values shown in Table A.3 in the Appendix.

DATA COLLECTION SHEET DENSITY AND UNIT WEIGHT OF SOIL SAMPLES BY DIRECT MEASUREMENT	
Lead Experimenter	
Date of Experiment	
Type of Sample	
Shape of Sample	
Source of Sample	
Mass of Sample (Step 4)	
Mass of Mold Plus Sample (Step 4a)	
Height of Cylinder (First) (Step 4b)	
Height of Cylinder (Second) (Step 4b)	
Height of Cylinder (Third) (Step 4b)	
Average Height of Cylinder (Step 4b)	
Diameter of Cylindrical Sample (First) (Step 4b)	
Diameter of Cylindrical Sample (Second) (Step 4b)	
Diameter of Cylindrical Sample (Third) (Step 4b)	
Average Diameter of Cylindrical Sample (Step 4b)	
Calculated Volume of Cylindrical Sample (Step 4b)	
Mass of Cylinder (Step 4c)	
Calculated Mass of Cylindrical Sample (Step 4d)	
Mass of Cuboidal Sample Plus Mold (Step a)	
Height of Cuboidal Sample (First) (Step 4e)	
Width of Cuboidal Sample (First) (Step 4e)	
Length of Cuboidal Sample (First) (Step 4e)	
Height of Cuboidal Sample (Second) (Step 4e)	

DATA COLLECTION SHEET DENSITY AND UNIT WEIGHT OF SOIL SAMPLES BY DIRECT MEASUREMENT	
Width of Cuboidal Sample (Second) (Step 4e)	
Length of Cuboidal Sample (Second) (Step 4e)	
Height of Cuboidal Sample (Third) (Step 4e)	
Width of Cuboidal Sample (Third) (Step 4e)	
Length of Cuboidal Sample (Third) (Step 4e)	
Average Height of Cuboidal Sample (Step 4e)	
Average Width of Cuboidal Sample (Step 4e)	
Average Length of Cuboidal Sample (Step 4e)	
Calculated Volume of Cuboidal Sample (Step 4e)	
Mass of Cuboidal Mold (Step 4c)	
Mass of Cuboidal Sample (Step 4f)	
Calculated Water Content of Sample from Data Analysis Section Step 1, w	
Calculated Moist Density from Data Analysis Section Step 5, ρ_t	
Calculated Dry Density from Data Analysis Section Step 6, ρ_d	
Calculated Moist Unit Weight from Data Analysis Section Step 7, γ_t	
Calculated Dry Unit Weight from Data Analysis Section Step 8, γ_d	

EXPERIMENT 11.6

DENSITY AND UNIT WEIGHT OF SOIL IN PLACE BY SAND-CONE METHOD

(Consistent with ASTM Method D 1556–15)

Date: ______________________________

Principal Investigator:

Name: ______________________________

Email: ______________________________

Phone: ______________________________

Collaborators:

Name: ______________________________

Email: ______________________________

Phone: ______________________________

Name: ______________________________

Email: ______________________________

Phone: ______________________________

Name: ______________________________

Email: ______________________________

Phone: ______________________________

Name: ______________________________

Email: ______________________________

Phone: ______________________________

Objective or Question to Be Addressed

This experiment is designed to determine the in-place density and unit weight of soils without appreciable amounts of rock or coarse materials in excess of 1½ inches (38 mm) in diameter and intact or in situ soils, provided the natural void or pore openings in the soil are small enough to prevent the sand used in the test from entering the voids.

Theory behind Experiment

This experiment requires that a small hole be hand excavated in the soil to be tested. All the material from the hole is saved in a container. The hole is filled with free-flowing sand of a known density, and the volume of the hole is determined from the volume of sand used. The in-place wet density of the soil is determined by dividing the wet mass of the removed material by the volume of the hole. The water content of the material from the hole is determined, and the dry mass and the dry density of the in-place material are calculated using the wet mass of the soil, the water content, and the volume of the hole.

In this experiment, the pound (lbf) represents a unit of force (weight). It is not uncommon in engineering practice to concurrently use units representing both mass and force unless dynamic calculations (such as F = Ma) are involved. This practice combines two separate systems within a single standard. This experiment has been designed using inch-pound units with conversions provided in the SI system.

Potential Interferences and Interference Management Plan

This experiment is not suitable for organic, saturated, or highly plastic soils that would deform or compress during excavation of the test hole and may not be suitable for soils consisting of unbound granular materials that will not maintain stable sides in the test hole, soils containing appreciable amounts of coarse material larger than 1 ½ inches (38 mm), and granular soils having high void ratios.

The units used in this experiment are unusual in that they are stated in both inch-pound units and equivalent SI units. The SI units are generally regarded as standard. The values stated in each system may not be exact equivalents; therefore, each system should be used independently of the other. Combining values from the two systems may result in nonconformance with the standard.

This is primarily a field experiment, although use in a laboratory for educational purposes is encouraged. Therefore, mass measurements should be recorded and reported to the nearest 0.01 lbm (5 g), water-content calculations should be recorded and reported to the nearest 1 %, and density values should be recorded and reported to three significant digits.

Uniformly graded sand is needed to prevent segregation during handling, storage, and use. Sand free of fines and fine sand particles is required to prevent significant bulk density changes with normal daily changes in atmospheric humidity. Sand comprised of durable, natural subrounded, or rounded particles is desirable. Crushed sand or sand having angular particles may not be free flowing, a condition that can cause bridging, resulting in inaccurate density determinations.

Data to Be Collected with an Explanation of How They Will be Collected (Ex.: "Water Temperature Data for Three Days Using a Continuous Data Logger")

Using the attached data sheet, record at a minimum the following information:

- Names of all experimenters
- Date of experiment and date of sample collection
- Identification of the type of aggregate (granite, schist, etc.) and the source of the sample (crushed rock, river run, etc.)
- The visual classification of the soil being tested (group name and symbol in accordance with the Uniform Soil Classification System)
- The bulk density of the sand used in the test, to three significant digits (this should be provided by the provider of the sand)
- Test hole volume, to four significant digits
- In-place wet density, to three significant digits
- In-place dry density, to three significant digits
- In-place dry unit weight, to three significant digits
- In-place water content of the soil expressed as a percentage of dry mass to the nearest 1 %

Tools, Equipment, and Supplies Required

- Container of known volume, V_1, that is approximately the same size as the expected size of the test hole and that allows the sand to fall approximately the same distance into the hole as the hole is deep
- Sand-cone density apparatus, commercially available, consisting of sand container, sand cone, and base plate
- Clean, dry sand, uniform in density and grading, uncemented, durable, and free flowing. Any gradation may be used that has a coefficient of uniformity (Cu = D60/D10) less than 2.0, a maximum particle size smaller than the 2.0 mm (No. 10) sieve size, and less than 3 % by weight passing the 250 µm (No. 60) sieve size. A standard commercially available sand designed for this purpose is recommended
- Balances or scale with a minimum capacity of 44 lbf (20 kg) and 0.01 lbf (5 g) readability or better
- A vented, thermostatically controlled, forced-draft, drying oven, capable of maintaining a uniform temperature of 110 +/− 5 °C (230 +/− 9 °F) throughout the drying chamber. The oven shall have a means of indicating the oven drying chamber temperature when in operation
- Miscellaneous equipment, such as a small knife, small pick, chisel, small trowel, screwdriver, or spoons for digging test holes, large nails or spikes for securing the base plate; buckets with lids, plastic-lined cloth sacks, or other suitable containers for retaining the density samples, moisture sample, and density sand, respectively; small paint brush, calculator, notebook or test forms, etc.

Safety

Soil samples excavated from a hole in the ground can be difficult to reach and extract from the hole. Twisting sprains are common. Care in lifting and handling these materials requires consideration of bending and twisting that can cause muscle strains and back injury.

The drying oven used in this experiment will operate at a temperature of 110 +/– 5 °C (230 +/– 9 °F). Burns are a significant risk with this equipment. Heat-resistant gloves should be used when putting sample into the oven and when removing them.

In the event of a burn, immediately run the affected area under cold water and seek medical attention for swelling, blisters, or breaking of the skin.

As with all experiments, it is appropriate to wear gloves, eye protection, and a lab coat during this experiment. Most fine aggregate contains significant quantities of dust that may be released during handling. Specifically, eye protection and a dust mask are recommended during the sample preparation procedures.

Planned Experimental Procedures and Steps, in Order

1. Put on personal protective equipment and gather all necessary materials and supplies in a convenient location.
2. Inspect the cone apparatus for damage, free rotation of the valve, and properly matched baseplate. Measure and record the volume of the sand container, V_1.
3. Fill the cone container with sand for which the dry bulk density has been determined in advance or reported by the supplier. If that information is not available, calibrate the apparatus by the following procedure.
   a. Weigh the empty apparatus sand container, M_{sc}.
   b. Fill the apparatus sand container with sand that is dried to the same state anticipated during use in testing, M_a.
   c. Determine the mass of the sand container filled with sand, M_b.
   d. Calculate the mass of material to fill the container, as $M_b - M_a = M_c$.
   e. Place the base plate on a clean, level, plane surface. Invert the container/apparatus and seat the funnel in the flanged center hole in the base plate.
   f. Mark and identify the apparatus and base plate so that the same apparatus and plate can be matched and reseated in the same position during testing.
   g. Open the valve on the apparatus sand container fully until the sand flow stops, making sure the apparatus, base plate, or plane surface are not jarred or vibrated before the valve is closed.
   h. Close the valve sharply, remove the apparatus sand container and determine the mass of the container and remaining sand, M_d.
   i. Calculate the mass of sand used to fill the funnel and base plate as the difference between the initial and final mass, $M_b - M_d = M_t$.
   j. Repeat this procedure a minimum of three times. The maximum variation between any one determination and the average should not exceed 1 %. Use the average of the three determinations for this value in the test calculations.
   k. The average of these three measurements is used as the mass of sand, M_2, used to fill the cone and base plate.
4. Calculate the bulk density of the sand as follows:

$$\rho_1 = M_c V_1$$

where: ρ_1 = bulk-density of the sand, lbm/ft^3 (g/cm^3)
M_c = mass of the sand to fill the calibration sand container, lbm (g), (average value from Step 3d)
V_1 = volume of the calibration container, ft^3 (cm^3)

5. Calculate and record the total mass of the container and sand, lbm (g), as M_b. (average value from Step 3c).
6. Level the surface of the location to be tested. The base plate may be used as a tool for striking off the surface to a smooth, level plane.
7. Seat the base plate on the plane surface, making sure there is contact with the ground surface all the way around the edge of the flanged center hole.
8. Mark the outline of the base plate on the ground surface to check for movement during the test and, if needed, secure the plate against movement using nails pushed into the soil adjacent to the edge of the plate.
9. Excavate a small test hole in the surface of the soil through the center hole in the base plate. Avoid disturbance to the sides of the hole in the soil to the degree possible and carefully save all the excavated soil in a clean, dry, container.

   NOTE: The test hole volume will depend on the anticipated maximum particle size in the soil to be tested and the depth of the compacted layer. Test hole volumes should be as large as practical to minimize errors and should not be less than the volumes indicated in Table 11.6.1. A hole depth should be selected that will provide a representative sample of the soil. The sides of the hole should slope slightly inward, and the bottom should be reasonably flat or concave. The hole should be kept as free as possible of pockets, overhangs, and sharp obtrusions since these affect the accuracy of the test. Soils that are essentially granular require extreme care and may require digging a conical-shaped test hole.
10. Place all excavated soil, and any soil loosened during digging, in a moisture-tight container that is marked to identify the test number. Take care to avoid losing any materials. Protect this material from any loss of moisture until the mass has been determined and a specimen has been obtained for a water-content determination.
11. Clean the flange of the base plate hole, invert the sand-cone apparatus and seat the sand-cone funnel into the flanged hole. Eliminate or minimize vibrations in the test area due to personnel or equipment.
12. Open the valve and allow the sand to fill the hole, funnel, and base plate. Take care to avoid jarring or vibrating the apparatus while the sand is running.
13. When the sand stops flowing, close the valve.
14. Determine the mass of the apparatus sand container with the remaining sand, M_p.
15. Calculate and record the mass of sand used to fill the test hole, funnel, and base plate as:

$$M_1 = M_b - M_p$$

    Where: M_1 = mass of sand used to fill the test hole, funnel, and base plate
    M_b = mass of container full of sand (Step 3c)
    M_p = mass of the apparatus sand container with the remaining sand (Step 14)
16. Thoroughly mix the material removed from the hole and either obtain a representative specimen for water-content determination or use the entire sample.
17. Determine the water content of the soil excavated from the hole using the following procedure. (See Experiment 11.2 for details on this procedure.)
    a. Weigh and record the mass of a clean and dry sample container and its lid, M_6.
    b. Place the moist sample in the container and set the lid securely in place.

Weigh the container and sample, M_7. Record this value to four decimal places.

c. Calculate the mass of the wet soil from the test hole as, $M_7 - M_6 = M_w$.
d. Remove the lid and place the container with the moist specimen in a drying oven. Dry the specimen to a constant mass. Maintain the drying oven at 110 +/− 5 °C (230 +/− 5 °F) during the drying process. The type of material tested, the size of the sample, the oven type and capacity, and other factors will influence the time required to obtain a constant mass. In most cases, drying a test sample overnight (about 12–16 hours) will normally be sufficient, when using forced draft ovens. Particularly moist samples and samples larger than 500 grams may require longer drying times.
e. Test for sample dryness by placing a small strip of torn paper on top of the material immediately upon removal from the oven. If the paper strip curls, the material is not dry.
f. After the specimen has dried to a constant mass, remove the container from the oven and immediately replace the lid or place the sample in a desiccator.
g. Allow the sample and container to cool to room temperature.
h. Weigh the container and oven-dried sample, M_8.
i. Calculate the moisture content of the original sample using the following equations. All three equations say the same thing but are provided to illustrate the evolution of the last, most simplistic rendering of the equations.

$$w = ((\text{mass of water})/(\text{mass of oven-dried sample})) \times 100$$
$$w = ((M_7 - M_8)/(M_8 - M_6)) \times 100$$
$$w = (M_{w/}M_s) \times 100$$

where: w = water content, %

M_7 = mass of container and moist sample, g (Step 17b)
M_8 = mass of container and oven-dried sample, g (Step 17h)
M_6 = mass of container, g (Step 17a)
M_w = mass of water, g
M_s = mass of oven-dried soil, g

18. Calculate the dry mass of the soil excavated from the test hole as $M_8 - M_6 = M_s$.
19. Calculate the wet mass of the soil excavated from the test hole as $M_7 - M_6 = M_3$.

Data Collection Format and Data Collection Sheets Tailored to Experiment

See attached data collection sheet.

Data Analysis Plan

Calculate the bulk density of the sand as follows:

$$\rho_1 = M_3/V_1$$

where: ρ_1 = bulk-density of the sand, lbm/ft³ (g/cm³)
M_b = mass of the sand to fill the sand container, lbm (g) (Step 3 c)

V_1 = volume of the sand container, ft^3 (cm^3) (provided by container manufacturer or measured in advance) (Step 2)

Calculate the volume of the test hole as follows:

$$V = (M_1 - M_2)/\rho_1$$

where: V = volume of the test hole, ft^3 (cm^3)
M_1 = mass of the sand used to fill the test hole, funnel, and base plate, lbm (g) (Step 15)
M_2= mass of the sand used to fill the funnel and base plate, lbm (g), (Step 3 k)
ρ_1 = bulk density of the sand, lbm/ft^3 (g/cm^3)

Calculate the dry mass of material removed from the test hole as follows:

$$M_4 = 100\, M_3/(w + 100)$$

where: w = water content of the material removed from test hole, % (Step 17)
M_3 = moist mass of the material from test hole, lbm (g) (Step 19)
M_4 = dry mass of material from test hole, lbm (g) (Step 18)

Calculate the in-place wet and dry density of the material tested as follows:

$$\rho_m = M_3/V$$
$$\rho_d = M_4/V$$

where: V = volume of the test hole, ft^3 (cm^3)
M_3 = moist mass of the material from the test hole, lbm (g) (Step 19)
M_4 = dry mass of the material from the test hole, lbm (g) (Step 18)
ρ_m = wet density of the tested material, lbm/ft^3 (g/cm^3)
ρ_d = dry density of the tested material, lbm/ft^3 (g/cm^3)

Expected Outcomes

The expected outcomes from this experiment should be reasonably consistent with the values shown in Table A.3 in the Appendix.

TABLE 11.6.1
Minimum Test Hole Volumes Based on Maximum Size of Included Particle

Maximum Particle Size		Minimum Test Hole Volumes	
inches	**millimeters**	**cubic centimeters**	**cubic feet**
1/2	12.7	1415	0.05
1	25.4	2125	0.075
1 1/2	38	2830	0.1

Source: From ASTM Standard D1556/D1556M – 15.

DATA COLLECTION SHEET DENSITY AND UNIT WEIGHT OF SOIL IN PLACE BY SAND-CONE METHOD	
Lead Experimenter	
Date of Experiment	
Rock Type Sampled	
Source of Sample	
Visual USC Classification of Sample	
Volume of Apparatus Sand Container (Step 2), V_1	
Mass of Empty Sand Container (Step 3a), M_{sc}	
Mass of Filled Apparatus Sand Container, First Try (Step 3b), M_{a1}	
Mass of Filled Apparatus Sand Container, First Try (Step 3c), M_{b1}	
Mass of Sand to Fill Apparatus Sand Container, First Try (Step 3d), M_{c1}	
Mass of Sand Container and Remaining Sand First Try (Step 3h), M_{d1}	
Mass of Sand to Fill Funnel and Base First Try (Step 3i), M_{t1}	
Mass of Filled Apparatus Sand Container, Second Try (Step 3b), M_{a2}	
Mass of Filled Apparatus Sand Container, Second Try (Step 3c), M_{b2}	
Mass of Sand to Fill Apparatus Sand Container, Second Try (Step 3d), M_{c2}	
Mass of Sand Container and Remaining Sand Second Try (Step 3h), M_{d2}	
Mass of Sand to Fill Funnel and Base Second Try (Step 3i), M_{t2}	
Mass of Filled Apparatus Sand Container, Third Try (Step 3a), M_{a3}	
Mass of Filled Apparatus Sand Container, Third Try (Step 3c), M_{b3}	
Mass of Sand to Fill Apparatus Sand Container, Third Try (Step 3d), M_{c3}	
Mass of Sand Container and Remaining Sand Third Try (Step 3h), M_{d3}	

DATA COLLECTION SHEET DENSITY AND UNIT WEIGHT OF SOIL IN PLACE BY SAND-CONE METHOD	
Mass of Sand to Fill Funnel and Base Third Try (Step 3i), M_{t3}	
Average Value of Sand to Fill Cone and Base Plate (Step 3k), M_2	
Calculated Bulk Density of the Sand (Step 4)	
Calculated Mass of Container and Sand (Step 5)	
Mass of Partially Filled Apparatus Sand Container (Step 14), $M_{p.}$	
Calculated Mass of Sand to Fill the Test Hole, Cone, and Base Plate (Step 15), M_1	
Mass of Clean, Dry Container and Lid (Step 17a), M_6	
Mass of Container with Wet Soil from Test Hole (Step 17b), M_7	
Mass of Container with Oven-Dried Soil from Test Hole (Step 17h), M_8	
Mass of Moist Soil Removed from Test Hole (Step 17c), M_w	
Calculated Moisture Content of the Original Sample (Step 17i), w	
Calculated Dry Mass of Soil from Test Hole (Step 18), M_s	
Calculated Wet Mass of Soil from Test Hole (Step 19), M_3	
Calculated Volume of Test Hole, V	
Calculated In-situ Wet Density of the Original Material	
Calculated In-situ Dry Density of the Original Material	

EXPERIMENT 11.7

LIQUID LIMIT, PLASTIC LIMIT, AND PLASTICITY INDEX OF SOILS

(Consistent with ASTM Method D 4318–17)

Date: ______________________________

Principal Investigator:

Name: ______________________________

Email: ______________________________

Phone: ______________________________

Collaborators:

Name: ______________________________

Email: ______________________________

Phone: ______________________________

Name: ______________________________

Email: ______________________________

Phone: ______________________________

Name: ______________________________

Email: ______________________________

Phone: ______________________________

Name: ______________________________

Email: ______________________________

Phone: ______________________________

Objective or Question to Be Addressed

This experiment provides three test methods to determine the liquid limit, plastic limit, and the plasticity index of soils defined as *cohesive soils*, which generally means those soils finer than 425 μm (No. 40 sieve). The three methods provided describe the *liquid limit* (LL), which is defined as the water content, in percent, of a soil at the arbitrarily defined boundary between the semi-liquid and plastic states; the *plastic limit* (PL), which is defined as the water content, in percent, of a soil at the boundary between the plastic and semi-solid states; and the *plasticity index* (PI), which is defined as the range of water content over which a soil behaves plastically. Numerically, the PI is the difference between the water content at the liquid limit and water content at the plastic limit.

The LL and PL of soils (along with the shrinkage limit) are often referred to as the *Atterberg limits*.

Theory behind Experiment

Two procedures for preparing test samples are provided in the ASTM Method 4318: a wet preparation procedure and a dry preparation procedure. In most cases, the wet preparation procedure is preferred and that is the procedure used in this experiment. Similarly, two procedures are provided in ASTM Method 4318 for determining the liquid limit: a multipoint method and a one-point method. The multipoint method is generally the preferred method and that is the method used in this experiment.

The LL test is conducted first because the PL test is performed on material prepared for the LL test. When determining the PL, two procedures are provided in ASTM Method 4318 for rolling portions of the test sample as follows: hand rolling and using a specified rolling device. The hand method provides a better sense of the water content in the soil during the procedure and is, therefore, more useful for student work and is the procedure used in this experiment.

Procedurally, a sample is processed to remove any material larger than 425 μm (No. 40 sieve). The LL is then determined by performing trials in which a portion of the sample is spread in a brass cup, divided in two by a grooving tool, and then allowed to flow together from the shocks caused by repeatedly dropping the cup in a standard mechanical device.

The PL is determined by alternately pressing together and rolling a small portion of plastic soil into a 3.2 mm (1/8 inch) diameter thread until its water content is reduced to a point at which the thread crumbles and can no longer be pressed together and re-rolled. The water content of the soil at this point is reported as the PL.

The plasticity index is calculated as the difference between the LL and the PL.

These three characteristics of fine-grained soil are used to classify the soils within several soil classification systems. The LL, PL, and PI are also used with other soil behavioral characteristics to indicate the compressibility, hydraulic conductivity, compactability, shrink-swell probability, and shear strength of various soil matrixes.

Potential Interferences and Interference Management Plan

The LL and PL of many soils that have been allowed to dry before testing may be considerably different from values obtained on non-dried samples. For accurate characterization of soils, samples should not be permitted to dry before testing. For educational purposes, use of previously dried soils will yield useful data if thoroughly moistened and allowed to rest for a day or two prior to testing. The resulting data should not be assumed to accurately represent the actual source material. That is why the wet preparation procedure is used in this experiment.

The multipoint method for determination of the LL is generally more precise than the one-point method. For that reason, the multipoint method is used in this experiment.

The composition and concentration of soluble salts in a soil affect the values of the LL and PL as well as the water content values of soils. Special consideration should therefore be given to soils from a marine environment or other sources where high soluble salt concentrations may be present. The degree to which the salts present in these soils are diluted or concentrated must be given careful consideration.

Where distilled water is referred to in this test method, either distilled or demineralized water may be used. As with soil water containing salts, the cations of salts present in tap water may also exchange with the natural cations in the soil and significantly alter the test results if tap water is used in the soaking and washing operations. Unless it is known that such cations are not present in the tap water, the use of distilled or demineralized water is required.

Data to Be Collected with an Explanation of How They Will Be Collected (Ex.: "Water Temperature Data for Three Days Using a Continuous Data Logger")

Using the attached data sheet, record at a minimum the following information:

- Name of all experimenters
- Date of the test
- Sample type and source-identifying information
- Description of sample, such as approximate maximum grain size, estimate of the percentage of sample retained on the 425 μm (No. 40) sieve, and as-received water content
- Equipment used, such as hand rolled for PL, manual or mechanical LL device, metal or plastic grooving tool
- LL, PL, and PI to the nearest whole number
- If the liquid limit or plastic limit tests could not be performed, or if the plastic limit is equal to or greater than the liquid limit, report the soil as non-plastic, NP

Tools, Equipment, and Supplies Required

- Liquid limit device
  - A commercially available hand-operated or electrically driven mechanical device consisting of a brass cup suspended from a carriage designed to control its drop onto the surface of a block of resilient material that serves as the base of the device.
- Counter (optional)
  - A separate hand-activated device or automatic mechanical mechanism attached to the LL device to count the number of drops of the cup during operation of the LL device

- Flat grooving tool
  - A commercially available tool made of plastic or noncorroding metal for creating a groove in the soil during the LL test and usually incorporating a height gauge for adjusting the height-of-drop of the LL device.NOTE: Curved tools were typically used for this test in the past but were found to be less accurate than the flat tool because the round tool does not control the depth of the soil as well as a flat tool does
- Height gauge
  - A metal gauge block for adjusting the height-of-drop of the cup, if not provided as an integral part of the LL device
- Water content containers
  - Small corrosion-resistant containers with snug-fitting lids to prevent samples from drying before testing
- Balance, with a readability of 0.01 g
- Mixing/storage container or dish
  - A container in which to mix the soil sample prior to testing and to store the prepared material
- Ground glass plate
  - A ground glass plate of sufficient size for rolling plastic limit threads. Spatula: spatula or small knife for handling samples
- Wash bottle
  - For adding controlled amounts of water to soil and washing fines from coarse particles
- Drying oven
- Distilled or demineralized water for preparing samples

Safety

Containers will be hot when removed from the drying oven. Handle hot containers with heat-resistant gloves or tongs. Some soil types can retain considerable heat, and serious burns could result from improper handling.

In the event of a burn, run the affected area under cold water immediately and seek medical assistance for pain and/or broken skin or blisters.

Suitable eye protection is recommended due to the possibility of particle shattering during the heating, mixing, or mass determinations.

Highly organic soils and soils containing oil or other contaminates may ignite into flames during drying. Means for smothering flames to prevent operator injury or oven damage should be available during testing.

Fumes given off from contaminated soils or wastes may be toxic, and the oven should be vented accordingly.

Due to the possibility of steam explosions, or thermal stress shattering porous or brittle aggregates, a loose or vented covering over the sample container during drying may be appropriate to prevent operator injury or oven damage. A cover also prevents scattering of the test sample in the oven during the drying cycle. Paper or cardboard covers may ignite if overheated and should be avoided.

Planned Experimental Procedures and Steps, in Order

1. Put on appropriate personal protective equipment and gather the required tools and samples in a convenient location.
2. Verify by visual and manual methods that the sample to be tested contains very little or no material retained on a 425 μm (No. 40) sieve.
3. Prepare 150–200 grams of material by mixing the sample thoroughly with distilled or demineralized water on the glass plate or in the mixing dish using the spatula. If necessary, soak the material in a mixing or storage dish with a small amount of water to soften the material before the start of mixing.
4. Adjust the water content of the material to bring it to a consistency that would require about 25–35 drops of the cup of the LL device (called *blows*) to close the groove. This will take some experience to judge well. The soil at this moisture content may have the consistency of soft cookie dough.
5. Place the prepared material in a mixing/storage dish, check its consistency (adjust if required), cover to prevent loss of moisture, and allow to stand (cure) for at least 16 hours (overnight).
6. After the standing period and immediately before starting the test, thoroughly remix the soil.

Liquid Limit Procedure:

1. Using a spatula, place a portion of the prepared soil in the cup of the LL device at the point where the cup rests on the base, squeeze it down, and spread it into the cup to a depth of about 10 mm at its deepest point, tapering to form an approximately horizontal surface. Take care to eliminate air bubbles from the soil, but form the soil with as few strokes as possible. Keep the unused soil in the mixing/storage dish. Cover the dish with a wet towel (or use other means) to retain the moisture in the soil.
2. Form a groove in the soil mass in the bowl by drawing the flat grooving tool, beveled edge forward, through the soil mass on a central line, joining the highest point to the lowest point on the rim of the cup. When cutting the groove, hold the grooving tool against the surface of the cup and draw in an arc, maintaining the tool perpendicular to the surface of the cup throughout its movement. Take care to prevent sliding the soil mass along the inside surface of the cup.
3. Verify that the base and the underside of the cup are clean of any particles of soil or debris.
4. Lift and drop the cup by turning the crank at a rate of 1.9 to 2.1 drops per second until the two halves of the soil mass come in contact at the bottom of the groove along a distance of 13 mm (1/2 inch). Only the soil at the surface of the cup should touch, not the entire depth of the soil mass. Do not hold the base of the machine while the crank is turned.
5. Verify that an air bubble has not caused premature closing of the groove by observing that both sides of the groove have flowed together with approximately the same shape. If a bubble has caused premature closing of the groove, reform the soil in the cup, adding a small amount of soil to make

up for that lost in the grooving operation and repeat Steps 2–4. If the soil slides on the surface of the cup, repeat 2–4 at a higher water content. If, after several trials at successively higher water contents, the soil mass continues to slide in the cup or if the number of drops required to close the groove is always less than 25, record that the LL could not be determined and report the soil as nonplastic without performing the PL test.

6. Record the number of drops, N, required to close the groove.
7. Obtain a water content sample by removing a slice of soil approximately the width of the spatula, extending from edge to edge of the soil cake at right angles to the groove and including that portion of the groove in which the soil flowed together, place that aliquot of the sample from the cup in a container of known mass, and cover.
8. Return the soil remaining in the cup to the mixing/storage dish. Wash and dry the cup and grooving tool and reattach the cup to the carriage in preparation for the next trial.
9. Remix the entire soil sample in the mixing/storage dish, adding distilled water to increase the water content of the soil and decrease the number of drops required to close the groove. Repeat Steps 2–9 for at least two additional trials, producing successively lower numbers of drops to close the groove. One of the trials shall be for a closure requiring 25–35 blows, one for closure between 20 and 30 blows, and one trial for a closure requiring 15–25 blows.
10. Determine the water content, w_n, of the soil water content sample from each trial, n, in accordance with Experiment 11.2 or 11.3.

Plastic Limit Procedure:

1. Select a 20 grams or more portion of soil from the material prepared for the LL test either after the second mixing before the test or from the soil remaining after completion of the LL test.
2. Reduce the water content of the soil to a consistency at which it can be rolled without sticking to the hands by spreading or mixing continuously on the glass plate or in the mixing/storage dish. The drying process may be accelerated by exposing the soil to the air current from an electric fan.
3. From the prepared 20 grams sample, select a 1.5–2 grams portion. Form the selected portion into an ellipsoidal mass.
4. Roll the soil mass between the palm or fingers and the ground-glass plate with just sufficient pressure to roll the mass into a thread of uniform diameter throughout its length. The thread should be further deformed on each stroke so that its diameter reaches 3.2 mm (1/8 inch), taking no more than 2 minutes. The amount of hand or finger pressure required will vary greatly according to the soil being tested, that is, the required pressure typically increases with increasing plasticity. Fragile soils of low plasticity are best rolled under the outer edge of the palm or at the base of the thumb. A normal rate of rolling for most soils should be 80–90 strokes per minute, counting a stroke as one complete motion of the hand forward and back to the starting position. This

rate of rolling may have to be decreased for very fragile soils. A 3.2 mm (1/8 inch) diameter rod or tube is useful for frequent comparison with the soil thread to ascertain when the thread has reached the proper diameter.

5. When the diameter of the thread becomes 3.2 mm, break the thread into several pieces. Squeeze the pieces together, kneading the soil between the thumb and first finger of each hand, reforming the soil into an ellipsoidal mass, and re-roll.
6. Continue this alternate rolling to a thread 3.2 mm in diameter, gathering together, kneading, and re-rolling until the thread crumbles under the pressure required for rolling and the soil can no longer be rolled into a 3.2 mm diameter thread. It does not matter if the thread breaks into threads of shorter length. Roll each of these shorter threads to 3.2 mm in diameter. The only requirement for continuing the test is that these threads can be reformed into an ellipsoidal mass and rolled out again.
7. If crumbling occurs when the thread has a diameter greater than 3.2 mm, this shall be considered a satisfactory end point, provided the soil has been previously rolled into a thread 3.2 mm in diameter. Crumbling of the thread will manifest itself differently with the various types of soil. Some soils fall apart in numerous small aggregations of particles, others may form an outside tubular layer that starts splitting at both ends. The splitting progresses toward the middle, and finally, the thread falls apart in many small platy particles. Fat clay soils require much pressure to deform the thread, particularly as they approach the plastic limit. With these soils, the thread breaks into a series of barrel-shaped segments about 3.2–9.5 mm (1/8 to 3/8 inch) in length.
8. Gather the portions of the crumbled thread together and place them in a container of known mass. Immediately cover the container.
9. Select another 1.5–2 grams portion of soil from the PL sample. Form the selected portion into an ellipsoidal mass and repeat Steps 4–9 until the container has at least 6 grams of soil.
10. Repeat Steps 3–9 to make another container holding at least 6 grams of soil.
11. Separately determine the water content of the soil in each container in accordance with Experiment 11.2 or 11.3.

Data Collection Format and Data Collection Sheets Tailored to Experiment

See attached data collection sheet.

Data Analysis Plan

Calculation for Liquid Limit:

1. Plot the relationship between the water content, w_n, and the corresponding numbers of drops of the cup, N_n, on a semilogarithmic graph with the water content as the ordinates on the arithmetical scale, and the number of drops as the abscissas on a logarithmic scale.

2. Draw the straight line of best fit through the (three or more) plotted points or add a trend line.
3. Take the water content corresponding to the intersection of the line with the 25 drop abscissa as the LL of the soil and round to the nearest whole number.

Calculation for Plastic Limit:

1. Compute the average of the two water contents (trial PLs) and round to the nearest whole number. This value is the PL.
2. Repeat the test if the difference between the two trial PLs is greater than the acceptable range of two results listed in Table 11.7.1.

Calculation for Plasticity Index:
Calculate the PI as follows:

$$PI = LL - PL$$

where: LL = liquid limit (whole number)
PL = plastic limit (whole number)

Both the LL and PL are reported as whole numbers. If either the LL or PL could not be determined, or if the PL is equal to or greater than the LL, report the soil as nonplastic, NP.

Expected Outcomes

These limits will vary widely for different soil types and matrices. Consistency with the values shown in Table 11.7.1 is expected.

TABLE 11.7.1
Acceptable Results

	AVERAGE VALUES OF RESULTS, % POINTS			ACCEPTABLE RANGE OF TWO RESULTS, +/− %		
SOIL TYPE	LL	PL	PI	LL	PL	PI
CH	60	21	39	2	1	2
CL	33	20	13	1	1	1
ML	27	23	4	2	2	2

Soil Type Based on Uniform Soil Classification System

Source: Table adapted from ASTM Method D 4318–17.

DATA COLLECTION SHEET LIQUID LIMIT, PLASTIC LIMIT, AND PLASTICITY INDEX OF SOILS	
Lead Experimenter	
Date of Experiment	
Type of Sample	
Description of Sample	
Source of Sample	
Equipment Used for Testing Sample	
Number of Drops to Close Groove (Step 6)	
Mass of Empty Liquid Limit Soil Container 1 (Step 7)	
Number of Drops to Close Groove (Step 6)	
Mass of Empty Liquid Limit Soil Container 2 (Step 7)	
Number of Drops to Close Groove (Step 6)	
Mass of Empty Liquid Limit Soil Container 3 (Step 7)	
Number of Drops to Close Groove (Step 6)	
Mass of Empty Liquid Limit Soil Container 4 (Step 7)	
Number of Drops to Close Groove (Step 6)	
Mass of Empty Liquid Limit Soil Container 5 (Step 7)	
Calculated Liquid Limit from Data Analysis Plan	
Water Content of Liquid Limit Sample 1 (Step 10), w_1	
Water Content of Liquid Limit Sample 2 (Step 10), w_2	

Water Content of Liquid Limit Sample 3 (Step 10), w_3	
Water Content of Liquid Limit Sample 4 (Step 10), w_4	
Water Content of Liquid Limit Sample 5 (Step 10), w_5	
Mass of Soil Container 1 for Plastic Limit (Step 8)	
Mass of Soil Container 2 for Plastic Limit (Step 8)	
Water Content of Plastic Limit Sample 1 (Step 11), w_1	
Water Content of Plastic Limit Sample 2 (Step 11), w_2	
Calculated Plastic Limit from Data Analysis Plan	
Calculated Plasticity Index from Data Analysis Plan	

EXPERIMENT 11.8

SPECIFIC GRAVITY OF SOIL SOLIDS BY WATER PYCNOMETER

(Consistent with ASTM Method D 854–14)

Date: ______________________________

Principal Investigator:

Name: ______________________________

Email: ______________________________

Phone: ______________________________

Collaborators:

Name: ______________________________

Email: ______________________________

Phone: ______________________________

Name: ______________________________

Email: ______________________________

Phone: ______________________________

Name: ______________________________

Email: ______________________________

Phone: ______________________________

Name: ______________________________

Email: ______________________________

Phone: ______________________________

Objective or Question to Be Addressed

This experiment uses a previously calibrated water pycnometer to determine the specific gravity of soil solids that pass the 4.75 mm (No. 4) sieve. When the soil sample contains particles larger than the 4.75 mm sieve, this test method is used only for the fraction of the soil sample passing the 4.75 mm sieve.

ASTM Method D 854–14 provides two methods for this experiment. The preferred method is described as the procedure for moist specimens, and that is the procedure on which this experiment is based. The alternative method uses oven-dried samples and yields less reliable results.

Theory behind Experiment

The specific gravity of soil solids is routinely used in civil engineering design to calculate various soil relationships, such as void ratio and degree of saturation, along with the density of the soil solids. Density is determined by multiplying the specific gravity by the density of water (at the proper temperature).

Potential Interferences and Interference Management Plan

The term "soil solids" refers to naturally occurring mineral particles or soil-like particles that are not readily soluble in water. Therefore, the specific gravity of soil solids containing extraneous water-soluble matter, such as sodium chloride, and soils containing contaminants with a specific gravity less than one, typically require special treatment or a qualified definition of their specific gravity. This experiment does not deal with those contaminated natural soils.

Data to Be Collected with an Explanation of How They Will Be Collected (Ex.: "Water Temperature Data for Three Days Using a Continuous Data Logger")

Using the attached data sheet, record at a minimum the following information:

- Name of all experimenters
- Date of the test
- Identification of the type of soil and source of the sample being tested
- The visual classification of the soil being tested (group name and symbol in accordance with the Uniform Soil Classification System)
- Percent of soil particles passing the 4.75 mm (No. 4) sieve
- If any soil or material was excluded from the test specimen, describe the excluded material
- All mass measurements (to the nearest 0.01 grams)
- Test temperature (to the nearest 0.1 °C)
- Specific gravity at 20 °C (G, Gs, $G_{20°C}$) to the nearest 0.01

Tools, Equipment, and Supplies Required

- Calibrated pycnometer
  - The water pycnometer shall be either a stoppered flask, stoppered iodine flask, or volumetric flask with a minimum capacity of 750–1,000 mL, and

properly calibrated in accordance with the manufacturer's instructions. The volume of the pycnometer must be 2–3 times greater than the volume of the soil–water mixture used during the de-airing portion of the test.

- Balance
    - A balance with a readability of 0.01 grams and a capacity of at least 1,000 grams
- Drying oven
    - A thermostatically controlled, forced-draft oven, capable of maintaining a uniform temperature of 110 +/– 5 °C (230 +/– 5 °F) throughout the drying chamber
- Thermometric device
    - Capable of measuring the temperature range from 0 to 150 °C (0–300 °F) with a readability of 0.1 °C (0.2 °F) and a maximum permissible error of +/– 0.5 °C (0.5 °F). The device must be capable of being immersed in the sample and calibration solutions to a depth ranging between 25 and 80 mm.
- Desiccator
    - A desiccator cabinet or large desiccator jar of suitable size containing silica gel or anhydrous calcium sulfate
- Entrapped air-removal apparatus: To remove entrapped air during the de-airing process, one of the following means should be utilized:
    - Hotplate or Bunsen burner capable of maintaining a temperature adequate to boil water
    - Vacuum system, vacuum pump, or water aspirator capable of producing a partial vacuum of 100 mm (26 inch) of mercury (Hg) or less absolute pressure
- Insulated container
    - A Styrofoam cooler and cover or equivalent container that can hold between three and six pycnometers plus a beaker (or bottle) of de-aired water, and a thermometer. This is required to maintain a controlled temperature environment where changes will be uniform and gradual.
- Funnel: A noncorrosive smooth-surface funnel with a stem that extends past the calibration mark on the volumetric flask or stoppered seal on the stoppered flasks. The diameter of the stem of the funnel must be large enough that soil solids will easily pass through.
- Pycnometer filling tube (optional): A device to assist in adding de-aired water to the pycnometer without disturbing the soil–water mixture. The device may be fabricated as follows. Plug a 6–10 mm (1/4 to 3/8 inch) diameter plastic tube at one end and cut two small vents (notches) just above the plug. The vents should be perpendicular to the axis of the tube and diametrically opposed. Connect a valve to the other end of the tube and run a line to the valve from a supply of de-aired water.
- Sieve: A 4.75 mm (No. 4) sieve
- Blender (optional): A blender with mixing blades built into the base of the mixing container
- Distilled water sufficient for mixing the soil sample
- Appropriate personal protective equipment, including heat-resistant gloves and tongs for removing hot items from the oven

Safety

Containers will be hot when removed from the drying oven. Handle hot containers with heat-resistant gloves or tongs. Some soil types can retain considerable heat, and serious burns could result from improper handling.

In the event of a burn, run the affected area under cold water immediately and seek medical assistance for pain and/or broken skin or blisters.

Suitable eye protection is recommended due to the possibility of particle shattering during the heating, mixing, or mass determinations.

Highly organic soils and soils containing oil or other contaminates may ignite into flames during drying. Means for smothering flames to prevent operator injury or oven damage should be available during testing.

Fumes given off from contaminated soils or wastes may be toxic, and the oven should be vented accordingly.

Due to the possibility of steam explosions, or thermal stress shattering porous or brittle aggregates, a loose or vented covering over the sample container during drying may be appropriate to prevent operator injury or oven damage. A cover also prevents scattering of the test sample in the oven during the drying cycle. Paper or cardboard covers may ignite if overheated and should be avoided.

Planned Experimental Procedures and Steps, in Order

1. Put on appropriate personal protective equipment and gather all necessary samples and supplies in a convenient location.
2. Determine the water content of a portion of the sample in accordance with Experiment 11.2 or 11.3.
3. Using this water content, select a wet mass for the specific gravity sample in accordance with Table 11.8.1. Obtain a sample within this range. Do not attempt to obtain an exact predetermined mass.

   NOTE: The test sample may be moist or oven-dried soil and representative of the soil solids that pass the 4.75 mm (No. 4) sieve in the total sample. Table 11.8.1 gives guidelines on recommended dry soil mass versus soil type and pycnometer size. Two important factors in determining the amount of soil solids to test are that the mass of the soil solids divided by its specific gravity should yield four significant figures and the mixture of soil solids and water should be a slurry, not a highly viscous fluid during the de-airing process.
4. Place about 100 mL of water into the mixing container of a blender or equivalent device.
5. Add the soil and blend.
6. Using the funnel, pour the slurry into the pycnometer. Rinse any soil particles remaining on the funnel into the pycnometer using a wash/spray squirt bottle.
7. Prepare the soil slurry by adding water until the water level is between 1/3 and 1/2 of the depth of the main body of the pycnometer. Agitate the water until slurry is formed. Rinse any soil adhering to the pycnometer into the slurry.
8. If a viscous paste, rather than a fluid slurry is formed, use a pycnometer having a larger volume.

9. De-air the soil slurry using either heat (boiling), vacuum, or combining heat and vacuum.
   a. When using the heat-only method (boiling), use a duration of at least 2 hours after the soil–water mixture comes to a full boil. Use only enough heat to keep the slurry boiling. Agitate the slurry as necessary to prevent any soil from sticking to or drying onto the glass above the slurry surface.
   b. If only a vacuum is used, the pycnometer must be continually agitated under the vacuum for at least 2 hours. Continually agitated means the silt/clay soil solids will remain in suspension, and the slurry will be in constant motion. The vacuum must remain relatively constant and be sufficient to cause bubbling at the beginning of the de-airing process.
   c. If a combination of heat and vacuum are used, the pycnometers can be placed in a warm water bath (not more than 40 °C [104 °F]) while applying the vacuum. The water level in the bath should be slightly below the water level in the pycnometer because if the pycnometer glass becomes hot, the soil will typically stick to or dry onto the glass. The duration of vacuum and heat must be at least 1 hour after the initiation of boiling. During the process, the slurry should be agitated as necessary to maintain boiling and prevent soil from drying onto the pycnometer.
10. Fill the pycnometer with de-aired water by introducing the water through a piece of small-diameter flexible tubing with its outlet end kept just below the surface of the slurry in the pycnometer. If the added water becomes cloudy, do not add water above the calibration mark or into the stopper seal area. Add the remaining water the next day.
11. If using the stoppered iodine flask, fill the flask so that the base of the stopper is submerged in water. Then rest the stopper at an angle on the flared neck to prevent air entrapment under the stopper.
12. If using a volumetric or stoppered flask, fill the flask to above or below the calibration mark, depending on preference.
13. If heat has been used, allow the specimen to cool to approximately room temperature before proceeding.
14. Put the pycnometer(s) into a covered insulated container along with the thermometric device (or the temperature-sensing portion of the thermometric device), a beaker (or bottle) of de-aired water, stopper(s) (if a stoppered pycnometer is being used), and either an eyedropper or pipette. Keep these items in the closed container overnight to achieve thermal equilibrium.
15. If the insulated container is not positioned near a balance, move the insulated container near the balance or vice versa. Open the container and remove the pycnometer. Only touch the rim of the pycnometer, using a gloved hand, because the heat from hands can change the thermal equilibrium.
16. Place the pycnometer on an insulated block (Styrofoam or equivalent).
17. If using a volumetric flask, adjust the water to the calibration mark following the procedure in Step 10. If a stoppered flask is used, adjust the water to prevent entrapment of any air bubbles below the stopper during its placement. If

water has to be added, use the thermally equilibrated water from the insulated container. Then place the stopper in the bottle. If water has to be removed, before or after inserting the stopper, use an eyedropper from the insulated container.

18. Dry the rim of the pycnometer using a paper towel. Be sure the entire exterior of the flask is dry.
19. Measure and record the mass of the pycnometer containing the soil and water to the nearest 0.01 grams using the same balance used for pycnometer calibration.
20. Measure and record the test temperature, T_t, of the slurry/soil–water mixture to the nearest 0.1 °C using the thermometric device and method used during the pycnometer calibration.
21. Determine the mass, P_1, of a clean, empty, pan to the nearest 0.01 grams.
22. Transfer the soil slurry to the pan. It is imperative that all of the soil be transferred. Water can be added as needed to ensure this.
23. Dry the sample slurry to a constant mass in an oven maintained at 110 +/– 5 °C (230 +/– 5 °F) and then cool it in a desiccator. If the pan can be sealed so that the soil cannot absorb moisture during cooling, a desiccator is not required.
24. Measure the dry mass of soil solids plus the pan, P_2, to the nearest 0.01 grams using the designated balance.
25. Calculate and record the mass of dry soil solids (P_2–P_1) to the nearest 0.01 grams.

Data Collection Format and Data Collection Sheets Tailored to Experiment

See attached data collection sheet.

Data Analysis Plan

The mass of the pycnometer and water at the test temperature is calculated by the following equation:

$$M_{pw,t} = M_p + (V_p \times \rho_{w,t})$$

where: $M_{pw,t}$ = mass of the pycnometer and water at the test temperature (T_t), grams
M_p = average calibrated mass of the dry pycnometer, grams
V_p = average calibrated volume of the pycnometer, mL
$\rho_{w,t}$ = density of water at the test temperature (Tt), g/mL from Table A.5

Calculate the specific gravity of soil solids at the test temperature, Gt, using the following equations:

$$G_t = p_s/\rho_{w,t}$$
$$M_s = (M_{pw,t} - (M_{pws,t} - M_s))$$

where: G_t = specific gravity of the soil solids at the test temperature
p_s = density of the soil solids, Mg/m^3 or g/cm^3
$\rho_{w,t}$ = density of water at the test temperature, T_t, from Table A.5 g/mL or g/cm^3
M_s = mass of the oven dry soil solids (g)
$M_{pws,t}$ = mass of pycnometer, water, and soil solids at the test temperature, T_t, grams

Calculate the specific gravity of the soil solids at 20 °C using the following equation:

$$G_{20\,°C} = K \times G_t$$

where: K = the temperature coefficient given in Table A.5 in the Appendix for the test temperature used for the test

Expected Outcomes

The outcomes from this experiment will depend on the actual soil samples tested.

TABLE 11.8.1
Suggested Mass of Samples When Using a 500 mL Pycnometer

USC SOIL TYPE	DRY MASS, grams
SP, SP-SM	100 +/− 10
SP-SC, SM SC	75 +/− 10
Silt or Clay	50 +/− 10

DATA COLLECTION SHEET SPECIFIC GRAVITY OF SOIL SOLIDS BY WATER PYCNOMETER	
Lead Experimenter	
Date of Experiment	
Source of Sample	
Visual USC Classification of Sample	
Percent of Soil Particles Passing the 4.75 mm (No. 4) Sieve	
Calculated Water Content of Sample (Step 2)	
Mass of the Pycnometer with Soil and Water (Step 19)	
Temperature of Slurry (Step 20), T_1	
Mass of Clean, Empty Pan (Step 21), P_1	
Mass of Oven-Dried Soil and Pan (Step 24), P_2	
Mass of Oven-Dried Solids (Step 25)	
Calculated Mass of Pycnometer and Water at the Test Temperature	
Calculated Specific Gravity of Soil Solids at Test Temperature	
Calculated Specific Gravity of Soil Solids at 20 °C	

EXPERIMENT 11.9[1]

DETERMINATION OF POROSITY AND VOID RATIO IN SOIL

Date: ______________________________

Principal Investigator:

Name: ______________________________

Email: ______________________________

Phone: ______________________________

Collaborators:

Name: ______________________________

Email: ______________________________

Phone: ______________________________

Name: ______________________________

Email: ______________________________

Phone: ______________________________

Name: ______________________________

Email: ______________________________

Phone: ______________________________

Name: ______________________________

Email: ______________________________

Phone: ______________________________

Name: ______________________________

Email: ______________________________

Phone: ______________________________

Objective or Question to Be Addressed

This experiment assesses various soil types by demonstrating the variation and magnitude of porosity and void ratio and allowing comparison of those characteristics.

Theory behind Experiment

The characteristics associated with a specific soil type factor into the stability of the soil, transport of water through the soil, and ability of contamination to spread throughout the soil. Void ratio and porosity are commonly used soil characteristics. The *void ratio* is the ratio of the volume of voids (open spaces, i.e., air and water) in a soil to the volume of solids. *Porosity* is the ratio of void volume to *total* volume.

Because it is a ratio between two components of a composite volume of material, the void ratio can be greater than 1. It can also be expressed as a fraction. The porosity is a ratio between a component of a given volume of material and the total volume of the composite material. Therefore, the porosity cannot be greater than 1, but it is expressed as a percentage or a fraction.

The void ratio and porosity are slightly different. The void ratio is the ratio of *void volume* to *solids volume*; porosity is the ratio of *void volume* to *total volume.* This creates a situation where the porosity can never be equal to or greater than 1, but the void ratio can, and does, exceed 1 in certain soil types in which the volume of the soil particles is less than the volume of the voids.

The generally accepted equation for the calculation of the void ratio is:

$$E = \frac{V_v}{V_s} = \frac{\left(V_a + V_w\right)}{V_s}$$

Where:

E = void ratio
V_v = volume of voids, m^3 or ft^3
V_s = volume of solids, m^3 or ft^3
V_a = volume of air, m^3 or ft^3
V_w = volume of water, m^3 or ft^3

The generally accepted equation for the calculation of porosity is:

$$\eta = \frac{V_v}{V_s}$$

Where:

η = porosity
V_v = volume of voids, m^3 or ft^3
V_s = volume of total sample, m^3 or ft^3

It is assumed that filling the voids in the soil with water will not cause an increase in the total volume of the saturated soil. It is also assumed that any void not filled with water will be filled with air. However, due to the inherent buoyancy of soil particles in water, it is possible for some swelling to occur as water is added. An adjustment needs to be made to the output data to account for any swelling that may occur. Hence, the test procedure requires a careful measurement of the volume of the soil prior to the addition of water and again at the end of the water addition phase.

Temperature can also affect the volume of the saturated soil. The water will expand in volume as the temperature rises. Therefore, it is important to ensure that the entire procedure is done as rapidly as possible and that all components are at the essentially same room temperature throughout the procedure.

Data to Be Collected with an Explanation of How They Will Be Collected (Ex.: "Water Temperature Data for Three Days Using a Continuous Data Logger")

The information to be gathered in this experiment are those data required to properly calculate the void ratio and porosity of the various soil types being tested, the appropriate soil classification of the soil being tested by the USCS (Uniform Soil Classification System), the volume of water used to fill the voids, and the total volume of soil and voids.

The soil volume is calculated in four ways, using the:

1. Volume of water required to fill the voids, subtracted from the total volume of saturated soil
2. Mass of the soil used multiplied by standard soil mass values found from standard references
3. Subtracting of the mass of water used to fill the voids, divided by the standard mass of water at the test temperature, from the total mass of saturated soil
4. Measurement of the volume used as indicated by the graduations on the side of each beaker

A reasonable average volume of soil is then calculated. The amount of compaction of the soil samples is important to these calculations. It is instructive, therefore, to test samples that are tamped lightly to represent "normal" soil conditions and to simultaneously test compacted samples to represent compacted conditions.

Potential Interferences and Interference Management Plan

The accuracy of the results in this experiment is particularly important because small changes in the data can have a statistically significant influence on the results. The sensitivity of data to specific independent variable changes is not uniform. It is important to consider the issue of variable sensitivity when evaluating the dependent variable changes associated with multiple independent variable adjustments. See Section 2.6 in Chapter 2 for a discussion of variable sensitivity.

The masses of soil and water used are sufficiently small that minor changes in the measurements can generate significant errors in the final calculations. Therefore, it is important to weigh masses accurately, measure volumes carefully, and avoid disruption of the samples when adding water to the tamped soils.

The temperature of all the materials should be essentially the same (+/– 1 °C or 2 °F) at the start and conclusion of the experiment. Changes in temperature can change the volume of the water or soil by a minor amount, but that amount will become significant when extrapolated to a large area or volume of material. If additional water is needed beyond that previously set aside, it will be important to ensure that the temperature of the new water is essentially the same (+/– 1 °C or 2 °F) as the original water being used.

Tools, Equipment, and Supplies Required

This experiment anticipates that soil from three different soil classifications will be tested. The materials and tools needed for more than three classifications or samples must be adjusted accordingly.

- Approximately 2 kg (4.4 lb) of each of three different soil types
- Six 1 L (0.26 gal) clean, graduated, glass beakers
- Three clean spatulas
- Approximately 4 L (1.2 gal) of tap water

Safety

Safety issues with this experiment include glass breakage, leading to cuts or puncture wounds and splashing of dirty water into eyes. Safety gloves, eye goggles, and a lab coat should be worn during conduct of these experiments.

Planned Experimental Procedures and Steps, in Order

1. Assemble all required tools and supplies in a convenient and handy location and put on the appropriate personal protection equipment.
2. Classify or verify prior classifications of all soil types to be tested.
3. Arrange and label all beakers for the type of soil and compaction status to be tested.
4. Weigh each beaker and record the weights.
5. Slowly scoop approximately 500 mL (0.13 gal) of the appropriate soil type into each labeled beaker.
6. Properly tamp the contents of each beaker either lightly or completely, as the particular beaker requires.
7. Weigh each beaker and record the mass of soil as the difference between the filled mass of the beaker and the empty mass of the beaker.
8. Measure and record the volume of soil in each beaker based on the graduations shown on the side of the beakers.
9. Record the temperature of the water to be used to saturate the soil and ensure that the soil samples are at normal room temperature (approximately 22 °C or 72 °F).
10. Slowly introduce tap water into each sample from a graduated cylinder or volumetric pipet. Care needs to be taken to avoid significant disruption of

the tamped soil surface during addition of the water and to distribute the added water across the entire surface of the sample. Allow the soil to fully saturate as the water is added. Saturation has been attained when the water level in the beaker and the top of the soil in the beaker are identical and no further absorption is occurring.

11. Carefully record the volume of water added to each beaker.
12. Carefully weigh each beaker and record the masses.
13. Note any change in soil volume that has occurred, based on the graduations on the side of the beaker, as a result of water addition, leading to swelling of the soil, and record the changed total volume.
14. Calculate the volume of water used by subtracting the mass of dry soil from the mass of saturated soil and dividing by the unit weight of water at the measured temperature (see Table A.5).
15. Calculate the volume of soil used by each of the four methods discussed in the experiment theory section and establish the volume to be used for the porosity and void ratio calculations.
16. Calculate the porosity and void ratios for each soil type using the equations provided in the experiment theory section.

Data Collection Format and Data Collection Sheets Tailored to Experiment

See attached data collection sheet.

Data Analysis Plan

Review of the measured and calculated data will determine the outcomes of this experiment. Since the experiment is designed to identify specific soil characteristics, reporting of those measured characteristics for each of the tested soils, being careful not to expand the data to soil classifications or types not measured, is the desired outcome of the experiment. Discussing any differences between the characteristics measured and the standard value provided will be instructive as an outcome of this experiment. Students may also consider the different characteristics and how they may impact engineering decisions, such as structural design or contaminant treatment processes.

Anticipated Outcomes

Soils classified by the Uniform Soil Classification System typically demonstrate void ratios and porosities as shown in Table A.4 in the Appendix. Density estimates for various soil types are provided in Table A.3 in the Appendix. Table A.5 provides the density of water at various temperatures. Intermediate temperatures may be interpolated linearly.

DATA COLLECTION SHEET SOIL POROSITY AND VOID RATIO DETERMINATIONS								
DATE SAMPLES COLLECTED:								
DATE SAMPLES TESTED:								
PRINCIPAL INVESTIGATOR:								
Soil Type and Compaction Status	Sample 1		Sample 2		Sample 3		Sample 4	
	Tamped	Compacted	Tamped	Compacted	Tamped	Compacted	Tamped	Compacted
Empty Beaker Weight, g								
Filled Beaker Weight, g								
Weight of Sample in Beaker, g								
Volume of Sample Based on Beaker Markings, mL								
Temperature of Water to be Added to Sample								
Volume of Water Added to Sample, mL								
Weight of Saturated Sample and Beaker, g								
Volume of Saturated Sample Based on Beaker Markings, mL								
Change in Volume as a Result of Saturation, mL								
Calculated Volume of Water Used Based on Subtracting the Mass of Dry Sample from the Mass of Saturated Sample and Dividing by the Unit Mass of Water at Test Temperature, mL								
Calculated Sample Volume Based on the Volume of Water Required to Fill the Voids, Subtracted from the Total Volume of Saturated Sample, mL								
Calculated Sample Volume Based on the Mass of the Sample Used Multiplied by Standard Sample Mass Values Found from Standard References, mL								
Calculated Sample Volume Based on Subtracting the Mass of Water used to fill the Voids, Divided by the Standard Density of Water at the Test Temperature, from the Total Mass of Saturated Sample, mL								
Calculated Sample Volume Based on Carefully Measuring the Volume of Sample Used as Indicated by the Graduations on the side of Each Beaker, mL								
Calculated Porosity of the Sample, %								
Calculated Void Ratio of the Sample, %								

EXPERIMENT 11.10

BULK DENSITY AND VOIDS IN AGGREGATE

(Consistent with ASTM Method C 29)

Date: ______________________________

Principal Investigator:

Name: ______________________________

Email: ______________________________

Phone: ______________________________

Collaborators:

Name: ______________________________

Email: ______________________________

Phone: ______________________________

Name: ______________________________

Email: ______________________________

Phone: ______________________________

Name: ______________________________

Email: ______________________________

Phone: ______________________________

Name: ______________________________

Email: ______________________________

Phone: ______________________________

Objective or Question to Be Addressed

This experiment is used to characterize aggregate, a construction material that includes rock, stone, soil, and particles. *Unit weight* is the traditional terminology used to describe the property determined by this experiment, while *dry bulk density* is the term used in a lab setting, but both terms refer to the weight per unit volume, or density, of the material. The dry bulk density of aggregate will be determined for a compacted or loose condition and to calculate the ratio of voids between particles in fine, coarse, or mixed aggregates not exceeding 125 mm (5 inches) in nominal maximum size.

Theory behind Experiment

Bulk density values are used to make decisions about construction materials and methods. They are used for selecting the proportions of aggregates in concrete mixtures and for determining the mass of aggregates needed to fill specific excavations and trenches. This procedure results in a calculation for dry bulk density, but water and surface moisture can impact construction and material transportation. This is why it is important to note that aggregates in hauling units and stockpiles usually contain absorbed and surface moisture, which can affect bulking. This procedure provides a method for computing the percentage of voids between the aggregate particles based on the bulk density of the dry aggregates.

Potential Interferences and Interference Management Plan

The size of the test sample should be between 125 and 200 % of the quantity required to fill the measure, as defined in the procedures section, in order to ensure that a truly representative sample is tested. Handling of the aggregates should be done in a way that avoids segregation of the grain sizes. During the experiment, the sample is dried to a constant mass in an oven at 110 +/– 5 °C (230 +/– 10 °F). The sample is then considered to be at constant mass. The state of a constant mass is defined as having been reached when the difference in mass between two consecutive measurements, taken one hour apart, is less than 0.1 % of the second weighing.

Data to Be Collected with an Explanation of How They Will Be Collected (Ex.: "Water Temperature Data for Three Days Using a Continuous Data Logger")

Record and report at a minimum the following information:

- Names of all experimenters
- Date of experiment and date of sample collection
- Identification of the type of aggregate (granite, schist, etc.) and the source of the sample (crushed rock, river run, etc.)
- Visual classification of the soil being tested (group name and symbol in accordance with the Uniform Soil Classification System)
- Maximum nominal size of the aggregate
- Description of any materials excluded from the test specimen

- All mass measurements (to the nearest 0.01 g)
- Test temperature (to the nearest 0.1 °C)
- Dry bulk density from this experiment to the nearest 10 kg/m^3 (1 lb/ft^3)
- Void content to the nearest 1 %

Tools, Equipment, and Supplies Required

- Balance or scale accurate to the greater of 0.05 kg (0.1 lb) or 0.1 % of the mass of material being tested, and a capacity of 2,000 kg/m^3 (125 lb/ft^3) times the volume of the measure being used
- Round, plain, steel tamping rod with a diameter of 16 +/– 2 mm (5/8 +/– 1/16 inches) and a length of at least 100 mm (4 inches) greater than the depth of the measuring container or mold in which rodding is being performed, but not greater than 750 mm (30 inches) in overall length
- Standard (commercially available) cylindrical measuring container made of steel or other suitable metal, watertight and sufficiently rigid to retain its form under rough usage, preferably with handles
- Shovel or scoop – a shovel or scoop of convenient size for filling the measuring container with aggregate
- Equipment for measuring volume of the measure, if not specified by the manufacturer

Safety

The volumes of aggregate being handled may be heavy, and care is needed when lifting or moving containers of these aggregates.

Dust associated with rodding or jigging of the samples will be hazardous if inhaled. A dust mask is recommended for use in this experiment.

Personal protection equipment required includes gloves, eye protection, and a lab coat.

Planned Experimental Procedures and Steps, in Order

1. Put on appropriate personal protective equipment and gather all materials and supplies in a convenient location.
2. Determine and record the mass and volume of the empty measuring container to the nearest 0.05 kg (0.1 lb). This is the value of the variable T.
3. Determine and record the volume of the empty measuring container, as provided by the manufacturer. This is the value of the variable V.
   a. Determine the mass of a glass plate at least 6 mm (1/4 inch) thick and at least 25 mm (1 inch) larger than the diameter of the measure to be calibrated, measured to the nearest 0.05 kg (0.1 lb).
   b. Place a thin layer of grease on the rim of the measuring container to prevent leakage of water from the container.
   c. Fill the container with water at room temperature and cover with the plate glass in such a way as to eliminate bubbles and excess water.
   d. Remove any water that may have overflowed onto the measure or plate glass.

e. Determine the mass of the water, plate glass, and measure to the nearest 0.05 kg (0.1 lb). This is the value of the variable W.
f. Measure the temperature of the water to the nearest 0.5 °C (1 °F) and determine its density from Table A.5 in the Appendix.
g. Determine the volume of the container using the procedure included in the data analysis plan.

4. Consolidate the sample to be tested.
   - For samples having a nominal maximum size of 37.5 mm (11/2 inches) or less, consolidate by rodding.
   - For samples having a nominal maximum size greater than 37.5 mm (11/2 inches) but less than 125 mm (5 inches), use jigging.

5a. Rodding:

a1. Fill the measure one-third full and level the surface with the fingers.
a2. Rod the layer of aggregate with 25 strokes of the tamping rod evenly distributed over the surface.
a3. Fill the measure two-thirds full and again level and rod, as previously described.
a4. Fill the measure to overflowing and rod again in the manner previously described.
a5. Level the surface of the aggregate with the fingers or a straightedge in such a way that any slight projections of the larger pieces of the coarse aggregate approximately balance the larger voids in the surface below the top of the measure.

5b. Jigging:

b1. Fill the measure in three approximately equal layers as described in 5a, compacting each layer by placing the measure on a firm base, such as a cement-concrete floor, raising the opposite sides alternately about 50 mm (2 inches), and allowing the measure to drop in such a manner as to hit with a sharp, slapping blow.
b2. Compact each layer by dropping the measure 50 times in the manner described in step 5b1, 25 times on each side.
b3. Level the surface of the aggregate with a straightedge (ruler) in such a way that any slight projections of the larger pieces of the coarse aggregate approximately balance the larger voids in the surface below the top of the measure.

6. Determine and record the mass of the measure plus its contents to the nearest 0.05 kg (0.1 lb) in accordance with the data analysis plan. This is the value of variable G.

Data Collection Format and Data Collection Sheets Tailored to Experiment

See attached data collection sheet.

Data Analysis Plan

If not provided by the manufacturer, determine and calculate the volume of the measuring container as follows:

$$V = (W - M_2)/D$$
$$F = D/(W - M_2)$$

where: V = volume of the measure, m^3 (ft^3)
W = mass of the water, plate glass, and measure, kg (lb)
M_2 = mass of the plate glass and measure, kg (lb)
D = density of the water for the measured temperature, kg/m^3 (lb/ft^3)
F = factor for the measure, $1/m^3$ ($1/ft^3$)

Calculate the bulk density from the following equations.

$$M = (G - T)/V$$

or:

$$M = (G - T) \times F$$

where: M = bulk density of the aggregate, kg/m^3 (lb/ft^3)
G = mass of the aggregate plus the measure, kg (lb) (Step 6)
T = mass of the measure, kg (lb) (Step 2)
V = volume of the measure, m^3 (ft^3) (Step 3)
F = factor for measure, m^{-3} (ft^{-3})

Calculate the void content in the aggregate using the bulk density from the following equations.

$$\% \text{ Voids} = 100 \times ((S \times W) - M)/(S \times W)$$

where: M = bulk density of the aggregate, kg/m^3 (lb/ft^3)
S = bulk specific gravity (dry basis) as determined by Experiment 10.8 or 10.9
W = density of water, 998 kg/m^3 (62.3 lb/ft^3) at 70 °F (21 °C). See Table A.5 in the appendix for the properties of water with density at various temperatures.

Expected Outcomes

The expected outcomes from this experiment should be consistent with the data provided in Table A.3 in the Appendix.

DATA COLLECTION SHEET BULK DENSITY AND VOIDS IN AGGREGATE	
Lead Experimenter	
Date of Experiment	
Date of Sample Collection	
Source of Aggregate Sample	
Aggregate Type (Granite, Schist, Sandstone, etc.)	
Visual USC Classification	
Maximum Nominal Size of Aggregate	
Describe Any Excluded Material	
Mass (T) of Empty Measuring Container to nearest 0.01 lb (0.05 kg) (Step 2)	
Volume (V) of Empty Measuring Container to nearest 0.01 pt (0.05 L) (Step 3)	
Mass of Glass Plate (Step 3a)	
Mass (W) of Glass Plate, Plus Measure and Water to 0.05 g (0.1 lb) (Step 3e)	
Temperature of Water to 0.5 °C (1.0 °F) (Step 3f)	
Calculate Mass (M_2) of Plate Glass Plus Measuring Container	
Mass (G) of the Measure Plus Its Contents to Nearest 0.05 kg (0.1 lb) (Step 6)	
Calculated Bulk Density (M) of the Aggregate to Nearest 0.05 kg (0.1 lb)	
Calculated Void %	

NOTE

1 Francis J. Hopcroft and Abigail J. Charest, *Experiment Design for Environmental Engineering, Methods and Examples* (CRC Press, 2022, Experiment 10.3): 140.

12 Environmental Assessment Experiments

This chapter provides completed designs for example experiments that can be used to demonstrate specific engineering phenomena in a university laboratory or field collection site. They may be used as presented or modified by the user to adapt to other desired outcomes. Sections on actual outcomes and data interpretation are necessarily left blank in these examples. They should be completed by the investigators upon collection of the data.

Note that where an experiment indicates that it is consistent with a specified ASTM method or other standard, this does not mean that the procedure is identical to, nor does it include all the details of, the stated standard method. Outcomes that require strict compliance with the stated standards must use that standard method to conduct the test. Therefore, the experiment procedures outlined in this book are not a substitute for proper compliance with the stated standards, nor are they intended to be.

The experiments in this chapter are:

- 12.1 Membrane Filter Technique for Bacteria Testing
- 12.2 Multiple Tube Fermentation (MPN) Test for Bacteria
- 12.3 Coagulation, Floculation, and Sedimentation for the Removal of Organic and Inorganic Water Contaminants
- 12.4 Jar Test Procedure
- 12.5 Gravimetric Determination of Particulate Matter in Air
- 12.6 Static Puncture Strength of Geotextiles and Geotextile-Related Products
- 12.7 Specific Heat Capacity of Materials
- 12.8 Percolation Test Procedure

DOI: 10.1201/9781003346685-12

EXPERIMENT 12.1

MEMBRANE FILTER TECHNIQUE FOR COLIFORM AND FECAL COLIFORM

(Consistent with EPA Method 9132)

Date: ______________________________

Principal Investigator:

Name: ______________________________

Email: ______________________________

Phone: ______________________________

Collaborators:

Name: ______________________________

Email: ______________________________

Phone: ______________________________

Name: ______________________________

Email: ______________________________

Phone: ______________________________

Name: ______________________________

Email: ______________________________

Phone: ______________________________

Name: ______________________________

Email: ______________________________

Phone: ______________________________

Objective or Question to Be Addressed

The concentration of total coliform and fecal coliform bacteria in water can be determined in a variety of ways. This experiment is used to quantify the concentrations of total coliform and fecal coliform bacteria in a sample of water. The source water is assessed using a series of dilutions and biological methods to enumerate the colonies. This experiment examines the use of a microfilter to capture the target bacteria, incubates them for 24 hours, and determines the concentration of bacteria based on the volume of sample used and the number of colonies that grow on a proprietary substrate during the 24 hours of incubation.

Theory behind Experiment

Bacteria are too small to observe with the naked eye and normally require a high-resolution microscope to observe them. If they are allowed to grow into masses of bacteria, called *colonies*, the colonies can be counted (enumerated). The number of colonies counted is assumed to be equal to the original concentration of bacteria in the source water. When a single proprietary growth medium is used to assist with the growth of the bacteria in the different petri dishes, a form identified as M-Endo for total coliform and MF-Endo for fecal coliform, the total coliform colonies and the fecal coliform colonies will each display a golden-green or slightly reddish metallic sheen. It is important, therefore, to know which growth medium was used so that the correct form of bacteria is reported. When the specific growth medium is used, it also depresses the growth of other forms of bacteria, allowing for easier counting of the target colonies. There is also available a mixed growth medium, which will allow both total coliform and fecal coliform colonies to grow in the same petri dish simultaneously. In this case, the total coliform colonies will display a red color and the fecal coliform will display a blue color, but they are both in the same area and need to be counted separately. Since fecal coliform are also a coliform bacteria, when both are grown together in the same petri dish, the two sums need to be added for a final total coliform count.

In general, the procedure utilizes biological methods. A predetermined amount of sample (ideally 100 mL since the count is reported as the number of colony-forming units per 100 mL) is filtered through a membrane filter, generally a 0.45 micron filter, which retains the bacteria in the sample, while allowing the water and other smaller bacteria to pass through the filter. The filters (containing the bacteria larger than 0.45 microns) are placed in a sterilized petri dish on an absorbent pad saturated with M-Endo media or MF-Endo media and incubated inside a dark incubator for 24 +/– 1 hr at 35 +/– 0.5 °C (95 +/– 1 °F). Colonies displaying the characteristic sheen are then counted, typically using a standard laboratory table-top microscope for ease of counting, and the results are reported as the number of colony-forming units per 100 mL of original sample.

Occasionally, the bacterial colonies are grown on an M-Endo agar, rather than on a saturated pad. This technique is done the same way as the one described in this experiment, except that the agar needs to be mixed, poured into the petri dishes, and allowed to gel in a sterile environment prior to inserting the membrane filter. In those cases, the membrane filter is rolled directly onto the agar surface.

Note that EPA Method 9132 provides an intermediary step in which the saturated filters are first placed on an absorbent pad saturated with lauryl tryptose broth and incubated at 35 +/− 0.5 °C (95 +/− 1 °F) for 2 hours, then moved to a second absorbent pad saturated with the M-Endo media, MF-Endo media, or M-Endo agar for an additional 21 +/− 1 hours at 35 +/− 0.5 °C (95 +/− 1 °F). This step ensures that sufficient nutrients are available for rapid initial growth of the bacteria. It is usually not an issue for students testing wastewater samples or contaminated water samples. Indeed, it is usually necessary to highly dilute those samples in order to find a sufficiently low number of colonies to actually count. If the counts seem unusually low relative to the expected concentrations, a second test using the lauryl tryptose broth step may yield more reliable results. Even with natural water sources, however, this step is not normally used.

A more detailed treatment of this method is presented in *Standard Methods for the Examination of Water and Wastewater* and in *Microbiological Methods for Monitoring the Environment.*

Potential Interferences and Interference Management Plan

The presence of residual chlorine or other halogens can prevent or stunt the continuation of bacterial growth. To prevent this, sodium thiosulfate may be added to the sample as soon as possible after sampling. In a student lab setting, it is generally more useful to wait for all disinfectant to dissipate prior to testing, but the sodium thiosulfate may be used to stop the disinfection action when intermediate tests are desired.

Water samples high in copper, zinc, or other heavy metals that can be toxic to bacteria may lead to incorrect counts. Chelating agents, such as ethylenediaminetetraacetic acid (EDTA), may be added to reduce the presence of those toxins but should only be used where it is not possible to conduct the bacterial count test immediately after sampling and only when heavy metals are suspected of being present.

Turbidity caused by the presence of algae, suspended silt or clay particles, or other interfering material will coat the filter and obscure accurate counts of the colonies. In this case, use of a different technique for the bacteria counts is recommended, such as EPA Method 9131, the Multiple Tube Fermentation Technique (see Experiment 12.2).

Data to Be Collected with an Explanation of How They Will Be Collected (Ex.: "Water Temperature Data for Three Days Using a Continuous Data Logger")

The data to be collected during this experiment are temperature data inside the incubator for a period of 24 hours using a temperature data logger. This is to ensure that the samples were properly incubated and that the resulting counts are reasonably accurate.

After incubation, the number of colony-forming units in each petri dish will be counted and recorded on the data collection sheet. Where multiple tests are conducted on various dilutions of the sample, the results will be averaged to yield a more reliable actual value. See the data analysis section of this experiment for how to do that averaging.

Using the attached data sheet, record at a minimum the following information:

- Lead experimenter
- Date of experiment

- Date of source water collection
- Storage conditions of source water
- Source of water
- Colony counts

Tools, Equipment, and Supplies Required

- 12 clean, sterilized, glass dilution bottles
- 12 clean, sterilized pipets of any convenient size, provided that they deliver the required volume accurately and quickly
- 12 clean, sterilized, graduated cylinders
- 12 commercially prepared, sealed vials of culture medium (the use of agar is not recommended due to the difficulties of sterilized preparation and the ready availability of sterile vials of liquid growth medium that are packaged in 1 mL vials for easy use of 1 vial per petri dish)
- 12 commercially prepared, sterilized petri dishes, 50 × 12 mm (2 inches × 0.4 inches), or other appropriate size, containing a sterilized absorbent pad of the same inside diameter
- 1 sterilized filtering apparatus (constructed of glass, plastic, porcelain, or any noncorrosive and bacteriologically inert metal) consisting of a seamless funnel fastened by a locking device to a porous plate on top of a tube inserted into the top of a 1 L Erlenmeyer filtering flask
- 1 electric vacuum pump, a filter pump operating on water pressure, a hand aspirator, or other means of securing a reliable and constant pressure differential. An additional flask may be connected between the filtering flask and the vacuum source to trap carryover water, if desired, to protect the pump
- 12 sterile, gridded, filter membranes (0.45 +/– 0.02 um)
- 1 pair of sterile forceps with rounded tips to minimize the risk of damaging the filter paper or sterile pads
- 1 incubator capable of providing and maintaining a temperature of 35 +/– 0.5 °C (95 +/– 1 °F) and a high level of humidity (approximately 90 % relative humidity).
- 1 microscope and light source with a magnification of 10–15 diameters and a light source adjusted to give maximum sheen discernment. Avoid using a microscope illuminator with an optical system for light concentration from an incandescent light source for coliform colony identification on Endo-type media
- Sodium sulfate or lauryl tryptose broth, if used, should be laboratory grade and sterile
- 1 L sample of water to be tested, to be diluted as necessary: an ideal diluted sample volume will yield a count of about 20–80 colony-forming units for total coliform or for fecal coliform bacteria, and not more than about 150 other miscellaneous bacteria colony-forming units of other types
- 1 L of sterile deionized water (lab water)
- 1 L of sterile stock phosphate buffer solution
    - Prepared by dissolving 34 g potassium dihydrogen phosphate (KH_2PO_4) in 500 mL lab water, adjusted to pH 7.2 +/– 0.5 with 1 N sodium hydroxide (NaOH), and diluted to 1 L with lab water

- 1 L of dilution water
  - Prepare by adding 1.25 mL stock phosphate buffer solution and 5 mL magnesium chloride solution (38 g $MgCl_2$/liter lab water or 81.1 g $MgCl_2 \cdot 6\ H_2O$/liter lab water) to 1 L lab water. Dispense in amounts that will provide 99 +/− 2 mL or 9 +/− 0.2 mL after autoclaving for 15 minutes

Safety

Bacteriological tests are done to determine the safety of the water sample. It is assumed at the outset that the sample contains a high concentration of bacteria, specifically coliform and/or fecal coliform bacteria, that are harmful to humans. Care needs to be taken to avoid contact with the samples, including accidental ingestion or contact with eyes. The use of eye protection, protective gloves, and a lab coat is required. Note that ingestion or eye contact can occur from touching the sample with a gloved hand and then touching the face, mouth, or eye with the contaminated glove.

In the event of physical contact, immediately wash the area thoroughly with hot water and soap. Change gloves.

In the event of eye contact, flush the eye with an eye station wash for up to 15 minutes and contact medical assistance for further advice.

Planned Experimental Procedures and Steps, in Order

1. Put on appropriate personal protective equipment and gather all materials and supplies together in a convenient location.
2. Prepare 12 sterile dilution bottles of at least 100 mL capacity and mark them for dilutions.
3. Using a clean, sterile pipet for each dilution, create dilutions of the source water by placing 100 mL, 50 mL, 25 mL. 10 mL, and 1 mL of source water in separate dilution bottles and adding sterile dilution water to create samples of at least 100 mL. Note that except for the 100 mL aliquot of source water, the final volume of the dilutions does not need to be precise, provided that it is at least 100 mL and not excessive beyond that because the volume of source water will still be the same so long as the total diluted volume is run through the filter membrane. Create two samples of each concentration and carefully cover or close the dilution bottles to prevent the introduction of random bacteria from the air. Also, prepare two dilution bottles with 100 mL each of the lab water (these will be used as negative controls).
4. Carefully mark each petri dish on the *bottom* of the dish with a permanent marker to indicate the dilution that is to be used for that dish and whether the test is for total coliform or fecal coliform. Keep the dishes closed during this step. Marking the bottom will prevent the markings from interfering with the colony counts later.
5. Prepare each petri dish one at a time, keeping the covers on as much as possible to avoid extraneous contamination. In each of the six dishes dedicated to total coliform testing, pipette the contents of a 1 mL vial of total coliform M-Endo broth on the sterile pad. Avoid touching the sides of the dish or the pad with the tip of the vial and spread the M-Endo around the pad as much as possible to provide uniform saturation of the pad. The M-Endo will

continue to spread throughout the pad while the filters are being prepared, so absolute saturation of the entire pad initially is not critical. Note that the nutrients are placed on the pad, not on the filter, because putting the nutrients onto the filter can cause the bacteria to be washed to the side or off the membrane altogether, rendering the count unreliable.

6. In each of the six dishes dedicated to fecal coliform testing, place the contents of a 1 mL vial of fecal coliform MF-Endo broth on the sterile pad. Avoid touching the sides of the dish or the pad with the vial and spread the Endo around the pad as much as possible to provide uniform saturation of the pad. The MF-Endo will continue to spread throughout the pad while the filters are being prepared, so absolute saturation of the entire pad initially is not critical. Note that the nutrients are placed on the pad, not on the filter, because putting the nutrients onto the filter can cause the bacteria to be washed to the side or off the membrane altogether, rendering the count unreliable.
7. Assemble the sterile filtering apparatus, being careful to avoid contamination of the inside of the funnel or the surface upon which the filter membrane will rest.
8. Using sterile forceps, carefully remove a sterile filter membrane from its container and place it carefully on the center of the filter holding surface of the filtering apparatus, gridded side up. Do not touch the filter in any way, except with the sterile forceps, or the results of the test will be compromised and inaccurate.
9. Turn on the vacuum pump and pour the undiluted 100 mL aliquot of source water through the filter membrane, being careful to pour the sample through slowly and to distribute the sample across the entire surface of the membrane, while avoiding any spillage on the sides of the funnel. If spillage on the sides of the funnel does occur, slowly rinse the sides of the funnel with sterile dilution water to ensure that the entire volume of source water has passed through the filter membrane.
10. When the sample has been fully drawn through the membrane, immediately shut of the vacuum pump and remove the funnel from the filtering apparatus.
11. Open the petri dish designated for the undiluted source water sample and total coliform. Then, using the sterile forceps, carefully remove the filter membrane from the filtering apparatus and roll it into the petri dish, keeping the gridded side up. The grids will make counting the colonies much easier later. Note that the membrane may still be "stuck" to the filter plate from the force of the vacuum and may need to be gently nudged at the edges to release the vacuum before it will come off. It is very fragile and will easily tear if care is not exercised. To roll the membrane into the dish, set one edge of the filter on the pad at the outside edge of the dish, then slowly lower the membrane into the dish to where releasing it will allow it to fall directly onto the center of the dish. The objective is to minimize or avoid trapping air pockets between the membrane and the pad. The membrane needs to be in complete contact with the pad for the growth medium to be effective. As soon as the membrane is in the dish, replace the cover and place the petri dish into the incubator.
12. Sterilize the filtering apparatus funnel and membrane holding plate and reassemble the filtering apparatus.
13. Repeat Steps 7–12 for each of the other five total coliform dilutions and the total coliform negative control, then the five fecal coliform dilutions and

the fecal coliform negative control. Be careful to make sure the filter membranes are placed in the correct petri dishes for dilution and coliform type.

14. Incubate all the tightly closed petri dishes for 24 hours, plus or minus 1 hour at 35 +/−0.5 °C (95 +/− 1 °F).
15. At the end of the incubation period, remove the petri dishes and count the colony-forming units using the microscope, recording the concentrations per the instructions in the data analysis section of the experiment. Note that the grid lines on the membrane will now be available to assist with the counting. It should not be necessary to open the petri dishes to do the counting, although there is no downside to doing so except that they have a strong, generally unpleasant odor after incubation. If the top of the dish is very moist on the inside, which obscures the filter surface, turning the dish on its side and gently tapping it on the workbench will usually cause the moisture to immediately drain off to the side of the pad, clearing the cover for reading the counts. If that does not work, removing the cover to do the counts is the only practicable solution.

Data Collection Format and Data Collection Sheets Tailored to Experiment

See attached data collection sheet.

Data Analysis Plan

Coliform concentrations are reported as total coliform colony-forming units/100 mL or fecal coliform colony-forming units/100 mL. The concentration is calculated using membrane filters with 20–80 coliform colonies and not more than 200 colonies of all types per membrane, using the following equation:

(Total) coliform colonies counted × 100 = coliform colonies/100 mL of filtered original sample (not the volume of diluted sample)

[(Total) coliform colonies counted × 100/mL of source water] = coliform colonies/100 mL of source water

With drinking water and other good-quality source water where 100 mL of undiluted source water is filtered and incubated, the number of total coliform or fecal coliform colonies will be fewer than 20 per membrane, and thus fewer than 20/100 mL. In that case, all the colonies are counted and the formula is used to obtain the total coliform or fecal coliform concentration. If more than one membrane has been filtered, using an average count/100 mL of all the membranes is appropriate.

If there are so many colonies formed that they cover the entire filtration area of the membrane, or the colonies are not discrete, the results are reported as "too numerous to count" (TNTC). If the total number of bacterial colonies, coliforms plus noncoliforms, exceeds 200 per membrane, or if the colonies are too indistinct for accurate counting, the results are also reported as TNTC. In either case, a new sample needs to be collected, appropriately diluted, and tested.

A sample of the original source water smaller than 100 mL may be filtered directly, if desired. Note, however, that evenly distributing the sample across the membrane surface becomes much more difficult as the sample size is decreased. Diluting the small original sample to 100 mL with sterile dilution water generally produces better results.

If, after testing, no filter has a coliform count falling in the ideal range, the coliform count total on all the filters combined is reported as the number per 100 mL. For example, if duplicate 50 mL portions were examined, (which implies 2–100 mL samples diluted 2:1 and thus each containing 50 mL of original source water), and the two membranes had 5 and 3 coliform colonies, respectively, a count of eight coliform colonies per 100 mL would be reported.

$$(5 + 3) \times 100/(50 + 50) = 8/100 \text{ mL}$$

Similarly, if 50, 25, and 10 mL portions were examined, (implying one sample of the original source water diluted 2:1, one sample diluted to 4:1, and one sample diluted to 10:1), and the counts were 15, 6, and 1 coliform colonies, respectively, the count would be reported as 23/100 mL based on the following equations:

$$(15 + 6 + 1) \times 100/(50 + 25 + 10) = 22.88 \text{ or } 23/100 \text{ mL}$$

Conversely, if 10, 1, and 0.1 mL portions of source water were filtered, resulting in counts of 40, 9, and 1 coliform colonies, respectively, only the 10 mL sample would be used for calculating the coliform density because this filter had a coliform count falling within the ideal range of 20–80. The result would be reported as 400/100 mL, even if the membrane in that sample also showed a total bacterial count greater than 200, so long as the coliform were actually countable.

$$(40 \times 100)/10 = 400/100 \text{ mL}$$

The statistical reliability of the membrane filter technique results is generally better than that of the multiple tube fermentation technique (Experiment 12.2). Nevertheless, membrane counts are not absolute numbers. Where decisions regarding human health, safety, and welfare are to be made based on these numbers, duplicate sampling and testing should be conducted for good data reliability.

Note that if the negative control dishes show any colonies, the number of colonies in the control dishes are subtracted from the count for all the other dilutions in that array. If the number of colonies in the negative control sample equals or exceeds the number of colonies in any of the dilution dishes, the dilution water should be considered contaminated and the test run again using a new batch of dilution water.

Expected Outcomes

Wastewater should show total and fecal coliform counts in the TNTC range, even when highly diluted, for samples that have not been disinfected. Disinfected samples should show counts in the ideal range or less. Untreated drinking water sources should show counts within the ideal range, but surface water may be contaminated from natural sources, such as animal wastes and airborne bacteria. They will ideally show counts within the ideal range or lower, but may not do so if contaminated. Treated drinking water sources should show counts of fewer than 1 coliform and 0 fecal coliform per 100 mL. Water collected at swimming beaches will generally show counts within the ideal range or lower, but normally may be considered safe with counts as high as 100 colony-forming units/100 mL.

DATA COLLECTION SHEET MEMBRANE FILTER TECHNIQUE FOR COLIFORM AND FECAL COLIFORM			
Lead Experimenter			
Date of Experiment			
Date of Source Water Collection			
Storage Conditions of Source Water			
Source of Water			
Colony Count – Dilution Water – 100 mL			
Colony Count – 100 mL, Total Coliform		Count per 100 mL of Source Water	
Colony Count – 50 mL, Total Coliform		Count per 100 mL of Source Water	
Colony Count – 25 mL, Total Coliform		Count per 100 mL of Source Water	
Colony Count – 10 mL, Total Coliform		Count per 100 mL of Source Water	
Colony Count – 1 mL, Total Coliform		Count per 100 mL of Source Water	
Colony Count – 100 mL, Fecal Coliform		Count per 100 mL of Source Water	
Colony Count – 50 mL, Fecal Coliform		Count per 100 mL of Source Water	
Colony Count – 25 mL, Fecal Coliform		Count per 100 mL of Source Water	
Colony Count – 10 mL, Fecal Coliform		Count per 100 mL of Source Water	
Colony Count – 1 mL, Fecal Coliform		Count per 100 mL of Source Water	

EXPERIMENT 12.2

MULTIPLE TUBE FERMENTATION (MOST PROBABLE NUMBER) TECHNIQUE FOR COLIFORM BACTERIA

(Consistent with EPA Method 9131)

Date: ________________________

Principal Investigator:

Name: ________________________

Email: ________________________

Phone: ________________________

Collaborators:

Name: ________________________

Email: ________________________

Phone: ________________________

Name: ________________________

Email: ________________________

Phone: ________________________

Name: ________________________

Email: ________________________

Phone: ________________________

Name: ________________________

Email: ________________________

Phone: ________________________

Objective or Question to Be Addressed

This experiment is used to quantify the coliform bacteria population in a sample of surface water or groundwater. The multiple-tube fermentation technique is a three-step test procedure. The results are expressed in terms of the most probable number (MPN), a statistically valid value derived from the results of the testing.

Theory behind Experiment

The three steps of the multiple-tube fermentation technique are referred to as the presumptive step, the confirming step, and the completed test. The experiment requires the simultaneous testing of at least five aliquots of the same sample of source water using special test tube arrangements, as described in the experiment procedures section.

During the presumptive step, a series of lauryl tryptose broth primary fermentation tubes are inoculated with various dilutions of the source water. The inoculated tubes are incubated at 35 +/– 0.5 °C (95 +/– 1 °F) for 24 +/– 2 hours, at which time the tubes are examined for gas formation. For the tubes in which no gas is formed, incubation is continued for an additional 24 +/– 2 hours and examined for gas formation at the end of 48 +/– 3 hours. Formation of gas in any amount within 48 +/– 3 hours is considered a positive and presumptive test.

During the second, or confirming step, all fermentation tubes showing gas formation during the 24-hour and 48-hour periods are subjected to further fermentation. In Step 2, fermentation tubes containing brilliant green lactose bile broth are inoculated with medium from the tubes showing a positive result in Step 1. Inoculation should be performed as soon as possible after gas formation occurs. The newly inoculated tubes are incubated for an additional 48 +/– 3 hours at 35 +/– 0.5 °C (95 +/– 1 °F). Formation of gas at any time during the 48-hour incubation period is considered a positive confirmed test.

During Step 3, the completed test, all samples showing a positive result in Step 2 are analyzed. This step can also be used as a quality-control measure on 20 % of all samples analyzed. For this step, one or more plates of eosin methylene blue agar gel are streaked with the sample to be analyzed. A separate set of plates is used for each dilution showing a positive result in Step 2. The streaked plates are incubated for 24 +/– 2 hours at 35 +/– 0.5 °C (95 +/– 1 °F). After incubation, one or more of the typical colonies growing on the agar (nucleated, with or without a metallic sheen) is transferred to a lauryl tryptose broth fermentation tube and a nutrient agar slant. The fermentation tubes and agar slants are incubated at 35 +/– 0.5 °C (95 +/– 1 °F) for 24 +/– 2 hours, or for 48 +/– 3 hours if gas is not produced. From the agar slants corresponding to the fermentation tubes in which gas formation occurs, gram-stained samples are examined microscopically. The formation of gas in the fermentation tube and the presence of gram-negative, non-spore-forming, rod-shaped bacteria in the agar culture is considered a completed positive test, demonstrating the presence of coliform bacteria in the analyzed sample.

Note that the entire procedure requires up to 48 +/– 3 hours for Step 1, a second 48 +/–3 hours for Step 2, and an additional 48 +/– 3 hours for Step 3. It is further noted that by that time, the conditions in any natural water source will likely have changed, perhaps dramatically, and the results may not be relevant.

A more detailed treatment of this method is presented in *Standard Methods for the Examination of Water and Wastewater* and in *Microbiological Methods for Monitoring the Environment.*

Potential Interferences and Interference Management Plan

The presence of residual chlorine or other halogens can interfere with or stop the continuation of bacterial growth. To prevent this, sodium thiosulfate may be added to the sterile sample collection container at the source-water collection site.

Water samples high in copper, zinc, or other heavy metals can be toxic to bacteria. Chelating agents such as ethylenediaminetetraacetic acid (EDTA) may be added at the source-water collection site to overcome these concerns, but this step should only be included when heavy metals are known or highly suspected of being present.

It is important to keep in mind that MPN tables are probability calculations and inherently have poor precision. They include a 23 % positive bias that generally results in high value. The precision of the MPN can be improved by increasing the number of sample portions examined and the number of samples analyzed from the same source water and sampling point.

Do not suspend bacteria in any dilution water for more than 30 minutes at room temperature because death or multiplication may occur, depending on the species.

Data to Be Collected with an Explanation of How They Will Be Collected (Ex.: "Water Temperature Data for Three Days Using a Continuous Data Logger")

Data to be collected during this three-step experiment include the number of aliquots of source water tested, the dilutions of those aliquots used, and the results of testing at each step of the procedure. Time is measured and recorded for the appearance of positive results at each step.

Using the attached data sheet, record at a minimum the following information:

- Lead experimenter
- Date of experiment
- Date of source water collection
- Storage conditions of source water
- Source of water
- Tube observations

Tools, Equipment, and Supplies Required

- Incubator capable of maintaining a uniform and constant temperature of 35 +/− 0.5 °C (95 +/− 1 °F) for the duration of the testing procedures
- Water bath sufficient in size and depth to immerse the tubes to the upper level of the media
- Microscope of sufficient magnification (1.5 to 2.0, minimum)
- pH meter accurate to 0.1 pH units, or better

- Analytical balance with a sensitivity of at least 0.1 g at a load of 150 g
- 10 clean, sterilized, glass dilution bottles
- 10 clean, sterilized pipets of any convenient size, provided that they deliver the required volume accurately and quickly
- 10 clean, sterilized graduated cylinders
- 5 sterile dilution bottles (minimum) or tubes with glass stoppers or screw caps equipped with liners that do not produce toxic or bacteriostatic compounds on sterilization. Mark gradation levels on the side of the dilution bottles or tubes
- 10 sterile petri dishes (minimum)
- 10 sterile 10 mm × 75 mm fermentation tubes (minimum) with shell vials of such size that they will be filled completely with medium and at least partly submerged in the fermentation tubes
- 10 sterile 3 mm diameter wire inoculation loops (minimum) made of 22- or 24-gauge nickel alloy or platinum-iridium suitable for flame sterilization or single-service transfer loops of aluminum or stainless steel
- 1 L of sterile deionized water (lab water)
- 1 L of sterile stock phosphate buffer solution
    - Prepared by dissolving 34 g potassium dihydrogen phosphate (KH_2PO_4) in 500 mL lab water, adjusted to pH 7.2 +/– 0.5 with 1 N sodium hydroxide (NaOH), and diluted to 1 L with lab water
- 1 L of dilution water
    - Prepare by adding 1.25 mL stock phosphate buffer solution and 5 mL magnesium chloride solution (38 g $MgCl_2$/liter lab water or 81.1 g $MgCl_2 \cdot 6\ H_2O$/liter lab water) to 1 L lab water. Dispense in amounts that will provide 99 +/– 2 mL or 9 +/– 0.2 mL after autoclaving for 15 minutes
- 1 L of a 10 % solution of peptone in lab water. Dilute a measured volume to provide a final 0.1 % solution with a final pH of 6.8. Dispense in amounts to provide 99 +/– 2 mL or 9 +/– 0.2 mL after autoclaving for 15 minutes
- 1 L lauryl tryptose broth prepared by dissolving 20 g tryptose; 5 g lactose; 5 g diphosphate hydrogen phosphate; 2.75 g KH_2PO_4, 2.75 g potassium dihydrogen phosphate, KH_2PO_4; 5 g sodium chloride, NaCl; and 0.1 g sodium lauryl sulfate in 1 L of lab water. (Lauryl tryptose broth is also available in a prepackaged dry powder form, which may be prepared to replicate the component concentrations cited here.) Dispense the broth into fermentation tubes that contain inverted vials. Add an amount sufficient to cover the inverted vial, at least partially, after sterilization has taken place. Sterilize at 121 °C (250 °F) for 12–15 minutes. The pH should be 6.8 +/– 0.2 after sterilization
- 1 L brilliant green lactose bile broth consisting of 10 g peptone, 10 g lactose, 20 g oxgall, and 0.0133 g brilliant green dissolved in 1 L lab water. This broth is also available in a prepackaged dry powder form. Dispense the broth into fermentation tubes that contain inverted vials. Add an amount sufficient to cover the inverted vial, at least partially, after sterilization has taken place. Sterilize at 121 °C (250 °F) for 12–15 minutes. The pH should be 7.2 +/– 0.2 after sterilization

- 100 mL ammonium oxalate-crystal violet (Hucker's solution) prepared by dissolving 2 g crystal violet (90 % dye content) in 20 mL 95 % ethyl alcohol, dissolving 0.8 g $(NH_4)_2C_2O_4 \cdot H2O$ in 80 mL lab water, mix the two solutions, and age for 24 hours before use; filter through paper into a staining bottle
- 300 mL Lugol's solution, Gram's modification: prepared by grinding 1 g iodine crystals and 2 g potassium iodine (KI) in a mortar. Add lab water, a few milliliters at a time, and grind thoroughly after each addition until the solution is complete. Rinse the solution into an amber glass bottle with the remaining water (using a total of 300 mL of lab water)
- 110 mL counterstain
  - Prepared by dissolving 2.5 g safranin dye in 100 mL 95 % ethyl alcohol and add 10 mL of that solution to 100 mL lab water
- 500 mL of acetone alcohol prepared by mixing equal volumes of ethyl alcohol, 95 %, with acetone

Commercially available Gram staining kits may be substituted for the ammonium oxalate-crystal violet solution, the Lugol's solution, the counterstain, and the acetone alcohol.

Safety

Samples tested for the presence or absence of coliform bacteria are presumed to be contaminated with those bacteria until proven otherwise. Coliform bacteria are indicator organisms that are highly associated with fecal contamination and the presence of *E. coli* bacteria. Ingestion of the sample could result in serious digestive system upset and health problems. Care needs to be taken to avoid ingestion or contact with eyes. Personal protective equipment consisting of eye protection, gloves, and a lab coat are required during all stages of this experiment. Ingestion or eye contact most commonly occurs when a gloved hand is touched to the face or eye, transferring a minute amount of sample to the body.

In the event of exposure to the skin, immediately wash the affected area with hot water and soap. In the event of ingestion, contact medical professionals for advice. In the event of eye contact, immediately flush the eye with an eye wash station or bottle and contact medical professionals for further advice.

Planned Experimental Procedures and Steps, in Order

Step 1 – Presumptive Step

1. Put on personal protective equipment and gather all tools, supplies, and samples in a convenient location.
2. Using a clean, sterile pipet for each dilution, create dilutions of the source water by placing 100 mL, 50 mL, 25 mL, 10 mL, and 1 mL of source water in separate dilution bottles and adding sterile dilution water to create samples of at least 100 mL. Note that except for the 100 mL aliquot of source

water, the final volume of the dilutions does not need to be precise, provided that it is at least 100 mL and not excessive beyond that because the volume of source water will still be the same so long as the total diluted volume is run through the filter membrane. Create two samples of each concentration and carefully cover or close the dilution bottles to prevent the introduction of random bacteria from the air. Also, add two dilution bottles with 100 mL each of the lab water (these will be used as negative controls).

3. Prepare a series of five sterile fermentation tubes by marking each tube with the dilution of source water to be fermented in that tube.
4. Using sterile pipets, inoculate a series of fermentation tubes ("primary" fermentation tubes) with appropriate diluted aliquots of sample. Use a separate sterile pipet for each different dilution. Do not prepare dilutions in direct sunlight. Use caution to maintain sterility of the pipets by not touching any surface or glassware with the tip. Insert pipets no more than 2.5 cm (1 inch) below the surface of the sample or dilution. After adding the sample, mix thoroughly by shaking the test tube rack. Do not invert the tubes.
5. Incubate inoculated fermentation tubes at 35 +/– 0.5 °C (95 +/– 1 °F). After 24 +/– 2 hours, shake each tube gently and examine it, and if no gas has formed and been trapped in the inverted vial, reincubate and reexamine at the end of 48 +/– 3 hours. Record the presence or absence of gas formation, regardless of amount, at each examination of the tubes.
6. Record the formation of gas in any amount in the inner fermentation tubes or vials within 48 +/– 3 hours as a positive presumptive test. Note that if gas is formed as a result of fermentation, as opposed to air bubbles, the broth medium will become cloudy.
7. Record the absence of gas formation at the end of 48 +/– 3 hours of incubation as a negative test.

Step 2 – Confirming Step

1. Submit all primary fermentation tubes showing any amount of gas within 48 hours of incubation to the confirming step. Transfer medium from each confirmed tube immediately upon the appearance of gas formation; do not wait until the entire 48 hours have expired.
2. Gently shake or rotate each primary fermentation tube showing gas in order and, using a sterile metal loop, 3 mm in diameter, transfer one loopful of culture to a fermentation tube containing brilliant green lactose bile broth.
3. Incubate the inoculated brilliant green lactose bile broth tube for 48 +/– 3 hours at 35 +/– 0.5 °C (95 +/– 1 °F). Record the formation of gas in any amount in the inverted vial of the brilliant green lactose bile broth fermentation tube at any time within 48 +/– 3 hours as a positive confirming test.

Step 3 – Completed Test

1. Submit all positive tubes from Step 2 to Step 3 testing to definitively establish the presence of coliform bacteria.

2. Streak one or more eosin methylene blue plates from each tube of brilliant green lactose bile broth showing gas as soon as possible after the appearance of gas. Streaks should be separated by at least 0.5 cm.
3. Incubate plates (inverted) at 35 +/– 0.5 °C (95 +/– 1 °F) for 24 +/– 2 hours.
4. Note (but it is not necessary to count) the colonies developing on the eosin methylene blue agar. The colonies are recorded as *typical* if they are nucleated, with or without a metallic sheen; *atypical* if they are opaque, unnucleated, mucoid, and pink after 24 hours of incubation, or *negative* if they display any other characteristics.
5. From each plate, select one or more *typical*, well-isolated coliform colonies, or if no typical colonies are present, select two or more colonies considered most likely to consist of organisms of the coliform group (this requires significant knowledge of the nature of coliform bacteria and may be ignored in undergraduate laboratory experiments) and transfer growth from each isolate to a lauryl tryptose broth fermentation tube (a secondary broth tube) and to a nutrient agar slant. NOTE: to the extent possible, when transferring colonies, choose well-isolated colonies and barely touch the surface of the colony with a flame-sterilized, air-cooled transfer needle to minimize the danger of transferring a mixed culture.
6. Incubate the secondary broth tubes at 35 +/– 0.5 °C (95 +/– 1 °F) for 24 +/– 2 hours. If gas is not produced within 24 +/– 2 hours, reincubate and examine again at 48 +/– 3 hours. Microscopically examine Gram-stained preparations (see the Gram-staining procedure next) from the 24-hour agar slant cultures corresponding to the secondary tubes that show gas.
7. Record the formation of gas in the secondary tube of lauryl tryptose broth within 48 +/– 3 hours and the presence of gram-negative, non-spore-forming, rod-shaped bacteria in the agar culture as a positive completed test, demonstrating the presence of coliform bacteria.

Gram-Stain Procedure

1. Prepare a light emulsion of the bacterial growth from an agar slant in a drop of lab water on a glass slide.
2. Air-dry or fix the emulsion by passing the slide through a flame and stain for 1 minute with the ammonium oxalate-crystal violet solution.
3. Rinse the slide in tap water and then apply Lugol's solution for 1 minute.
4. Rinse the stained slide in tap water, then decolorize for approximately 15–30 seconds with acetone alcohol by holding the slide between the fingers and letting acetone alcohol flow across the stained smear until no more stain is removed. Take care to not over-decolorize.
5. Counterstain with safranin for 15 seconds, then rinse with tap water, blot dry with bibulous paper, and examine microscopically.

6. Record cells that decolorize and accept the safranin stain, thereby turning pink, as *gram-negative*. Record cells that do not decolorize but do retain the crystal violet stain, thereby retaining a deep blue color, as *gram-positive*.

Computing and Recording of MPN

1. A reasonable estimate of the density of coliform bacteria in the sample can be obtained from the most probable number (MPN) table, based on the number of positive tubes in each dilution of the confirmed or completed test.

   Calculation of the MPN by multiplying the MPN Index per 100 mL by the lowest dilution factor used. For example, using Table 12.2.1, if two of the five 10 mL samples of source water showed positive results, the MPN would be reported as 5.1/(the MPN Index) times 1 (the lowest, and in this case only, dilution factor) yielding an MPN of 5.1/100 mL with a probability of the actual value being between 0.5 and 19.2/100 mL.

   If less than 10 mL of source water are tested, the MPN Index value would be multiplied by the dilution factor to yield the MPN. Thus, if only 1 mL of source water was tested in each sample, the dilution factor would be 10, the MPN Index would become 51, and the MPN would be reported as 51 colony forming units/100 mL.

   Table 12.2.1 shows MPN indices and 95 % confidence limits for potable water testing, and Table 12.2.2 describes the MPN indices and 95 % confidence limits for general use.
2. Note that three dilutions are necessary for the determination of the MPN Index. For example (see Table 12.2.2), if five 10 mL, five 1 mL, and five 0.1 mL aliquots of a sample are tested and four of the 10 mL, two of the 1 mL, and none of the 0.1 mL aliquots give positive results, the coded result is 4–2–0 and the MPN Index is 22/100 mL.
3. All MPN values for water samples should be reported on the basis of a 100 mL sample.

Data Collection Format and Data Collection Sheets Tailored to Experiment

See attached data collection sheet.

Data Analysis Plan

The positive results from the Step 3 tests are recorded as the MPN of coliform present in the original sample of source water.

Expected Outcomes

The expected outcomes from this experiment will depend on the source water and the accuracy of the procedures used. Drinking-water sources and sources that have been disinfected, should show MPN in the 0–2/100 mL range. Contaminated sources will show higher MPN, perhaps in the millions or more per 100 mL.

TABLE 12.2.1
MPN Index and 95 % Confidence Limits for Positive Results from Five 10 mL Dilutions of Source Water

Number of Positive Tubes Out of 5	MPN Index/100 mL	Lower 95 % Confidence Limit	Upper 95 % Confidence Limit
0	<2.2	0	6.0
1	2.2	0.1	12.6
2	5.1	0.5	19.2
3	9.2	1.6	29.4
4	16	3.3	52.9
5	>16	8.0	infinite

Source: Adapted from HACH.com and EPA Method 9131.

TABLE 12.2.2
MPN Index and 95 % Confidence Limits for Positive and Negative Results from Five Serial Dilutions of Source Water

Number of Positive Tubes				95 % Confidence Limits	
5 of 10 mL Each	5 of 1 mL Each	5 of 0.1 mL Each	MPN Index per 100 mL	Lower	Upper
0	0	0	<2		
0	0	1	2	<0.5	7
0	1	0	2	<0.5	7
0	2	0	4	<0.5	11
1	0	0	2	<0.5	7
1	0	1	4	<0.5	11
1	1	0	4	<0.5	11
1	1	1	6	<0.5	15
1	2	0	6	<0.5	15
2	0	0	5	<0.5	13
2	0	1	7	1	17
2	1	0	7	1	17
2	1	1	9	2	21
2	2	0	9	2	21
2	3	0	12	3	28
3	0	0	8	1	19
3	0	1	11	2	25
3	1	0	11	2	25
3	1	1	14	4	34
3	2	0	14	4	34

(*Continued*)

TABLE 12.2.2 (Continued)

Number of Positive Tubes				95 % Confidence Limits	
5 of 10 mL Each	5 of 1 mL Each	5 of 0.1 mL Each	MPN Index per 100 mL	Lower	Upper
3	2	1	17	5	46
3	3	0	17	5	46
4	0	0	13	3	31
4	0	1	17	5	46
4	1	0	17	5	46
4	1	1	21	7	63
4	1	2	26	9	78
4	2	0	22	7	67
4	2	1	26	9	78
4	3	0	27	9	80
4	3	1	33	11	93
4	4	0	34	12	93
5	0	0	23	7	70
5	0	1	31	11	89
5	0	2	43	15	110
5	1	0	33	11	93
5	1	1	46	16	120
5	1	2	63	21	150
5	2	0	49	17	130
5	2	1	70	23	170
5	2	2	94	28	220
5	3	0	79	25	190
5	3	1	110	31	250
5	3	2	140	37	340
5	3	3	180	44	500
5	4	0	130	35	300
5	4	1	170	43	490
5	4	2	220	57	700
5	4	3	280	90	850
5	4	4	350	120	1000
5	5	0	240	68	750
5	5	1	350	12	1000
5	5	2	540	180	1400
5	5	3	920	300	3200
5	5	4	1600	640	5800
5	5	5	>2400		

Source: Adapted from HACH.com and EPA Method 9131.

DATA COLLECTION SHEET MULTIPLE TUBE FERMENTATION (MOST PROBABLE NUMBER) TECHNIQUE FOR COLIFORM BACTERIA					
Lead Experimenter					
Date of Experiment					
Date of Source Water Collection					
Source Water Storage Method Used					
Source of Water					
PRESUMPTIVE STEP – Step 1					
Fermentation Tube	1	2	3	4	5
Gas Formation?					
Time of Gas Observance					
Positive (P) or Negative (N)?					
CONFIRMING STEP – Step 2					
Fermentation Tube	1	2	3	4	5
Gas Formation?					
Time of Gas Observance					
Positive (P) or Negative (N)?					
COMPLETED TEST – Step 3					
Eosin Methylene Blue Plate	1	2	3	4	5
Typical (T), Atypical (A), or Negative (N)?					
Secondary Broth Tubes	1	2	3	4	5
Positive (P) or Negative (N)?					
Coliform Present? Yes (Y) or No (N)					
GRAM-STAIN PROCEDURE					
Slide Number	1	2	3	4	5
Gram-Positive (P) or Gram-Negative (N)?					
MPN CALCULATION					
Dilution (mL of source water)					
Calculated MPN					

EXPERIMENT 12.3[1]

COAGULATION, FLOCCULATION, AND SEDIMENTATION FOR THE REMOVAL OF ORGANIC AND INORGANIC WATER CONTAMINANTS

Date: ______________________________

Principal Investigator:

Name: ______________________________

Email: ______________________________

Phone: ______________________________

Collaborators:

Name: ______________________________

Email: ______________________________

Phone: ______________________________

Name: ______________________________

Email: ______________________________

Phone: ______________________________

Name: ______________________________

Email: ______________________________

Phone: ______________________________

Name: ______________________________

Email: ______________________________

Phone: ______________________________

Name: ______________________________

Email: ______________________________

Phone: ______________________________

Objective or Question to Be Addressed

The removal of dissolved contaminants from water is nuanced and can be as much an art as a science. The science has been well developed over time, but the application requires some patience and careful attention to the concentration of contaminants in the water. Organic contaminants and inorganic contaminants can be removed in essentially the same way, but the chemicals used are quite different.

The general process for removal of dissolved solids is to convert the dissolved solids to fine solid particles, called *pin floc*, in a process called *coagulation*, to then combine the fine pin floc particles together, in a process called *flocculation*, into large enough particles to settle out, and to then let the large particles settle out of the water through a process of *sedimentation*.

This experiment examines the general nature of the three separate steps to the removal of dissolved solids from water or wastewater.

Theory behind Experiment

As noted, the general process for removal of dissolved solids is to convert the dissolved solids to fine solid particles, called *pin floc*, in a process called *coagulation*, to then combine the fine pin floc particles together, in a process called *flocculation*, into large enough particles to settle out, and to then let the large particles settle out of the water through a process of *sedimentation*.

The process of coagulation requires the changing of the surface electric charge of the dissolved particles so that they can detach from the water molecules and merge together into fine pin floc particles. This is done through the addition of a *coagulant*, a chemical selected for that purpose. The objective of the coagulant is to neutralize the negative electrical charge of the dissolved solid by combining with those solids using the positive charge of the coagulant. Once that occurs, the dissolved solids can combine with each other to form the pin floc.

The selection of the coagulant depends on the nature of the contaminant. A coagulant that works well for organic contaminants generally will not work as well for inorganic contaminants. The opposite is also true. The coagulant most commonly used at water treatment plants in the United States is aluminum sulfate, also called alum, for organic contaminants and ferric sulfate, ferric chloride, or a commercial polymer for inorganic contaminants.

In the second step, the water containing the pin floc is slowly stirred to allow the fine particles to contact each other and stick together, without being so agitated as to break the floc up as soon as it forms. This is the process of flocculation. Note that sometimes the concepts of coagulation and flocculation are combined in discussion and referred to as either coagulation or flocculation. We believe they are two separate steps and can best be understood if viewed in that fashion.

The final step of the process is removal of the large floc from the treated water. That has historically been done using sedimentation principals. It can also be done effectively using filtration principles, generally with a sand filter or some form of microfiltration or ultrafiltration. The use of microfilters or ultrafiltration reduces the amount of time needed to form the larger floc and to allow those larger floc to settle out. The filters tend to clog faster, however, and require greater operating maintenance. A sand filter following sedimentation is recommended in most cases, in any event, and that is often seen as the justification for going straight to filtration from flocculation.

In a water treatment facility, the coagulant is added to the water and it is rapidly mixed so that the coagulant is circulated throughout the water. This step is often done with an inline mixer and is confined to about 30 seconds of mixing. The treated water is then discharged to a second reactor where it is slowly stirred for a period of 30 minutes to one hour. If sedimentation is used following the flocculation step, a third reactor is used to allow the water 1–4 hours to settle out. If filtration is used without prior sedimentation, the time to filter is based on the overflow rate of the filter media.

This experiment uses the sedimentation process and does not delve into the filtration process.

Data to Be Collected with an Explanation of How They Will Be Collected (Ex.: "Water Temperature Data for Three Days Using a Continuous Data Logger")

Using the attached data sheet, record at a minimum the following information:

- Name of all experimenters
- During this test, visual observations will be made and recorded every five minutes of the color, size, and volume of floc being developed during the stirring phase
- During the settling phase, measurement using the markings on the side of the beakers will be made every five minutes of the depth and color of sediments colleting at the bottom of the beaker

Tools, Equipment, and Supplies Required

- 1 6-paddle Phipps and Bird mixer, or equivalent
- 6 clean 2 L beakers that will fit the mixer and have volumetric markings on the side of the beaker
- Approximately 3,500 mg (0.1235 oz) of aluminum sulfate
- Approximately 3,500 mg (0.1235 oz) of ferric sulfate
- Testing equipment and supplies sufficient to test the settled water for the concentration of aluminum sulfate and ferric sulfate

Safety

The chemicals used in this experiment are toxic to human heath if ingested or if placed in contact with eyes. Safety gloves, eye protection, and a lab coat are recommended for personal protection. In the event of skin exposure, washing with warm water and soap is recommended. For eye contact, flush with an eye wash for 15 minutes and seek medical attention immediately.

Planned Experimental Procedures and Steps, in Order

1. Gather all required equipment and materials and don appropriate personal protective equipment.
2. Fill six 2 L (2.11 qt) beakers with tap water to the 2 L mark on the beakers.
3. Add 3 g (0.109 oz) of fine inorganic clay particles and 3 g (0.109 oz) of fine organic silt to each beaker
4. Place the six beakers on the 6-paddle mixer and set the bottom of the paddle blades at approximately the 0.5 L mark on the beakers for stirring.
5. Set the mixer to stir the six beakers at a high speed (approximately 100–150 rpm) and turn the mixer on. Let the mixer stir the samples for 5 minutes, then turn the mixer off.
6. After turning off the mixer, let the samples rest for 5 additional minutes.
7. At the end of the rest period, note and record the depth of silt and clay that has settled, if any.
8. Add aluminum sulfate (alum) and ferric sulfate to each of the beakers in the following concentrations. Note that the beakers contain 2 L of water and the concentrations shown are in mg/L. If alum is not added at exactly these concentrations, report the concentrations actually used.
   Beaker 1 50 mg/L (0.00167 oz/qt) of each chemical
   Beaker 2 100 mg/L (0.00334 oz/qt) of each chemical
   Beaker 3 200 mg/L (0.00668 oz/qt) of each chemical
   Beaker 4 300 mg/L (0.0100 oz/qt) of each chemical
   Beaker 5 400 mg/L (0.0134 oz/qt) of each chemical
   Beaker 6 500 mg/L (0.0167 oz/qt) of each chemical
9. Turn the mixer on for 15 seconds at very high speed – approximately 250–300 rpm – to simulate the rapid mix phase of water treatment.
10. At the end of the 15 second rapid mix phase, immediately turn the mixer speed down to a slow stirring rate of approximately 7–10 rpm for 30 minutes to simulate the stirring phase of water treatment. Every 5 minutes during the stirring period, note and record observations regarding the development of floc, including color, size, and number. Note that these parameters are observed as anecdotal observations, rather than precise measurements.
11. At the end of the 30 minute stirring period, turn off the mixer and let the beakers settle for an additional 30. Every 5 minutes during this settling period, note and record observations regarding the depth and color of accumulated floc in each beaker. Note that these parameters are based on the volume markers on the side of the beakers. The settlements may not be uniform, and it may be necessary to estimate an average depth of sediment in specific beakers. The accuracy of these measurements is important for data analysis.
12. Test the settled water in each beaker for the concentration of aluminum sulfate and ferric sulfate and record those values.
13. Calculate and record the amount of aluminum sulfate and the amount of ferric sulfate used in each beaker during this test in mg/L.
13. Plot the concentration of aluminum sulfate used in each beaker against the depth of settled sediments in that beaker and the concentration of ferric

sulfate used in each beaker against the depth of sediments in that beaker. Both lines should be graphed on the same chart for easy comparison. Determine and report the optimum concentrations of aluminum sulfate and ferric sulfate needed to adequately treat these samples.

14. Dispose of the contents of all beakers in accordance with laboratory waste disposal practices.

Data Collection Format and Data Collection Sheets Tailored to Experiment

See attached data collection form.

Data Analysis Plan

The observations specified will be made during the conduct of the experiment. The amount of each chemical used will be converted to mg/L and plotted against the volume of sediment generated in each beaker. Based on the two lines on the chart, the optimal concentration of aluminum sulfate and ferric sulfate will be determined and reported.

Expected Outcomes

It is expected that a fine pin floc will rapidly develop as soon as the stirring phase begins and that the larger floc will develop rapidly thereafter. Settling of the large floc will occur rapidly and the accumulations of sediments will stabilize in each beaker quickly. The volume of sediment in each beaker is expected to increase based on the amount of each chemical utilized. It is expected that essentially all the chemicals will be used in the first two or three beakers and that only a portion of the chemicals will be utilized in the other three beakers. The two graph lines should flatten out as the optimal concentration of each chemical is achieved.

DATA COLLECTION FORM COAGULATION, FLOCCULATION, SEDIMENTATION EXPERIMENT												
Beaker	Mass of $Al_2(SO_4)_3$ Added, mg	Mass of $Fe_2(SO_4)_3$ Added, mg	Final Mass of $Al_2(SO_4)_3$ in Settled Water, mg	Final Mass of $Fe_2(SO_4)_3$ in Settled Water, mg		Mass of $Al_2(SO_4)_3$ Utilized, mg	Mass of $Fe_2(SO_4)_3$ Utilized, mg					
1												
2												
3												
4												
5												
6												
OBSERVATION OF FLOC DURING STIRRING PHASE							OBSERVATION OF FLOC DURING SETTLING PHASE					
OBSERVATIONS OF FLOC COLOR							OBSERVATIONS OF FLOC COLOR					
BEAKER	1	2	3	4	5	6	1	2	3	4	5	6
TIME, min												
5												
10												
15												
20												
25												
30												
OBSERVATIONS OF FLOC SIZE							OBSERVATIONS OF FLOC VOLUME, mL					
BEAKER	1	2	3	4	5	6	1	2	3	4	5	6
TIME, min												
5												
10												
15												
20												
25												
30												
OBSERVATIONS OF FLOC DENSITY												
BEAKER	1	2	3	4	5	6						
TIME, min												
5												
10												
15												
20												
25												
30												

EXPERIMENT 12.4[2]

JAR TEST PROCEDURE

Date: ______________________________

Principal Investigator:

Name: ______________________________

Email: ______________________________

Phone: ______________________________

Collaborators:

Name: ______________________________

Email: ______________________________

Phone: ______________________________

Name: ______________________________

Email: ______________________________

Phone: ______________________________

Name: ______________________________

Email: ______________________________

Phone: ______________________________

Name: ______________________________

Email: ______________________________

Phone: ______________________________

Name: ______________________________

Email: ______________________________

Phone: ______________________________

Objective or Question to Be Addressed

Jar tests are used for a variety of reasons in environmental engineering. A common use is to determine the best reactant to use for a particular purpose or to determine the optimal dose of a reactant to achieve a specific objective. For example, a jar test may be used to determine which of two or more coagulants will work best within a given water source to remove a mix of organic and inorganic constituents. A follow-up test may then be conducted to determine the optimal concentration of the selected coagulant to use.

This experiment demonstrates the use of a jar test to determine the optimal mix of coagulants for an artificially prepared wastewater and to select the optimal concentration of the selected mix of coagulants to use.

Theory behind Experiment

Most organic compounds and many inorganic compounds may be coagulated and settled out to be removed from wastewater with aluminum hydroxide, sodium hydroxide, ferric chloride, or ferrous chloride. These chemicals form insoluble precipitates in the water that collect other particles of organic and inorganic origin as the insoluble particles precipitate out of the treated water. Identifying which chemical will work best for which organics or inorganics is often a guess and may depend on a lot of complicating factors, such as what else is in the water, what the various concentrations are, and what affinity any given contaminant has as for a particular coagulant versus any other.

By applying a series of concentrations of coagulant to a sample of wastewater and observing the results, optimal mixes of the best available coagulants can be determined empirically rather than through complicated calculation or extensive research. This can be very helpful when water-quality parameters are changing constantly and the time to do the research work is far longer than the time of change in the water quality.

Potential Interferences and Interference Management Plan

The type of coagulants used is dependent on a variety of factors, some of which may negatively impact each other. Some may work for several of the contaminants with varying degrees of efficiency. When one is added, some of everything may be removed, but not all of anything. When a second coagulant is added, it is possible to undo some of the work of the first additive and put some of the contaminants back into the solution while removing more of the others. It is also possible that less of one coagulant, yielding less than ideal results with many contaminants but excellent results for some of the contaminants and slightly more of a much less costly second coagulant, may yield far greater results for the contaminants treated poorly with the first coagulant.

To minimize these cross-purpose results, it is useful to do a separate jar test with each coagulant, to determine which of the contaminants that coagulant is best at removing, then preparing a mix of coagulants based on the optimal concentration of each. The final jar test with the mix of contaminants will determine whether there are any interferences among the coagulants. If interferences do occur, planning to separate the precipitates after each addition can often provide exceptional treatment of the target water.

Data to Be Collected with an Explanation of How They Will Be Collected (Ex.: "Water Temperature Data for Three Days Using a Continuous Data Logger")

The data to be collected are the concentrations of various contaminants introduced into an artificial wastewater. Samples will be collected from the wastewater before treatment and after treatment at various coagulant concentrations. Testing will be done with a Hach Company test kit, or equivalent, following the manufacturer's instructions. In most cases, a powder packet of reactant will be utilized in these tests and the concentration determined from a spectrophotometer programmed for the specific test and calibrated for each test in accordance with the manufacturer's instructions.

A separate jar test will be set up for each of the coagulants being tested, and the wastewater will be tested for concentrations of the target contaminants in each "jar" of each test for the concentration of each contaminant.

Using the attached data sheet, record at a minimum the following information:

- Name of all experimenters
- Date of the test
- Concentrations and names of chemicals added
- Results from each addition of chemical

Tools, Equipment, and Supplies Required

- Hach Company programmable spectrophotometer, or equivalent
- Phipps and Bird 6-paddle mixer, or equivalent
- Six 2 L (0.53 gal) glass or plastic beakers, round or square, that fit the mixer
- 55 L (15 gal) glass or plastic container to mix the artificial wastewater in
- Clean spatulas to measure contaminants and coagulants
- A supply of approximately 100 g or more of each of the following chemicals:
    - Calcium nitrate ($Ca(NO_3)_2$)
    - Ferrous carbonate ($FeCO_3$)
    - Ferrous sulfate ($FeSO_4$)
    - Manganese sulfate ($MnSO_4$)
    - Aluminum hydroxide ($Al(OH)_3$)
    - Ferric chloride ($FeCl_3$)
- Several 25–50 mL pipettes for sampling the reactors during the test
- Several sets of optically matched test tubes or sample jars for the spectrophotometer
- Sufficient powder pillows and other reagents needed for the testing of 30 samples, as required by the spectrophotometer manufacturer

Safety

The chemicals used in this experiment are toxic and caustic. Ingestion, inhalation, and skin contact should be avoided. Drying and irritation of the skin, serious damage to vision, or illness may result from contact.

In the event of skin contact, wash immediately with warm water and soap. In the event of eye contact, wash immediately with an eye flush station for up to 15 minutes and seek immediate medical assistance. In the event of inhalation, seek medical assistance immediately.

Gloves, eye protection, an inhalation mask, and a lab coat are required for this experiment.

Planned Experimental Procedures and Steps, in Order

1. Gather all material and equipment together and don the appropriate personal protective equipment.
2. Collect 50 L of tap water in a 55 L container and add the following chemicals in the amounts indicated and stir until all the materials are completely dissolved. The ferric carbonate may dissolve slowly. This will be the artificial wastewater.

   Calcium nitrate ($Ca(NO_3)_2$) 100 g
   Ferrous carbonate ($FeCO_3$) 2 g
   Ferrous sulfate ($FeSO_4$) 50 g
   Manganese sulfate ($MnSO_4$) 50 g
3. Sample and test the contents of this container for total iron, total calcium, total nitrate, total nitrite, and total manganese and record those values.
4. Collect 2 L of the artificial wastewater in each of six 2 L beakers and place them on the 6-paddle mixer.
5. Add the following amounts of aluminum hydroxide (alum) to the indicated beakers and record those amounts on the data sheet.

   Beaker mass of alum added

   1. 200 mg
   2. 400 mg
   3. 600 mg
   4. 800 mg
   5. 1,000 mg
   6. 1,200 mg
6. Turn the mixer on at a moderate rate (approximately 150 rpm) until all of the chemical is dissolved in all of the containers and then turn off the mixer. Allow the contents to settle, any floc to form, and the floc to settle. This should take about 30 minutes.
7. Sample the liquid above the floc and test for total iron, total calcium, total nitrate, total nitrite, and total manganese and record those values.
8. Plot the resulting concentrations of the recorded contaminants using a scatter plot with the concentration of alum on the y-axis and the percent removal of each tested contaminant on the x-axis.
9. Dispose of the contents of the six beakers in accordance with the hazardous waste disposal protocols of the laboratory.
10. Clean the six beakers and refill each with 2 L of the artificial wastewater.
11. Place the six beakers on the mixer and add the following amounts of ferric chloride to the designated beakers.

Beaker mass of ferric chloride added

1. 200 mg
2. 400 mg
3. 600 mg
4. 800 mg
5. 1,000 mg
6. 1,200 mg

12. Turn the mixer on at a moderate rate (approximately 150 rpm) until all of the chemical is dissolved in all of the containers and then turn off the mixer. Allow the contents to settle, any floc to form, and the floc to settle. This should take about 30 minutes.
13. Sample the liquid above the floc and test for total iron, total calcium, total nitrate, total nitrite, and total manganese and record those values.
14. Dispose of the contents of the six beakers in accordance with the hazardous waste disposal protocols of the laboratory.
15. Plot the resulting concentrations of the recorded contaminants using a scatter plot with the concentration of ferric chloride on the y-axis and the percent removal of each tested contaminant on the x-axis.
16. Create a table with the total removal percentage of each contaminant versus the concentration of each coagulant (see the data analysis section).
17. Select the concentration of each coagulant that optimizes the removal of the most contaminants.
18. Fill one clean 2 L beaker with artificial wastewater and place it on the mixer.
19. Add the calculated dose of each coagulant to the beaker, turn on the mixer until the coagulants are dissolved, then turn off the mixer and allow the contents to settle. This could take up to 30 minutes or so.
20. Sample the water above the floc and test for total iron, total calcium, total nitrate, total nitrite, and total manganese and record those values.
21. Compare the resulting values to the predicted values from the 10 graphs and the chart and discuss as to why there may be differences in the actual values versus the predicted values. Also determine whether a sedimentation step between the addition of the two coagulants might result in a better overall contaminant removal efficiency and justify that recommendation.

Data Collection Format and Data Collection Sheets Tailored to Experiment

See attached data collection form.

Data Analysis Plan

The data collected in this experiment will be plotted using scatter plots and the optimal concentrations for each coagulant listed on a chart showing percent removal efficiencies versus coagulant concentrations. The table should look something like the following:

Percent Contaminant Removal					
Contaminant	**Alum**		**Ferric Chloride**		**Aggregate**
Calcium					
Iron					
Nitrate					
Nitrite					
Manganese					

The optimal dosages will be those that together exhibit the most effective aggregate contaminant removal efficiency.

Expected Outcomes

It is expected that neither coagulant will remove all of any of the contaminants, but that each will be better for some and less effective on others. The actual percent removal efficiencies to be expected could be calculated from the chemical equations, but the complexity of the mixture is such that some of the molecules could easily combine with more than one of the coagulants or byproducts and a calculated determination is more easily done after a more thorough chemical analysis of the resulting solution concentrations.

DATA COLLECTION SHEET JAR TEST PROCEDURE						
Chemicals Added to Create Artificial Wastewater	Amount Added in mg/L	Resulting Concentrations in mg/L				
		Calcium	Iron	Nitrate	Nitrite	Manganese
Calcium Nitrate ($Ca(NO_3)_2$)						
Ferrous Carbonate ($FeCO_3$)						
Ferrous Sulfate ($FeSO_4$)						
Manganese Sulfate ($MnSO_4$)						
Coagulant Added in First Test						
Aluminum Hydroxide						
Coagulant Added in Second Test						
Ferric Chloride						
		Percent Removal				
		Calcium	Iron	Nitrate	Nitrite	Manganese
Aluminum Hydroxide						
Ferric Chloride						
Aggregate						

EXPERIMENT 12.5

GRAVIMETRIC DETERMINATION OF PARTICULATE MATTER IN AIR

Date: ______________________________

Principal Investigator:

Name: ______________________________

Email: ______________________________

Phone: ______________________________

Collaborators:

Name: ______________________________

Email: ______________________________

Phone: ______________________________

Name: ______________________________

Email: ______________________________

Phone: ______________________________

Name: ______________________________

Email: ______________________________

Phone: ______________________________

Name: ______________________________

Email: ______________________________

Phone: ______________________________

Name: ______________________________

Email: ______________________________

Phone: ______________________________

Objective or Question to Be Addressed

This experiment provides a method for determining the particulate matter concentration in air of specific particle size. The size of the particulates measured is dependent on the size of the filter used.

Theory behind Experiment

Particulate matter (PM) generally includes a complex mixture of extremely small particles and liquid droplets comprised of various pollutants, including acid vapors, organic chemicals, metals, and ubiquitous soil or dust particles. The size of particles is directly linked to their potential for causing health problems. Particles that are 10 micrometers in diameter or smaller are the particles that generally pass through the throat and nose and enter the lungs, causing the most concern from a health standpoint. The U.S. EPA groups particle pollution into two categories. The first is identified as inhalable coarse particles, which are larger than 2.5 microns and smaller than 10 microns in diameter, called PM_{10} *particles.* Those are typically found near roadways, transportation sources, coal-burning power plants, steel mills, mining operations, and dusty industries. The second category is identified as fine particles, which are 2.5 microns in diameter and smaller. These are called $PM_{2.5}$ *particles.* These particles are typically found in smoke and haze from forest fires and industrial air emissions.

A commonly used method for particulate matter analysis is the use of filter collection and gravimetric analysis. Size-selective devices are used to collect PM onto a filter, which is then analyzed gravimetrically. This experiment examines the PM_{10} and $PM_{2.5}$ concentrations in indoor air.

Potential Interferences and Interference Management Plan

Filter weighing is preferable done in a cleanroom to avoid potential contamination from the surrounding air as well as to have more control over temperature and humidity. A glove box can be used, if available, but a glove box is awkward to use effectively. For most university teaching laboratories, care in handling the filters will not produce such significant errors as to dilute the educational value of the experiment.

Handling filters carefully requires consideration of the following things that can adversely affect performance.

1. Most filter materials are soft and difficult to handle. The use of rubber-tipped tongs to grasp the filters as gently as possible and as close to the edge as possible, without otherwise touching the filter at all, will minimize damage to the filter surface and minimize the addition of normal oils from hands impacting the mass of the filter.
2. One of the biggest challenges when weighing filters is their very high sensitivity to humidity in their surroundings. That means that filters need to be acclimatized before use and before final testing. To minimize this effect, allow the filters to sit for 24 hours in a lightly covered container that will allow the filters to respond to ambient temperature and humidity, but exclude

dust particles, before placing them inside a filter holder immediately prior to use. Alternatively, commercially available, pre-loaded filter cartridges can be used after acclimatization to temperature while still sealed in their shipping containers, opened for use, immediately resealed after use, and stored at the same room temperature to which they were originally acclimatized prior to final weighing.

3. Filters are prone to electrostatic charging, which can attract additional particles during weighing procedures, and that can lead to inaccurate measurements. Touching a grounded metallic object prior to handling filters and touching plastic filter holders to a grounded metal object prior to opening can minimize these effects.
4. Filters tend to have a low density, which makes them easily susceptible to air buoyancy, and they appear to be lighter than they actually are. Air buoyancy influence differs with changing temperature, air pressure, and humidity. Weighing the filters in the same environment, ideally inside a closed balance inside a cleanroom, before and after PM sampling, can minimize this effect.
5. The mass of the clean filter and the mass of the exposed filter will usually be very close to the same value even when the procedure is conducted with absolute precision. Consequently, use of an enclosed balance, to minimize air current effects, and with an accuracy of 0.01 µg is important.

Data to Be Collected with an Explanation of How They Will Be Collected (Ex.: "Water Temperature Data for Three Days Using a Continuous Data Logger")

The data to be collected are restricted to the mass of filters (or filter assemblies) in µg using a delicate analytical balance.

Using the attached data sheet, record at a minimum the following information:

- Name of all experimenters
- Date of the test
- Mass of filters

Tools, Equipment, and Supplies Required

- Analytical balance accurate to 0.01 µg
- Supply of 2.5 µ filters, 10 µ filters, and an equal number of filter holders (or an equivalent supply of commercially available pre-loaded filters and holders of the same size ratings)
- Small, commercially available personal air pumps designed to be worn on the experimenter's coat or shirt (one for each filter size and one set of two for each experimenter)
- Static-free tubing of the same size as the filters and the air pump, approximately two feet of tubing per filter holder
- Rubber-tipped tongs to handle the filters

Safety

It is a reasonable assumption that there is nothing hazardous in the air to be tested inside a university laboratory, absent any direct knowledge otherwise. In a commercial setting, that assumption may not be correct.

The filters and procedures are not inherently dangerous to the experimenter, but care should be taken to avoid snagging the air tubing while walking about with the filters on.

As with all laboratory experiments, gloves, eye protection, and a lab coat are recommended for use during the laboratory portions of this experiment.

Planned Experimental Procedures and Steps, in Order

1. Put on appropriate personal protective equipment and gather all necessary tools and supplies in a convenient location.
2. For each filter to be tested (typically one 2.5 μ filter and one 10 μ filter for each experimenter), the following procedures should be very carefully followed.
   a. Mark a clean filter holder (or pre-loaded filter assembly) to indicate the size of the filter and a unique identifying name or number.
   b. Weigh a clean, blank filter on the scale (if using pre-loaded filters, weigh the entire assembly), M_o.
   c. As soon as the balance has stabilized, remove the filter and place it into a filter holder according to the instructions of the filter holder manufacturer.
   d. Calibrate an air pump to a convenient, stable air flow rate, typically at 3 L/min (0.1 ft^3/min) or lower, flow rate, F.
   e. Attach the filter assembly to the *intake* of the air pump with a suitable length of tubing, keeping the tubing as short as possible. The pump needs to draw air through the filter assembly, and through the filter before it is discharged from the exhaust side of the air pump. Be sure the filter assembly is oriented so that the air entering the filter pushes the filter against the filter support inside the filter assembly.
   f. Attach the filter and pump assembly securely to the coat or shirt of the experimenter at approximately waist height to avoid influences from the breathing of the experimenter.
   g. Turn on the air pump and carefully note the time, T_1.
   h. Allow the experimenter to walk about the building and campus for 1 hour.
   i. Turn off the air pump at one hour. Carefully note the time. The precise time of exposure is not important, but knowing the actual time of exposure is important, time, T_2.
   j. Immediately remove the filter from the assembly and re-weigh the filter (or the entire pre-loaded assembly), to obtain the filter weight, M_F.
3. Calculate the mass of particulates in the air using the equations in the data analysis plan section.

Data Collection Format and Data Collection Sheets Tailored to Experiment

See attached data collection sheet.

Data Analysis Plan

This experiment relies on the careful weighing of the filters. The mass and concentration of PM in the air is calculated from the following equations.

$$PM = M_F - M_o$$

Where: PM = mass of particulates collected on the filter, µg

M_F = mass of filter after exposure, µg
M_o = mass of filter prior to exposure, µg

$$T_T = T_2 - T_1$$

Where: T_T = total elapsed time of the test, min

T_1 = time that the air pump was turned on
T_2 = time when the air pump was turned off

$$PM_C = PM/(T_T \times F)$$

Where: PM_C = PM concentration in the air, µg/ft^3 or µg/m^3

PM = mass of particulate collected on the filter, µg
T_T = total elapsed time of the test, min
F = air pump flow rate in ft^3/min or m^3/min

Expected Outcomes

Indoor air is generally low in PM. Outdoor air may be much higher depending on the conditions of temperature, wind, and humidity and depending on the time of year. Experiments during the spring and fall, in particular, may exhibit much higher pollen counts and humidity and, therefore, much higher PM concentrations.

DATA COLLECTION SHEET GRAVIMETRIC DETERMINATION OF PARTICULATE MATTER IN AIR	
Lead Experimenter	
Date of Experiment	
Mass of Clean Blank Filter (Step 3b), M_o	
Flow Rate of Air from Calibrated Air Pump, (Step 3d), F	
Time Air Pump Is Turned On (Step 3g), T_1	
Time Air Pump Is Turned Off (Step 3i), T_2	
Mass of Exposed Filter (Step 3j), M_F	
Calculated Mass of Particulates per Data Analysis Plan (Step 4)	
Calculated Value of PM	
Calculated Value of T_T	
Calculated Value of PM_c	

EXPERIMENT 12.6

STATIC PUNCTURE STRENGTH OF GEOTEXTILE FABRICS

(Consistent with ASTM Method D 6241–14)

Date: ______________________________

Principal Investigator:

Name: ______________________________

Email: ______________________________

Phone: ______________________________

Collaborators:

Name: ______________________________

Email: ______________________________

Phone: ______________________________

Name: ______________________________

Email: ______________________________

Phone: ______________________________

Name: ______________________________

Email: ______________________________

Phone: ______________________________

Name: ______________________________

Email: ______________________________

Phone: ______________________________

Objective or Question to Be Addressed

This experiment is used to measure the force required to puncture a textile or other fabric. While initially designed to measure the puncture resistance of geotextiles and geotextile-related products, it is also applicable to any other fabric that can be properly secured in the testing device.

Theory behind Experiment

A test sample is clamped, tightly, but without tension, between circular plates and secured in a tensile or compression testing machine. A force is exerted against the center of the unsupported portion of the sample by a 2 +/– 0.1 inches (50 +/– 1 mm) diameter plunger attached to a load indicator until rupture occurs. The maximum force is defined as the puncture strength.

Potential Interferences and Interference Management Plan

Samples collected from a roll of material should be taken no closer to the edge of the fabric than 1/20 of the fabric width or 6 inches (150 mm), whichever is smaller. The inner and outer wraps of a roll, or any material containing folds, crushed areas, or other distortions will not be representative of the sample and should be excluded from the test sample. From each laboratory sample, cut the sample from the available material so that the edge of the specimen will extend beyond the edge of the clamp by at least 0.39 inches (10 mm) in all directions.

Separating multilayer samples and adding the results of each layer is not appropriate and will not yield reliable results.

Data to Be Collected with an Explanation of How They Will Be Collected (Ex.: "Water Temperature Data for Three Days Using a Continuous Data Logger")

Using the attached data sheet, record at a minimum the following information:

- Names of all experimenters
- Date of the experiment
- Type and source of the material tested
- Description of the testing apparatus and clamp system, including the means of holding the sample in the clamp
- Number of samples tested from the same material
- Average puncture strength of the material
- Variation of data within each group of specimens
- Variation, if any, from the described test method

Tools, Equipment, and Supplies Required

- Compression machine
    - Ideally a constant-rate-of extension (CRE) type, with autographic recorder, capable of holding a 2 +/– 0.1 inches (50 +/– 1 mm) steel rod

vertically and slowly pushing that rod vertically downward onto the center of a fabric holding device while recording the maximum load on the rod at failure of the material. The material should fail at a load between 10 and 90 % of the selected load range of the compression machine. If it fails at a load outside this range, select a new range to keep the failure load within this range.

- Plunger attached to the compression machine
  - Flat diameter of 2 +/– 0.1 inches (50 +/– 1 mm) with a radial edge of 0.1 inches (2.5 +/– 0.5 mm) suitable for mounting in the compression device
- Circular clamp
  - Inside diameter of 6 +/– 0.2 inches (150 +/– 1 mm) and an outside diameter sufficient to provide adequate clamping of the fabric without tearing of the fabric at the clamp restraints and tall enough to allow the fabric to stretch prior to failure. A clamp with an outside diameter of about 9.8 inches (250 mm) and approximately 6 inches (150 mm) clearance below the bottom of the clamp will generally be adequate.
- Several samples of the fabric for testing with the number of samples determined as follows:

  - When there exists a reliable estimate available of the coefficient of variation, v, for samples of similar materials determine the required number of specimens as follows:

$$n = (tv/A)^2$$

where: n = number of test specimens (rounded upward to a whole number)
v = reliable estimate of the coefficient of variation, %
t = value of Student's t one-sided limits, at the 95 % probability level, and the degrees of freedom, associated with the estimate of v
A = 5 % of the average value of the allowable variation

  - When there is no reliable estimate of v available, use 10 or more samples to develop a reliable average value for the penetration strength.

Safety

Cutting samples of material from a larger specimen creates a risk of injury to the cutter. The use of cut-resistant gloves is recommended during this procedure. In the event of a cut of any size, seek medical advice for infection control or stitching, if necessary.

The placement of the sample under the compression machine plunger creates a risk of premature activation of the machine, leading to a finger or hand injury. The machine should not be turned on until the sample is properly placed in the machine, ready for testing to avoid this risk.

This experiment requires repetitive testing of sequential samples. Repetition often creates complacency, and complacency leads to accidents. At least two people need to be present at all times during this experiment to minimize this risk.

Planned Experimental Procedures and Steps, in Order

1. Put on appropriate personal protective equipment and gather all materials and supplies together in a convenient location.
2. Center and secure the test sample in the circular clamp, ensuring that the sample extends beyond the outer edges of the clamping plates.
3. Mark the test specimen along the inside circumference of the clamp so that a measurement can be made of any potential slippage of the sample in the clamp. If slippage greater than 5 mm is observed, the test on that sample should be discarded.
4. Test the fabric at a machine speed of about 2 inches/min (50 mm/min) until the puncture rod completely ruptures the sample.
5. Read the maximum puncture strength and displacement from the greatest force registered on the recording instrument during the test. Note that with composite materials there may be a double peak. If so, the initial value should be reported even if the second peak is higher than the first one.
6. Calculate the average of the puncture strength for all samples tested as read directly from the recording instrument.

Data Collection Format and Data Collection Sheets Tailored to Experiment

See attached data collection sheet.

Data Analysis Plan

The data collected from this experiment are direct readings from the instrument used. The readings are averaged, and the result reported as the puncture resistance strength of the fabric. No further analyses are anticipated as part of the experiment.

Expected Outcomes

The outcome will be dependent on the actual material tested, and there is no standard outcome to which those data may be compared.

DATA COLLECTION SHEET STATIC PUNCTURE STRENGTH OF GEOTEXTILE FABRICS	
Lead Experimenter	
Date of Experiment	
Type of Material Tested	
Source of Test Material	
Penetration Strength for Sample 1	
Penetration Strength for Sample 2	
Penetration Strength for Sample 3	
Penetration Strength for Sample 4	
Penetration Strength for Sample 5	
Penetration Strength for Sample 6	
Penetration Strength for Sample 7	
Penetration Strength for Sample 8	
Penetration Strength for Sample 9	
Penetration Strength for Sample 10	
Average Penetration Strength	

EXPERIMENT 12.7[3]

SPECIFIC HEAT CAPACITY OF MATERIALS

Date: ____________________

Principal Investigator:

Name: ____________________

Email: ____________________

Phone: ____________________

Collaborators:

Name: ____________________

Email: ____________________

Phone: ____________________

Name: ____________________

Email: ____________________

Phone: ____________________

Name: ____________________

Email: ____________________

Phone: ____________________

Name: ____________________

Email: ____________________

Phone: ____________________

Objective or Question to Be Addressed

The objective of this experiment is to determine the specific heat capacity of various materials.

Theory behind Experiment

Specific heat capacity is a physical characteristic of a material or substance. It is an indication of the energy required to heat or cool an object. These factors are important in assessing energy consumption and assessing climate change. The *specific heat capacity* of a substance used for this experiment, indicated by the symbol C, is the heat capacity of a sample of the substance divided by the mass of the sample. In essence, it is the amount of energy that must be added to one unit of mass of the material, generally in the form of heat, to cause an increase of one unit in its temperature, generally measured in °C or °K. Note that °C need to be converted to °K for consistency with the definition of specific heat. Most thermometers do not measure directly in °K, however, so °C are usually used for measurement because a change of one degree C is equivalent to a change of one degree F and the actual temperature is not important; only the change in temperature is important in the experiment.

The SI units of specific heat are *joules per kelvin per kilogram*, J/(°K/kg). The English units of specific heat are BTU/°F/lb (1 BTU/°F/lb = 4,177.6 J/°K/kg). The English measurement units were originally defined such that the specific heat of water would be equal to 1 BTU/°F/lb.

Note that the specific heat often varies with temperature, however, and is different for each state of matter. For example, liquid water has a specific heat of about 4,182 J/(°K/kg) at 20 °C; ice just below 0 °C (32 °F) has a specific heat of only 2,093 J/(°K/kg). When a substance is undergoing a phase transition, such as melting or boiling, its specific heat is technically infinite because the heat goes into changing its state rather than raising its temperature. When discussing specific heat capacity remember that the word "specific" denotes an intrinsic characteristic and in this case means that the property is not dependent on the mass of the substance, while the term "heat capacity" is an extensive property and varies with the amount of a substance.

Note, too, that the specific heat of some substances, especially gases, is often much higher when they are allowed to expand as the heating occurs (referred to as the specific heat *at constant pressure*, c_p) than when is heated in a closed vessel that prevents expansion (referred to as the specific heat *at constant volume*, c_v). The term "heat capacity ratio" is applied to the quotient of these two values and denoted by the symbol γ. Doing such an experiment under constant volume conditions requires extensive procedures to guarantee no change in volume. That also dramatically increases internal pressures that need to be controlled. That procedure being dangerous, use of constant pressure calculations is most commonly done. If a constant volume value is needed, calculation of that value from the constant volume value and the various physical properties of the material, such as its coefficient of heat expansion, can be done. The procedure described here is a constant pressure procedure.

There are other definitions of specific heat used in esoteric applications that can be identified, if ever needed.

The specific heat capacity of each material is calculated from the formula:

$$C = \frac{Q}{m * T}$$

Where:
C = specific heat capacity
Q = amount of heat added, J
M = mass of material being heated, kg
T = change in temperature of the material, °C

The value of Q for this experiment design can be reasonably approximated by the formula:

Q = (Temperature of the hotplate) × (Time of heating in minutes)

Table A.6 in the Appendix provides standard values of specific heat for a variety of materials suitable for testing by the experimental procedure. Results from this experiment should be evaluated by comparison with the standard values shown. Note that the reported values for some materials differ significantly based on different sources. That suggests that calculation of these values is not a precise science for many materials. Experimental results from this experiment should be within reasonable concurrence with the range of values shown.

Conversion between units of measurement are based on the following equations:

1 BTU/(lb-°F) = 1.8 BTU/(lb-°C)
1 BTU/(lb-°F) = 4177.6 J/(kg-°K)
1 J/(kg-°K) = 1000 J/(g-°C)

Data to Be Collected with an Explanation of How They Will Be Collected (Ex.: "Water Temperature Data for Three Days Using a Continuous Data Logger")

Using the attached data sheet, record at a minimum the following information:

- Name of all experimenters
- Date of the test
- Mass of material being tested will be determined from weighing on a scale to the nearest 0.01 g (0.035 oz)
- Temperature of the hotplates will be measured using the preset temperature control guides on the units, verified by careful calibration of the units, to set them at 100 +/– 0.1 °C or 212 +/– 0.25 °F)
- Temperature of the mass will be measured initially and at the end of the test period using a thermometer inserted through a stopper on the top of the heating flask capable of reading to 0.01 °C or 0.25 °F
- Time of heating will be measured with a stopwatch, wristwatch, or other suitably accurate clock

Potential Interferences and Interference Management Plan

Interferences in this test procedure include impurities in the test samples, inaccuracies in the calibration of the hotplates and thermometers, improper calibration of scales used to mass the samples, inaccurate reading of heating times, and inaccurate calculation of results. To avoid these complications, care must be taken to calibrate the equipment properly and to check calculations carefully. Impurities are difficult to see in most samples, but it is unlikely that a pure substance will be available, suggesting that a variation, perhaps significant, from the stated values in the table may result from the experimental procedure.

Tools, Equipment, and Supplies Required

- 1 hotplate capable of temperature control to within 0.1 °C at 100 °C (0.02 °F at 212 °F) for each sample to be tested
- 1 250 mL (8.5 oz) Erlenmeyer flask for each sample to be tested
- 1 stopper (cork or rubber) for each flask sufficient to cover the top of the flask and containing a hole for the insertion of the selected thermometer or temperature probe.
- 1 digital thermometer or electronic thermometer with a long enough probe to be inserted through the stopper of a 250 mL (8.5 oz) Erlenmeyer flask. The thermometer must be capable of being inserted into the test sample contained within the flask and extend far enough out of the flask to read the temperature of the sample before and after heating to within 0.01 °C (0.02 °F).

Safety

Wear lab coats, safety glasses, and use heat-resistant gloves. This experiment uses hotplates set at a temperature of 100 °C (212 °F), and glass flasks are placed on top of those hotplates to heat the flasks and their contents. Everything will be very hot, and care must be used to handle all equipment with heat-resistant gloves at all times. Burns are the biggest risk faced with this experiment.

In addition to the heat risks, touching a hot surface generally elicits an uncontrolled reaction in people, which can dislodge glassware and cause breakage, yielding flying glass shards and burning of people, laboratory tables, floors, and other materials nearby. Cuts from hot glass are a risk in that scenario.

Depending on the material being tested, heating of the material may cause off-gassing of hazardous or dangerous gasses. Being knowledgeable about the potential emissions from hot samples is important to safeguard experimenters. Use of a fume hood in which to conduct the heating of the flasks will minimize the risks of accidental breakage and off-gassing.

Heating of material is being done inside a covered Erlenmeyer flask. The cork or stopper is to be placed *over* the mouth of the flask *but not inserted into the mouth of the flask, to avoid an increase in pressure inside the flask which could cause an explosion.* An increase in the internal pressure inside the flask would also change the constant pressure value of the calculated specific heat, rendering the experimental results inaccurate.

Planned Experimental Procedures and Steps, in Order

1. Preheat all hot plates to 100 °C (212 °F).
2. Arrange 250 mL (8.5 oz) Erlenmeyer flasks with a label to indicate the material to be tested in that beaker.

3. Place exactly 100 g (3.2 oz) of each material to be tested in separate beakers (any other mass is also acceptable provided the actual mass is recorded precisely.)
4. Place a cork or rubber stopper on the top of each beaker and insert a thermometer through the hole in the center of each cork or stopper so that the thermometer is inserted into the mass of material being tested.
   Do not insert the cork or stopper into the mouth of the flask to avoid overpressure during heating.
5. Record the initial temperature of each material.
6. Set each beaker on a separate hotplate sequentially at 2-minute intervals so that end results can be read without haste at approximate intervals of 2 minutes.
7. Allow the materials to heat for 10 minutes. Record the actual time of heating.
8. At the end of the 10-minute heating time, record the temperature of the material from the thermometer.
9. Calculate the change in temperature of the material being tested.
10. Calculate the value of Q using the formula:
    Q = (Temperature of the hotplate) × (Time of heating in minutes)
11. Calculate the specific heat capacity of each material using the formula:

$$C = \left(\frac{Q}{M * T} \right)$$

Where:
C = specific heat capacity
Q = amount of heat added
M = mass of material being heated
T = change in temperature of the material

Data Collection Format and Data Collection Sheets Tailored to Experiment

See attached data collection sheet.

Data Analysis Plan

A comparison will be made between the specific heat calculated for the test materials and the standard values for specific heat of that material as published elsewhere. In the event there is no identified standard value published for the material tested, a comparison will be made with similar materials demonstrating specific heat values slightly higher and slightly lower than that measured to determine whether the measured value is reasonable.

Anticipated Outcomes

The specific heat values calculated are very different for each material tested. The actual results will vary depending upon how accurately the hotplates are calibrated, how accurately they hold the set temperature, the starting and ending temperature of the samples tested and the accuracy of the thermometer measuring the temperature of the test samples. Calculated values should be consistent with the values shown in Table A.6.

DATA COLLECTION SHEET SPECIFIC HEAT CAPACITY OF MATERIALS					
DATE SAMPLES COLLECTED:					
DATE SAMPLES TESTED:					
PRINCIPAL INVESTIGATOR:					
Material Being Tested	SAMPLE 1	SAMPLE 2	SAMPLE 3	SAMPLE 4	SAMPLE 5
Mass of Sample Placed into Beaker					
Initial Temperature of Sample, °C					
Actual Time of Heating, mins					
Temperature of Sample after Heating, °C					
Change in Temperature during Heating, °C					
Calculated Value of Q					
Calculate Specific Heat of Each Sample					

EXPERIMENT 12.8

PERCOLATION TEST PROCEDURE[4]

Date: ______________________________

Principal Investigator:

Name: ______________________________

Email: ______________________________

Phone: ______________________________

Collaborators:

Name: ______________________________

Email: ______________________________

Phone: ______________________________

Name: ______________________________

Email: ______________________________

Phone: ______________________________

Name: ______________________________

Email: ______________________________

Phone: ______________________________

Name: ______________________________

Email: ______________________________

Phone: ______________________________

Objective or Question to Be Addressed

This experiment demonstrates the procedure for conducting a percolation test in soil. When a *subsurface wastewater treatment system* (sometimes referred to as a *leaching field* or *septic system*) is required, the design of the field is based on the ability of the subsurface soils to filter the volume of water generally expected from the structure or structures connected to it. This soil characteristic is determined by a *percolation test*, or *perc test*.

Theory behind Experiment

The percolation test method varies slightly depending on site conditions. The steps outlined next provide for that variation. Regardless of the method used, the rate developed from the test, measured in minutes per inch, is then used to go into a table, such as Table 12.8.1, to determine the area of leaching trench needed for the expected average daily flow and the soil type found at the site. If the percolation rate is less than 5 minutes per inch, a septic tank and leaching field will generally prove to be a functional disposal system. At percolation rates above that, the type of soil will generally dictate the loading rates. Silty and clayey soils will require a much lower loading rate than sandier soils, as indicated in Table 12.8.1.

The disposal field is typically constructed of a 4 inch (10 cm) diameter, perforated PVC pipe set at a slope of 0.005 ft/ft (0.0165 m/m) with individual lines generally less than 100 feet (30.54 m) long. The trenches are excavated to a depth of at least 18 inches (0.46 m) or to the top of the layer in which the percolation test was run. The bottom of the trench must be at least 5 feet (1.5 m) above the seasonal high groundwater table. The trench is generally constructed from 3 feet to 4 feet (0.9 to 1.2 m) wide and backfilled to a depth of 12–14 inches (30.5–35.6 cm) with gravel before the pipe is placed. The pipe is then backfilled with gravel to 2 inches (10 cm) above the pipe, and site soils are used to backfill the rest of the way with a layer of geotextile often placed on top of the gravel, first, to minimize the seepage of soil into the gravel layer.

Lateral trenches are placed 6 feet (1.8 m) or more on center. The total length of pipe required depends on the trench width, since the product of the trench width times the pipe length must equal the area requirement determined from the table. The sides of the trench are not counted when calculating the area or length of laterals required.

Potential Interferences and Interference Management Plan

The hole diameter and depth are important characteristics of this test due to the standardization of loading rates for different types of soil being dependent on the seepage rate out of the hole. The seepage rate out of the hole depends on the surface area of the hole, including the sides and the bottom. If those areas are significantly different from the defined hole, the rate developed will not accurately reflect the size requirements for the field.

Scratching the walls and bottom of a hole to reduce smearing during excavation is an area where interferences may occur. Care is needed to ensure that the scratches do not significantly alter the dimensions of the hole.

This test is predicated on the use of clean water in the percolation test hole, even though settled sewage is expected to be discharged through the field later. Using dirty water, containing suspended or floating debris that can clog the pores of the hole soils, will significantly alter the test results.

Maintaining the water level in the hole during the saturation step is a key element of the outcome. The level of water in the hole must be maintained at the designated depth, +/– 1 inch (2.5 cm), during the entire saturation period.

Data to Be Collected with an Explanation of How They Will Be Collected (Ex.: "Water Temperature Data for Three Days Using a Continuous Data Logger")

Using the attached data sheet, record at a minimum the following information:

- Name of all experimenters
- Date of the test
- This experiment requires the recording of precise time intervals and precise water depths. A measuring stick, which measures down from a fixed board across the top of the hole or measuring stick that measures up from the bottom hole are equally acceptable
- A watch or stopwatch is used to record the time after the start of the test that specific events occur

Tools, Equipment, and Supplies Required

- Posthole digger with a 12 inch (30.54 cm) diameter
- Bucket to carry water from a nearby source and an adequate supply of clean, potable water
- Measuring device to measure the depth of water in the hole
- Watch or stopwatch to measure time of seepage

Safety

Care walking around the site of a percolation test is needed to minimize the risk of collapsing the hole during the test or accidentally stepping into the hole, causing a foot or angle injury.

Soil can be abrasive on skin. Heavy work gloves are required when using the posthole digger.

Planned Experimental Procedures and Steps, in Order

1. Prepare a test hole located within the proposed disposal area that, in the judgment of the investigator, appears to be the most limiting. The test hole should have a diameter of 12 inches (30.5 cm), with vertical sides, and be 18 inches (45.7 cm) deep, not including any allowable liners or filter layers on either the bottom or sides.
2. Establish a fixed point at the top or bottom of the test hole from which all measurements will be made.

3. Scratch the bottom and sides of the test hole to remove any smeared soil surfaces, taking care not to significantly change the hole dimensions. Add 2 inches (5.1 cm) of coarse sand to protect the bottom from scouring or insert a board or other impervious object in the hole so that water may be poured down or on it during the filling operation. A mesh or perforated liner designed to maintain the test hole dimensions in extremely loose soils, while allowing essentially unrestricted flow of water, may also be used.
4. Carefully fill the hole with clear water to a minimum depth of 12 inches (30.5cm) from the bottom. Maintain this minimum 12 inches (30.5 cm) or greater water level by adding water as necessary in order to saturate surrounding soils for a period of no less than 15 minutes after first filling the hole.
5. After saturation, let the water level drop to a depth of 9 inches (22.9 cm) above the bottom of the hole and then measure the length of time in minutes for the water level to drop from a depth of 9 inches (22.9 cm) to a depth of 6 inches (15.2 cm). If the rate is erratic, the hole should be refilled and soaked until the drop per increment of time is steady. The time for the level to drop from a depth of 9 inches (22.9 cm) to a depth of 6 inches (15.2 cm), divided by three, is the *percolation rate* in *minutes per inch.*
6. In certain soils, particularly coarse sands, the soil may be so pervious as to make a percolation test difficult, impractical, and meaningless. In this case, the percolation test may generally be discontinued and a rate of 2 minutes per inch or less can be assumed provided that at least 24 gallons of water has been added to the percolation hole within 15 minutes and it is impossible to obtain a liquid depth of 9 inches (22.9 cm).

A variation of this procedure provides for the following modifications after the hole is excavated in accordance with the preceding Step 1.

A 2-inch (5.1 cm) layer of fine gravel or coarse sand is placed in the bottom of the hole and smoothed out. The hole is then filled with clean water to a depth of 12 inches (30.5 cm) above the gravel or sand. That depth is maintained for a period of at least 4 hours, and preferably overnight, by continually refilling it. If the hole holds water overnight, the depth is adjusted to 6 inches (15.2 cm) above the sand, then the depth the water level drops in 30 minutes is recorded. If the hole is empty in the morning, it is refilled to 6 inches (15.2 cm) and the change in depth is recorded at 30-minute intervals for 4 hours. The drop during the last 30-minute interval is recorded as the percolation rate in terms of minutes per 1 inch (2.5 cm) of drop.

Data Collection Format and Data Collection Sheets Tailored to Experiment

See attached data collection sheet.

Data Analysis Plan

The data are recorded and reported as measured. There is no additional analysis required. The results are then used to determine the size of the required distribution field as previously noted.

TABLE 12.8.1
Recommended Soil Loading Rates for Various Soil Types in gpd/ft^2 (L/d/m^2)

Percolation Rate in min/in	Soil Type: Sand, Loamy Sand	Loams, Sandy Loam	Silt Loam, Sandy Clay, Loams with Less Than 27 % Clay, Silt	Clays, Silty Clay Loam, Sandy Clay Loam with 27 % or More Clay, Clay Loams, Silty Clays
5 or less	0.74 (30.1)	0.60 (24.4)		
6	0.70 (28.5)	0.60 (24.4)		
7	0.68 (27.7)	0.60 (24.4)		
8	0.66 (26.9)	0.60 (24.4)		
10		0.60 (24.4)		
15		0.56 (22.8)	0.37 (15.1)	
20		0.53 (21.6)	0.34 (13.8)	
25		0.40 (16.3)	0.33 (13.4)	
30		0.33 (13.4)	0.29 (11.8)	
40			0.25 (10.2)	
50			0.20 (8.1)	0.20 (8.1)
60			0.15 (6.1)	0.15 (6.1)

Expected Outcomes

The rate observed should be consistent with the rate shown for the type of soil identified on the site and noted in Table 12.8.1.

DATA COLLECTION SHEET PERCOLATION TEST PROCEDURE		
Lead Investigator		
Date of Test		
Location of Test		
		NOTES
Dimensions of Finished Hole		
Diameter		
Depth		
Depth of Water in Initial Fill (Step 4)		
Time of Initial Filling		
Time at End of Saturation Period		
Time for Water Level to drop 3 Inches (7.62 cm)		
Time for Water Level to Drop 3 Additional Inches (7.62 cm)		
Calculated Percolation Rate (Step 5)		

NOTES

1 Francis J. Hopcroft and Abigail J. Charest, *Experiment Design for Environmental Engineering, Methods and Examples* (CRC Press, 2022, Experiment 14.5): 282.
2 Francis J. Hopcroft and Abigail J. Charest, *Experiment Design for Environmental Engineering, Methods and Examples* (CRC Press, 2022, Experiment 9.3): 75.
3 Francis J. Hopcroft and Abigail J. Charest, *Experiment Design for Environmental Engineering, Methods and Examples* (CRC Press, 2022, Experiment 9.5): 89.
4 Francis J. Hopcroft, *Wastewater Treatment Concepts and Practices* (Momentum Press, 2015).

13 Fluid Mechanics and Hydraulic Experiments

This chapter provides completed designs for example experiments that can be used to demonstrate specific engineering phenomena in a university laboratory or field collection site. They may be used as presented or modified by the user to adapt to other desired outcomes. Sections on actual outcomes and data interpretation are necessarily left blank in these examples. They should be completed by the investigators upon collection of the data.

Note that where an experiment indicates that it is consistent with a specified ASTM method or other standard, this does not mean that the procedure is identical to, nor does it include all the details of, the stated standard method. Outcomes that require strict compliance with the stated standards must use that standard method to conduct the test. Therefore, the experiment procedures outlined in this book are not a substitute for proper compliance with the stated standards, nor are they intended to be.

The experiments in this chapter are:

13.1 Properties of Fluids
13.2 Capillary Rise in Tubes and Soil
13.3 Buoyancy Forces on Submerged Surfaces
13.4 Siphons
13.5 Open Channel Flow Measurement by Velocity-Area Method
13.6 Open Channel Flow Measurement by Thin Plate Weir
13.7 Determination of the Horizontal Flow Rate through a Geosynthetic Screening Material

DOI: 10.1201/9781003346685-13

EXPERIMENT 13.1

PROPERTIES OF FLUIDS

Date: ______________________________

Principal Investigator:

Name: ______________________________

Email: ______________________________

Phone: ______________________________

Collaborators:

Name: ______________________________

Email: ______________________________

Phone: ______________________________

Name: ______________________________

Email: ______________________________

Phone: ______________________________

Name: ______________________________

Email: ______________________________

Phone: ______________________________

Name: ______________________________

Email: ______________________________

Phone: ______________________________

Objective or Question to Be Addressed

The definition of fluids includes a variety of materials such as water, other liquids, and also gases. Energy waves, such as heat, may also act a lot like a fluid. This experiment examines some of those fluid properties using water as the medium.

Theory behind Experiment

Water, and most other fluids, exhibit a variety of properties. They are generally recognized as density, specific gravity (for fluids other than water, since specific gravity relates the density of any other fluid to the density of water), dynamic and kinematic viscosity, specific heat, conductivity, Prandtl number, thermal diffusivity, coefficient of thermal expansion, and coefficient of volumetric expansion. The measurement of these properties is complex.

The first part of this experiment involves heating a known quantity of water by a finite and measurable amount and then measuring the density (mass/unit of volume), coefficient of thermal expansion, and coefficient of volumetric expansion. The change in density and the expansion coefficients are linear functions of temperature change for most fluids, *but not for water*. This experiment as designed uses water, but it may also be conducted using any other fluid. Table A.7 in the Appendix provides values for a selection of fluids for which there are constants. The density and the coefficients of expansion for water are temperature dependent, and the results will be curvilinear when plotted. As the temperature rises, density decreases, while the coefficients of thermal expansion increase. Similarly, as the temperature declines, density increases while the coefficients of thermal expansion decrease.

In the second part of the experiment, the concepts of fluid viscosity are examined. Viscosity is a measure of the resistance of a fluid to flowing. It depends on the size and shape of the particles of the liquid, as well as the attraction between those particles and the temperature of the fluid. Liquids that have a low viscosity flow quickly – such as water, rubbing alcohol, and vegetable oil. Liquids that have a high viscosity flow slowly – such as liquid honey, corn syrup, and molasses. The more viscous (or thick) a fluid is, the longer it will take for an object to move through that fluid.

The kinematic viscosity of a fluid is the ratio of the dynamic viscosity of the fluid to the density of that fluid. Mathematically, it is expressed as:

$$\nu = \mu/\rho$$

where: ν = kinematic viscosity, $\nu = \mu/\rho$, (m^2/s)
μ = dynamic viscosity ($N{\cdot}s/m^2$)
ρ = density (kg/m^3)

A conversion from absolute to kinematic viscosity in imperial units can be expressed as:

$$\nu = 6.7197\ 10^{-4}\ \mu/\gamma$$

where: ν = kinematic viscosity (ft^2/s)
μ = absolute or dynamic viscosity (cP)
γ = specific weight (lb/ft^3)

Note that since the density is a variable function of the temperature, so is the kinematic viscosity.

The dynamic viscosity is the kinematic viscosity of the liquid at a particular temperature multiplied by its density at that same temperature. Dynamic viscosity, also called absolute viscosity, is defined in physics as the tangential force per unit area required to move one horizontal plane with respect to another plane, at a unit velocity, while maintaining the planes a unit distance apart in the fluid. It is functionally a measure of internal resistance in the fluid.

The measurement of viscosity in difficult and beyond the scope of this experiment. What is possible is to understand the concept better and to then be able to apply the data from published tables of absolute viscosity to determine the viscosity at any desired temperature. Table A.11 in the Appendix provides a list of absolute viscosity values for several common fluids, while Table A.12 provides similar data with respect to dynamic viscosity. A linear conversion between stated temperatures will yield values that are sufficiently close to the actual for all practical purposes.

In the third part of the experiment, the electrical conductivity of the water is examined. This characteristic is generally a function of dissolved salts and metallic ions. Experiments could be run for days using various concentrations of salts and ions, but for the purposes of this experiment, only the salt concentration, represented by table salt (sodium chloride, NaCl), will be utilized. Using other dissolved metals of interest in a manner similar to that shown in this experiment will provide interesting and similar results for those ion concentrations.

The Prandtl number (P_r) is a dimensionless number, named after the German physicist Ludwig Prandtl, defined as the ratio of momentum diffusivity to thermal diffusivity. The Prandtl number is given as:

$$
\begin{aligned}
P_r &= \nu/\alpha \\
&= \text{kinematic viscosity/thermal diffusivity} \\
&= [(\mu/\rho)/(k/(c_p \cdot \rho)] \\
&= [(c_p \cdot \mu)/k]
\end{aligned}
$$

where: ν kinematic viscosity = μ/ρ, (m^2/s)
α = thermal diffusivity, $k/\rho c_p$, (m^2/s)
μ = dynamic viscosity, ($Pa{\cdot}s = N{\cdot}s/m^2$)
k = thermal conductivity, ($W/(m{\cdot}K)$)
c_p = specific heat, ($J/(kg{\cdot}K)$)
ρ = density, (kg/m^3)

Note that the Prandtl number is dependent only on the fluid and the fluid state. The Prandtl number is often found in property tables alongside other properties, such as viscosity and thermal conductivity. Those values are not explored in this experiment.

Specific heat determination is the subject of Experiment 12.7 and is not further examined in this experiment.

Potential Interferences and Interference Management Plan

The measurement of the water characteristics is a function of the purity of the water, and the values obtained from this experiment may be slightly different from the data in published tables. Reasonable concurrence with those published data should be expected, and the variation by temperature or ion concentrations should also be consistent with published data.

Data to Be Collected with an Explanation of How They Will Be Collected (Ex.: "Water Temperature Data for Three Days Using a Continuous Data Logger")

Using the attached data sheet, record at a minimum the following information:

- Names of all experimenters
- Date of experiment
- Mass of cylinders, empty and full, using a scale accurate to 0.01 g
- Mass, depth of fluid, and temperature of hot cylinders with hot contents
- Time to 0.01 s with a stopwatch
- Various calculated values developed from the data during the data analysis phase

Tools, Equipment, and Supplies Required

Part 1 – Density and Coefficient of Thermal Expansion

- 3 graduated Pyrex glass cylinders with etched markings for depth as close together as possible and a volume of at least 1.5 L and a depth of at least 15 inches (38 cm)
- 3 L of distilled water
- Refrigerator to chill water to 4 °C (39.2 °F)
- Hotplates to heat cylinders full of water to 90 °C (212 °F)
- 3 thermometers that will record liquid temperatures from 0 °C to 100 °C (32 °F to 212 °F)
- Scale to weigh the filled cylinders with a precision of 0.01 g

Part 2 – Viscosity

- 5 graduated glass cylinders with etched markings for depth as close together as possible and a volume of at least 1.5 L
- 2 L each of distilled water, acetone, olive oil, liquid honey, and Karo® corn syrup
- Stopwatch that records in hundredths of a second
- 5 small glass marbles of equal convenient size to fit into the cylinders without touching the sides
- Small plastic net for retrieving the marbles from the bottom of the filled cylinders

Part 3 – Conductivity

- 3 graduated Pyrex glass beakers with etched markings for depth and a volume of at least 1.5 L each
- 3 L of distilled water
- Approximately 0.5 kg of table salt
- Glass stirring rod at least 1.5 times as long as the cylinders are deep
- An approximately 400–500 mL sample of distilled water with an unknown (less than 5 mg/L) concentration of NaCl
- Commercially available and calibrated conductivity meter

Safety

This experiment does not use dangerous materials except in Part 2, where acetone is used. Acetone is volatile and may exhibit vaporization during the experiment. Conducting this phase of the experiment under a vented hood is recommended.

Heating of water to temperatures just below the boiling point is required. Measurement of the depth of fluid inside a hot, narrow, glass cylinder can cause burns. Heat-resistant gloves are recommended during the measurement of parameters of hot fluids during the experiment.

Planned Experimental Procedures and Steps, in Order

Part 1 – Density and Coefficients of Thermal Expansion

1. Put on appropriate personal protective equipment and gather all required tools, equipment, and supplies in a convenient location.
2. Measure and record the mass of each graduated cylinder being used.
3. Fill each of three graduated cylinders with 1 L of distilled water, insert a thermometer into each cylinder, then weigh and record the mass of each cylinder.
4. Cool all three cylinders to 4 °C (39.2 °F).
5. Quickly and carefully measure and record the depth of the water *by measuring along the outside of the cylinder* (do not put anything into the fluid to do the measuring) in each of the three cylinders.
6. Measure and record the mass of each cylinder.
7. Transfer all three cylinders to hotplates and heat them to 10 °C (50 °F).
8. Quickly and carefully measure and record the depth of the water, the temperature of the water, and the mass of each filled cylinder.
9. Repeat Steps 5 and 6 at temperatures of 20, 30, 40, 50, 60, 70, 80, and 90 °C (53.8, 86, 104, 122, 140, 158, and 176 °F).
10. Calculate and record the density and coefficients of thermal expansion for each cylinder at each temperature in accordance with the data analysis plan, then average the three values and report the average values as the density and coefficients of thermal expansion for the water at that temperature.
11. Separately plot the calculated average values on a normal-normal graph with temperature on the x-axis and the other three properties on the y-axis.
12. Compare the calculated average values to published data from the tables in the Appendix and elsewhere.

Part 2 – Viscosity

1. Put on appropriate personal protective equipment and gather all required tools, equipment, and supplies in a convenient location.
2. Fill five graduated cylinders with 1 L of fluid as follows:
   Number 1 – fill with acetone
   Number 2 – fill with distilled water
   Number 3 – fill with olive oil
   Number 4 – fill with liquid honey
   Number 5 – fill with Karo® corn syrup
3. Allow all five cylinders to stabilize to room temperature for a period of at least 24 hours.
4. Record the actual temperature of the fluids at the time of testing.
5. Using gloved hands (not vinyl or latex) or padded tongs (to avoid touching the sides of the cylinders, which could change the temperature slightly), place the cylinders in a convenient location where the depth of liquid in the cylinders can be accurately measured without further touching of the cylinders.
6. Carefully measure and record the depth of the liquid in each of the cylinders.
7. Drop the marble into the first cylinder by releasing it just above the surface of the liquid and record the time it takes the marble to just reach the bottom of the cylinder.
8. Repeat Step 6 for each of the other four cylinders in order.
9. Using the small net and gloved hands or padded tongs, retrieve the small marble from each cylinder and carefully clean and dry them.
10. Repeat Steps 5–8 four more times.
11. Average the values for all five trials and record those values as the values for the test liquids.
12. Calculate and record the dynamic viscosity of the five liquids in accordance with the data analysis plan.
13. Calculate and record the kinematic viscosity of each liquid in accordance with the data analysis plan.
14. Compare the results to published data such as that found in Tables A.11 and A.12 in the appendix. Table A.13 in the appendix provides a means of converting among various units used to express viscosity.
15. Repeat this experiment at different temperatures of the fluids and compare those results to published data.

Part 3 – Conductivity

1. Put on appropriate personal protective equipment and gather all required tools, equipment, and supplies in a convenient location.
2. Fill each of four graduated beakers with 1 L of distilled water.
3. Allow all four cylinders to stabilize to room temperature for a period of at least 24 hours.
4. Measure and record the conductivity of the water in each of the three cylinders using the conductivity meter.

5. Add 1 mg/L of NaCl to each of three cylinders and stir them with the glass rod until all the salt has been dissolved.
6. Measure and record the conductivity of the water in each of the three cylinders using the conductivity meter.
7. Add an additional 1 mg/L of NaCl to each of the cylinders and stir them with the glass rod until all the salt has been dissolved.
8. Measure and record the conductivity of the water in each of the three cylinders using the conductivity meter.
9. Repeat Steps 8 and 9 three more times until a total of 5 mg/L of NaCl have been added and the conductivity recorded.
10. Calculate the average value for the three cylinders at each concentration of NaCl and report that average value as the conductivity of the test water at the determined concentration of NaCl.
11. Plot the average conductivities on the x-axis vs. NaCl concentration on the y-axis of a normal-normal graph.
12. Create an unknown sample by adding approximately 1/8 teaspoon of NaCl to the fourth beaker and stir with a glass rod until all the salt is dissolved.
13. Measure the conductivity of the unknown sample of distilled water with NaCl and determine from the graph what the concentration of NaCl is in the unknown sample.

Data Collection Format and Data Collection Sheets Tailored to Experiment

See attached data collection sheet.

Data Analysis Plan

Part 1 – Density and Coefficients of Thermal Expansion

The density of fluids is a function of the temperature of the fluid and the coefficients of expansion. The density is defined as the mass per unit of volume, generally expressed in terms of kg/m^3 (lb/ft^3). For this experiment, subtract the mass of the empty cylinders from the mass of the filled cylinders at each temperature to determine the mass of the fluid inside the cylinder. Note that this mass may change due to evaporation as the cylinders are heated; thus, the need to re-measure each mass at each temperature.

Note that the density of a fluid when temperature changes can be expressed as:

$$\rho_1 = m/V_0\,(1 + \beta\,(t_1 - t_0))$$
$$= \rho_0/(1 + \beta\,(t_1 - t_0))$$

Where: m = mass of unit (kg, lb)
V = volume of unit (m^3, ft^3)
β = volumetric temperature expansion coefficient ($m^3/m^3/°C$, $ft^3/ft^3/°F$)
t_1 = final temperature (°C, °F)
t_0 = initial temperature (°C, °F)
ρ_1 = final density (kg/m^3, lb/ft^3)
ρ_0 = initial density (kg/m^3, lb/ft^3)

The coefficients of thermal expansion cause the volume of the fluid to change in direct relation to the change in temperature. The depth of the fluid in the cylinder (adjusted for any loss of mass due to heating) increases by a factor equal to the original volume multiplied by the linear coefficient of thermal expansion. The volume of the fluid in the cylinder will expand by a factor equal to the original mass multiplied by the volumetric coefficient of thermal expansion.

Calculate and record the volume of fluid in each cylinder at each temperature as the depth of the fluid times the cross-sectional area of the cylinder.

Calculate the linear coefficient of expansion as the depth of the fluid at each temperature divided by the depth of the fluid at the previous temperature, all divided by the change in temperature.

The linear thermal coefficient of expansion, α, can be expressed as:

$$dl = L_0\ \alpha\ (t_1 - t_0)$$

or

$$\alpha = dl/[L_0\ (t_1 - t_0)]$$

where: dl = change in depth (m, inches)
L_0 = initial depth (m, inches)
α = linear expansion coefficient (m/m/°C, inches/inch/°F)
t_0 = initial temperature (°C, °F)
t_1 = final temperature (°C, °F)

The volumetric coefficient of expansion, β, also called the cubic coefficient of expansion, can be expressed as:

$$dV = V_0\ \beta\ (t_1 - t_0)$$

or

$$\beta = [V_1 - V_0]/[V_0\ (t_1 - t_0)]$$

where: dV = change in volume (m^3, ft^3)
V_1 = new volume
V_0 = previous volume
β = volumetric coefficient of thermal expansion ($m^3/m^3/°C$, $ft^3/ft^3/°F$)
t_1 = final temperature (°C, °F)
t_0 = initial temperature (°C, °F)

Expected Outcomes

The values obtained from these experiments are expected to fall within reasonable conformity with published data such as that found in Tables A.5, A.11, and A.12 in the appendix.

DATA COLLECTION SHEET PROPERTIES OF FLUIDS								
Lead Experimenter								
Date of Experiment								
Part 1 – Density and Coefficients of Thermal Expansion								
Cylinder Number	1	2	3		Cylinder Number	1	2	3
At 10 °C					At 60 °C			
Mass of Empty Cylinders, Step 2					Mass of Empty Cylinders, Step 2			
Mass of Filled Cylinders, Step 3					Mass of Filled Cylinders, Step 3			
Depth of Fluid in Cooled Cylinders, Step 5					Depth of Fluid in Cooled Cylinders, Step 5			
Mass of Cooled Cylinders with Fluid, Step 6					Mass of Cooled Cylinders with Fluid, Step 6			
Mass of Heated Cylinder, Step 8					Mass of Heated Cylinder, Step 8			
Temperature of Heated Cylinders, Step 8					Temperature of Heated Cylinders, Step 8			
Depth of Fluid in Heated Cylinders, Step 8					Depth of Fluid in Heated Cylinders, Step 8			
Calculated Density					Calculated Density			
Calculated Coefficient of Thermal Expansion					Calculated Coefficient of Thermal Expansion			
Cylinder Number	1	2	3		Cylinder Number	1	2	3
At 20 °C					At 70 °C			
Mass of Empty Cylinders, Step 2					Mass of Empty Cylinders, Step 2			
Mass of Filled Cylinders, Step 3					Mass of Filled Cylinders, Step 3			
Depth of Fluid in Cooled Cylinders, Step 5					Depth of Fluid in Cooled Cylinders, Step 5			
Mass of Cooled Cylinders with Fluid, Step 6					Mass of Cooled Cylinders with Fluid, Step 6			
Mass of Heated Cylinder, Step 8					Mass of Heated Cylinder, Step 8			
Temperature of Heated Cylinders, Step 8					Temperature of Heated Cylinders, Step 8			

Depth of Fluid in Heated Cylinders, Step 8					Depth of Fluid in Heated Cylinders, Step 8			
Calculated Density					Calculated Density			
Calculated Coefficient of Thermal Expansion					Calculated Coefficient of Thermal Expansion			
At 30 °C					At 80 °C			
Mass of Empty Cylinders, Step 2					Mass of Empty Cylinders, Step 2			
Mass of Filled Cylinders, Step 3					Mass of Filled Cylinders, Step 3			
Depth of Fluid in Cooled Cylinders, Step 5					Depth of Fluid in Cooled Cylinders, Step 5			
Mass of Cooled Cylinders with Fluid, Step 6					Mass of Cooled Cylinders with Fluid, Step 6			
Mass of Heated Cylinder, Step 8					Mass of Heated Cylinder, Step 8			
Temperature of Heated Cylinders, Step 8					Temperature of Heated Cylinders, Step 8			
Depth of Fluid in Heated Cylinders, Step 8					Depth of Fluid in Heated Cylinders, Step 8			
Calculated Density					Calculated Density			
Calculated Coefficient of Thermal Expansion					Calculated Coefficient of Thermal Expansion			
At 40 °C					At 90 °C			
Mass of Empty Cylinders, Step 2					Mass of Empty Cylinders, Step 2			
Mass of Filled Cylinders, Step 3					Mass of Filled Cylinders, Step 3			
Depth of Fluid in Cooled Cylinders, Step 5					Depth of Fluid in Cooled Cylinders, Step 5			
Mass of Cooled Cylinders with Fluid, Step 6					Mass of Cooled Cylinders with Fluid, Step 6			
Mass of Heated Cylinder, Step 8					Mass of Heated Cylinder, Step 8			
Temperature of Heated Cylinders, Step 8					Temperature of Heated Cylinders, Step 8			
Depth of Fluid in Heated Cylinders, Step 8					Depth of Fluid in Heated Cylinders, Step 8			

Calculated Density					Calculated Density			
Calculated Coefficient of Thermal Expansion					Calculated Coefficient of Thermal Expansion			
At 40 °C					At 90 °C			
Mass of Empty Cylinders, Step 2					Mass of Empty Cylinders, Step 2			
Mass of Filled Cylinders, Step 3					Mass of Filled Cylinders, Step 3			
Depth of Fluid in Cooled Cylinders, Step 5					Depth of Fluid in Cooled Cylinders, Step 5			
Mass of Cooled Cylinders with Fluid, Step 6					Mass of Cooled Cylinders with Fluid, Step 6			
Mass of Heated Cylinder, Step 8					Mass of Heated Cylinder, Step 8			
Temperature of Heated Cylinders, Step 8					Temperature of Heated Cylinders, Step 8			
Depth of Fluid in Heated Cylinders, Step 8					Depth of Fluid in Heated Cylinders, Step 8			
Calculated Density					Calculated Density			
Calculated Coefficient of Thermal Expansion					Calculated Coefficient of Thermal Expansion			
Part 2 – Viscosity								
Cylinder Number	1	2	3	4	5			
Fluid in Cylinder								
Depth of Liquid, Step 6, Trial 1								
Time for Marble to Drop, Trial 1								
Depth of Liquid, Step 6, Trial 2								
Time for Marble to Drop, Trial 2								
Depth of Liquid, Step 6, Trial 3								
Time for Marble to Drop, Trial 3								

Depth of Liquid, Step 6, Trial 4								
Time for Marble to Drop, Trial 4								
Depth of Liquid, Step 6, Trial 5								
Time for Marble to Drop, Trial 5								
Average Depth, Step 11								
Average Time to Drop								
Calculated Dynamic Viscosity								
Calculated Kinematic Viscosity								
Part 3 – Conductivity								
Beaker Number	1	2	3			1	2	3
Measured Conductivity, 0 mg NaCl, Step 5					Measured Conductivity, 4 mg/L NaCl, Step 10			
Average Conductivity					Average Conductivity			
Measured Conductivity, 1 mg/L NaCl, Step 7								
Average Conductivity								
Measured Conductivity, 2 mg/L NaCl, Step 10					Measured Conductivity of Unknown Sample			
Average Conductivity					Predicted NaCl Concentration in Unknown Sample			
Measured Conductivity, 3 mg/L NaCl, Step 10								

EXPERIMENT 13.2

CAPILLARY RISE IN TUBES AND SOILS

Date: ______________________________

Principal Investigator:

Name: ______________________________

Email: ______________________________

Phone: ______________________________

Collaborators:

Name: ______________________________

Email: ______________________________

Phone: ______________________________

Name: ______________________________

Email: ______________________________

Phone: ______________________________

Name: ______________________________

Email: ______________________________

Phone: ______________________________

Name: ______________________________

Email: ______________________________

Phone: ______________________________

Name: ______________________________

Email: ______________________________

Phone: ______________________________

Objective or Question to Be Addressed

Fluids in general, particularly water, have the interesting characteristic of being able to flow upward (opposite the force of gravity) under the right conditions. This characteristic is called *capillarity*. This experiment examines the phenomenon of capillarity and the conditions that cause it.

Theory behind Experiment

Capillary action is defined by the U.S. Geologic Survey as the movement of water within the spaces of a porous material due to the forces of *adhesion*, *cohesion*, and *surface tension*. It is generally responsible for water rising into the interstitial spaces between soil particles and is responsible for the watering of plants after precipitation has stopped, among other things.

Capillary action occurs due to the forces of cohesion and adhesion. Cohesion is the propensity for the water molecules to stick together, and adhesion is the propensity of the water to "stick to" the walls of a vessel or the particles of soil in the ground. Adhesion causes an upward force on the liquid at the edges and results in an upward *meniscus* at the top of the water column. Surface tension in the water acts to hold the surface intact. Capillary action occurs when the adhesion to the walls is stronger than the cohesive forces between the liquid molecules. The height to which capillary action will take water in a uniform circular tube, such as a straw, is limited by surface tension and gravity.

The height to which the water will rise in a column is a result of the size of the opening (the diameter of the straw, for example) and the adhesion characteristics of the material and the water. When the force of gravity equals the adhesion forces drawing the water upward through the column, the rise stops.

The adhesion forces in a narrow tube are stronger than those in a wider tube relative to the associated force of gravity. That occurs because the surface area along which the adhesion forces are acting is relatively greater in a small diameter tube than in a larger diameter tube. Since the force of gravity acting on the column of water is a function of the volume of water in the column and the volume is a function of the tube or equivalent space diameter, the water will rise higher in a narrower tube or interstitial soil space.

Note that capillary action is not restricted to a vertical movement of water. Placing a paper towel adjacent to a spill of water, for example, will allow the water to be drawn horizontally into the spaces between the towel fibers. This form of capillarity is most often limited by the amount of water available, rather than the force of gravity stopping the flow. In addition, fluids that do not adhere to the tube will not demonstrate capillarity. For example, mercury will not adhere to glass and will, therefore, not exhibit capillarity in a glass tube. Mercury will adhere to various other metals, however, and will demonstrate capillarity in those materials.

Potential Interferences and Interference Management Plan

The success of this experiment relies on the use of clean glass tubes of uniform diameter. Dirty or distorted tubing will yield poor results. In addition, the material from which the tubes are made will be important. The use of mixed tube materials will yield inconsistent data that is not likely to plot well. The plot of the data will drive the answer to the final question of predicting the rise in soil, and that plot must be accurate for the data to be useful.

Data to Be Collected with an Explanation of How They Will Be Collected (Ex.: "Water Temperature Data for Three Days Using a Continuous Data Logger")

Using the attached data sheet, record at a minimum the following information:

- Names of all experimenters
- Date of experiment
- Diameters of all tubes used
- Results of the experiment

Tools, Equipment, and Supplies Required

- 2 L (0.53 gal) square beaker
- Several glass tubes of varied diameters
- 1.5 L (0.4 gal) of tap water
- 0.06 L (0.25 cups) of food coloring of desired color
- Glass stirring rod to mix food coloring into the water
- Measuring device to measure the rise of fluid inside the tubes relative to the surface of the water in the beaker

Safety

There are no inherently dangerous components or activities associated with this experiment. Care needs to be taken when handling the tubes, particularly the thinner tubes, to avoid breakage. Food coloring will stain hands and clothing if spilled and cause eye damage if splashed into the eyes.

In the event of a cut, however minor, wash the affected area immediately and cover with a bandage to avoid infection. For more serious cuts or lacerations, apply pressure to the wound to control bleeding and seek medical attention. Report all injuries to the lab supervisor.

Gloves, eye protection, and a lab coat are recommended for this experiment.

Planned Experimental Procedures and Steps, in Order

1. Put on appropriate personal protective equipment and gather all necessary tools and supplies in a convenient location.
2. Fill a 2 L (0.53 gal) square beaker with 1.5 L (0.4 gal) of tap water to which 0.06 L (0.25 cups) of food coloring of any favorite color (bright red or deep blue are recommended) has been added.
3. Insert four glass tubes of different inside diameters into the beaker and allow them a few minutes to stabilize. Diameters of 0.25, 0.50, 0.75, 1.0, 1.25, and 1.5 cm (0.1, 0.2, 0.3, 0.4, 0.5, and 0.6 inches) are suggested.

4. Observe the level of the liquid inside each tube.
5. Measure and record the height to which the water has risen in each tube above the surface of the water in the beaker.
6. Plot the rise of the water level in the tubes as a function of the tube diameter on normal-normal graph paper.
7. Calculate the equation of the line on the graph and calculate the expected rise of the water in an equivalent tube of 0.01 cm (0.004 inches) diameter, which is more or less equivalent to the interstitial space between soil particles.
8. Fill a 500 mL (0.13 gal) glass beaker about half full with coarse sand and level it.
9. *SLOWLY* pour approximately 100 mL (0.03 gal) of colored water into the bottom of the beaker through a tube extending down one side. Do not disturb the soil.
10. Observe the level of water in the beaker as it fills and measure the height to which it rises immediately after it has all been poured into the sand.
11. Leave the beaker undisturbed for 30 minutes and again observe the level to which the colored water has risen in the sand.
12. Wait an additional 30 minutes and measure the height of the colored water in the sand for the third time.
13. Observe how closely the estimated rise calculated in Step 7 compares to the observed rise in Steps 11 and 12 and discuss why they may be slightly different.

Data Collection Format and Data Collection Sheets Tailored to Experiment

See attached data collection sheet.

Data Analysis Plan

The calculation of the expected rise of the water in soil is based on the equation of the line from the plot of tubular rise observed. The rise is defined by the equivalent space diameter. When that diameter is selected and inserted into the equation, the expected rise should be determined.

Expected Outcomes

The expected outcome will be a rise greater than any observed in the experiment. There is a presumption made that the rise is linear beyond the extremes of the measured data. This may not be a valid assumption, but the outcome should be reasonably consistent with that prediction in most soils.

DATA COLLECTION SHEET CAPILLARY RISE IN TUBES AND SOIL						
Lead Experimenter						
Date of Experiment						
Tube Number	1	2	3	4	5	6
Observed Water Height above Water in Beaker – Step 5						
Calculated Rise in Soil – Step 7						
Water Height in Beaker Immediately after Filling – Step 10						
Height of Water in Sand after 30 min – Step 11						
Height of Water in Sand after 60 min – Step 12						

EXPERIMENT 13.3

BUOYANCY FORCES ON SUBMERGED SURFACES

Date: ______________________

Principal Investigator:

Name: ______________________

Email: ______________________

Phone: ______________________

Collaborators:

Name: ______________________

Email: ______________________

Phone: ______________________

Name: ______________________

Email: ______________________

Phone: ______________________

Name: ______________________

Email: ______________________

Phone: ______________________

Name: ______________________

Email: ______________________

Phone: ______________________

Objective or Question to Be Addressed

This experiment examines the forces that act on a submerged object. Things that are less dense than water will tend to float on or near the surface of the water; conversely, those that are more dense than water will sink. However, objects that sink still retain a certain amount of buoyancy equal to the density of the water they displace.

Theory behind Experiment

The buoyant force on a submerged object is calculated from the equation:

$$F_b = \rho V g$$

where: F_b = buoyant force, N
ρ = density of the fluid, kg/m^3
V = volume of displaced fluid, m^3
g = acceleration due to gravity, m/s^2

The density of the fluid is a function of both depth and temperature. The deeper into a fluid the object goes, the closer the density of the water becomes to the density of the object. While unusual, it is possible, in sufficiently deep and cold water, for an object that sinks at the surface to stop sinking at some depth and hang in the water column.

The procedure for determining the forces acting on a submerged object is to first determine the density of the object and then the density of the fluid. Those factors are inserted into the equation provided earlier, along with the force of gravitational acceleration, and the force in newtons (N) is calculated, being cognizant of the volume of the object that is actually submerged if it is floating at or near the surface.

A newton is defined as 1 kg·m/s^2 (it is a derived unit defined in terms of the SI base units). One N is the force needed to accelerate one kilogram of mass at the rate of one meter per second squared in the direction of the applied force. The U.S. customary unit of force is the pound (symbol: lb_f). One pound of force is equivalent to 4.44822 newtons.

Potential Interferences and Interference Management Plan

There are several possible sources of error that can occur during this seemingly simple experiment. The forces acting on a submerged object are sensitive to the depth of submergence, relative density of the object and fluid, and the temperature of both the fluid and object. Allowing the fluid time to settle and the object to stabilize within the fluid if it is floating make a big difference in the accuracy of the outcome. The fluid may penetrate into the surface pores of the test object, changing the determination of the volume of the object (under Step 3 of the procedure) and the depth at which the object floats. Note that a boat, for example, floats because it displaces a larger mass of water than the volume of space the boat occupies. If the water is allowed to enter the boat, however, the boat settles further into the water even though the mass of the boat has not changed – only the volume of water it displaces.

Data to Be Collected with an Explanation of How They Will Be Collected (Ex.: "Water Temperature Data for Three Days Using a Continuous Data Logger")

Using the attached data sheet, record at a minimum the following information:

- Name of all experimenters
- Date of the experiment
- Description of the object being tested, including dimensions, volume, and density
- Description of the fluid being used
- Temperature of the fluid at each test
- Density of the fluid at each test
- Results of the tests

Tools, Equipment, and Supplies Required

- Small object or series of objects of different material and size for testing
- Scale accurate to 0.01 g for weighing the objects
- Basin or beaker of suitable size to allow each of the test objects to be submerged
- Tap water
- Thermometer that will read between 1 and 100 °C (32 and 212 °F) accurate to 0.1 degree

Safety

There are no inherent safety concerns or risks associated with this experiment. Safety gloves, eye protection, and a lab coat are recommended for this experiment, as for all laboratory experiments.

Planned Experimental Procedures and Steps, in Order

1. Put on the appropriate personal protective equipment and gather all necessary tools and supplies in a convenient location.
2. Weigh and record the mass of the object to be submerged.
3. Determine the volume of the object that is to be subjected to the submerged forces.
   a. Measure the dimensions of the object if it is uniform and calculate the volume directly.
   b. If the object is irregular, fill a suitable beaker or tub with a known volume of water and known dimensions.
   c. Carefully measure the depth of the water inside the beaker or basin.
   d. Place the object in the water and measure the depth of the water with the object completely submerged. It may be necessary to hold the object under the water while that depth measurement is made.
   e. Calculate the volume of the submerged object as the volume of the water displaced by it while submerged. Note that the depth of submergence is

irrelevant if the entire object is below the elevation of the original water surface. The change in depth multiplied by the area of the basin equals the volume of the submerged object.

4. Calculate the density of the object as the mass divided by the volume.
5. Fill a small tub or sink with cold tap water.
6. Measure and record the temperature of the water.
7. Insert the object into the water and allow it to stabilize if it is floating. Allow the water to stop moving as well.
8. Determine whether the object is floating and if so what percentage of the mass is submerged and what percentage is above the water surface.
9. Using the tables in the Appendix, determine the density of the water at the test temperature.
10. Calculate the force in newtons acting on the object by inserting the measured data and the force of gravity into the force equation.
11. Remove the object from the water and allow the water in the beaker or basin to stabilize at a temperature of about 60 °F (15 °C).
12. Repeat Steps 6–10, recording the calculated force at the new temperature.
13. Repeat Step 11 and allow the water to stabilize at room temperature.
14. Repeat Steps 6–10, recording the calculated force at the new temperature.
15. Covert all the calculated forces to American standard units.
16. Plot the three data points, in either system of units, to determine a reasonable line of best fit with which to predict the force on the object at any other temperature within that range.

Data Collection Format and Data Collection Sheets Tailored to Experiment

See attached data collection sheet.

Data Analysis Plan

The data collected during this experiment include the volume and mass of the test object and the mass of the fluid. Those data are inserted into the standard formula to determine the force in newtons on the submerged object.

Expected Outcomes

It is expected that the results will be reasonable from this experiment and that the force on the test object will vary only slightly as a function of the test temperature.

DATA COLLECTION SHEET BUOYANCY FORCES ON SUBMERGED OBJECTS	
Lead Experimenter	
Date of Experiment	
Mass of Object to Be Measured, Step 2	
Depth of Water in Basin, Step 3c	
Depth of Water in Basin with Object, Step 3d	
Volume of Displaced Water, Step 3e	
Calculated Density of Object, Step 4	
1a. Temperature of Cold Water, Step 6	
1b. Percentage of Object Submerged, Step 8	
1c. Density of Water, Step 9	
1d. Calculated Force Acting on Object, Step 10	
1e. Calculated Force, American Standard Units, Step 15	
2a. Temperature of 60 °F Water, Step 6	
2b. Percentage of Object Submerged, Step 8	
2c. Density of Water, Step 9	
2d. Calculated Force Acting on Object, Step 10	
2e. Calculated Force, American Standard Units, Step 15	
3a. Temperature of Room Temperature Water, Step 6	
3b. Percentage of Object Submerged, Step 8	
3c. Density of Water, Step 9	
3d. Calculated Force Acting on Object, Step 10	
3e. Calculated Force, American Standard Units, Step 15	

EXPERIMENT 13.4

SIPHONS

Date: ______________________________

Principal Investigator:

Name: ______________________________

Email: ______________________________

Phone: ______________________________

Collaborators:

Name: ______________________________

Email: ______________________________

Phone: ______________________________

Name: ______________________________

Email: ______________________________

Phone: ______________________________

Name: ______________________________

Email: ______________________________

Phone: ______________________________

Name: ______________________________

Email: ______________________________

Phone: ______________________________

Objective or Question to Be Addressed

This experiment examines the principles upon which siphons function. A siphon is designed to cause a flow of water to rise up and over an obstacle (a regular siphon) or to flow under an obstacle (an inverted siphon) and regain the original hydraulic gradient at the other end. With a regular siphon, the hydraulic gradient line is below the flow line. With an inverted siphon the hydraulic gradient line is above the flow line.

Sewers and storm drains are generally designed such that they will flow by gravity through a series of street pipes and interceptor lines to a convenient discharge point for storm drains or to a pump station for sewer lines. For the most part, this concept works well. Occasionally, however, the hydraulic gradient of the pipeline falls above the ground surface. Where that happens, it is customary to raise the pipe out of the ground to maintain a constant gradient, support the pipe on piers or piles, or bury the raised pipe inside a soil berm.

When roads, canals, railways, and other obstructions are in the way, it is not possible to lay the pipeline above the ground. In those cases, a siphon is used to go over the obstruction or an inverted siphon is used to go under the obstruction.

Theory behind Experiment

In practice, the siphon is a closed tube that must remain full at all times. The pipe leading up to a regular siphon is designed with no outlets below the highest point on the inside of the siphon. This causes the pipe and the siphon to fill with fluid until the fluid level in the siphon reaches the bottom inside elevation at the top of the siphon. As the fluid inside the inlet pipe continues to flow in, the hydraulic gradient line exceeds the top inside elevation of the siphon and the entire siphon section becomes filled with fluid. The top inside elevation at the outlet of the siphon is set below the bottom inside elevation of the inlet pipe. This means that the column of water inside the inlet leg of the siphon is shorter than the column of water in the outlet leg of the siphon. This difference does not need to be large because the force of gravity on the column of water in the outlet leg will be greater than the force of gravity on the column of water in the inlet leg. That excess force of gravity pulls the water down the outlet leg, but since there is no break in the water column flowing through the siphon, that falling column at the outlet continues to draw the column up the inlet side, over the top of the siphon, and down the outlet side. Only when the inlet flow is sufficiently low to allow air to pass over the siphon and break the continuity of the flow will the siphon stop flowing. At that point, it does continue to flow until the flow of water in the inlet gets so low that the gradient line falls below the bottom inside elevation at the top of the siphon. The siphon will then start again as soon as the flow conditions just described reestablish themselves.

In the case of an inverted siphon, the same phenomenon occurs except that the siphon goes under the obstruction instead of over it. Here, the bottom inside elevation of the outlet leg is lower than the bottom inside elevation of the inlet pipe so that the column of water on the inlet side is longer than the column of water on the outlet side. This allows the inlet water to push the flow through the siphon until the two columns equalize. Therefore, the lower section of the siphon pipe will always remain full below the bottom inside elevation at the end of the outlet leg.

This type of system acts as a trap when the flow is low because the siphon pipe stays full, preventing gasses in the pipe below the siphon from flowing back up the inlet pipe. It can also allow sedimentation of debris inside the lower section of the siphon, however, unless significant flow velocity can be maintained to scour the pipe clean during periods of normal flow. To prevent that, the inverted siphon is often designed to split the flow into two smaller pipes that will increase the velocity of the flow through the system sufficiently to prevent siltation inside the siphon pipes. For large pipes subjected to variable flows, two or more siphons may be installed to operate singly or in combination to maintain scouring velocities in all the pipes.

Potential Interferences and Interference Management Plan

This experiment relies on the construction of a demonstration siphon system. If the elevation differences shown between the inlet and outlet are not maintained, the siphon may not function properly. In addition, any solids in the system will accumulate at the bottom of the inverted siphon if adequate flow velocity is not maintained.

Data to Be Collected with an Explanation of How They Will Be Collected (Ex.: "Water Temperature Data for Three Days Using a Continuous Data Logger")

Using the attached data sheet, record at a minimum the following information:

- Names of all experimenters
- Date of experiment
- Water flow rates and water pressure are determined at several locations through reading of gauges and calculation of expected values

Tools, Equipment, and Supplies Required

- Two flow systems built in general accordance with Figures 13.4.1 and 13.4.2
- Stopwatch

Safety

Excessive flow through the system could cause catastrophic failure of connections, causing water to spray out in unpredictable directions with indeterminate force. Care is needed to bring flow rates up to desired magnitudes slowly enough to avoid sudden pressure changes on the system.

Planned Experimental Procedures and Steps, in Order

1. Put on appropriate personal protective equipment and gather all necessary tools and supplies in a convenient location.
2. Set up a water flow system in accordance with the diagrams in the appendix, Figure 13.4.1, for a regular siphon.
3. Start the flow of water and control that flow such that the flow through the system is laminar. Note that on the inlet side of the siphon the pipe is flowing full and that it remains full through the siphon.
4. Measure and record the flow rate and the pressure in the system.
5. Shut the water flow off, then observe and record how the flow in the pipe and the siphon is affected.

6. Repeat Step 2 using Figure 13.4.2 for an inverted siphon.
7. Start the flow of water and control that flow such that the flow through the system is laminar. Note that on the inlet side of the siphon the pipe may not be flowing full but that it does flow full through the siphon. The outlet pipe may also be less than full.
8. Measure and record the flow rate and pressures throughout the system.
9. Shut the water flow off, then observe and record how the flow in the pipe and the siphon is affected.

Based on a flow rate determined from Step 4, use Bernoulli's equation to calculate the expected flow rate and pressures at each of the designated pressure gauges and flow meters.

$$P_1 + 1/2\ \rho v_1^2 + \rho g h_1 = P_2 + 1/2\ \rho v_2^2 + \rho g h_2$$

Where: P_1 = pressure energy at the measuring point
ρ = density of the fluid
v = velocity of the fluid at the measurement point
g = force of gravity
h = elevation of the measurement point above a fixed horizon

NOTE: P is referred to as the pressure energy of the fluid, $1/2\ \rho v^2$ is referred to as the kinetic energy per unit volume of the fluid, and ρgh is referred to as the potential energy per unit volume of the fluid.

10. Start flow through the system at the stipulated flow rate and record the actual pressures and flow rates at each of the designated locations. Discuss any variations noted.
11. Based on a flow rate determined in Step 8, use Bernoulli's equation to calculate the expected flow rate and pressures at each of the designated pressure gauges and flow meters.
12. Start flow through the system at the stipulated flow rate and record the actual pressures and flow rates at each of the designated locations. Discuss any variations noted.

Data Collection Format and Data Collection Sheets Tailored to Experiment

See attached data collection sheet.

Data Analysis Plan

The data collected are not subject to analysis except by opinionated observation. No further analysis is required.

Expected Outcomes

It is expected that the siphons used in this demonstration will act as specified and properly demonstrate the phenomena of siphons.

<table>
<tr><td colspan="4">DATA COLLECTION SHEET
SIPHONS EXPERIMENT</td></tr>
<tr><td colspan="4"></td></tr>
<tr><td colspan="4">Lead Experimenter</td></tr>
<tr><td colspan="4">Date of Experiment</td></tr>
<tr><td colspan="4"></td></tr>
<tr><td>Set Flow Rate through System, Step 3</td><td></td><td>Set Flow Rate through System, Step 6</td><td></td></tr>
<tr><td>Measured Flow Rate at Point 1</td><td></td><td>Measured Flow Rate at Point 1</td><td></td></tr>
<tr><td>Measured Flow Rate at Point 2</td><td></td><td>Measured Flow Rate at Point 2</td><td></td></tr>
<tr><td>Measured Flow Rate at Point 3</td><td></td><td>Measured Flow Rate at Point 3</td><td></td></tr>
<tr><td>Measured Flow Rate at Point 4</td><td></td><td>Measured Flow Rate at Point 4</td><td></td></tr>
<tr><td>Measured Flow Rate at Point 5</td><td></td><td>Measured Flow Rate at Point 5</td><td></td></tr>
<tr><td>Measured Pressure at Point 1</td><td></td><td>Measured Pressure at Point 1</td><td></td></tr>
<tr><td>Measured Pressure at Point 2</td><td></td><td>Measured Pressure at Point 2</td><td></td></tr>
<tr><td>Measured Pressure at Point 3</td><td></td><td>Measured Pressure at Point 3</td><td></td></tr>
<tr><td>Measured Pressure at Point 4</td><td></td><td>Measured Pressure at Point 4</td><td></td></tr>
<tr><td>Measured Pressure at Point 5</td><td></td><td>Measured Pressure at Point 5</td><td></td></tr>
<tr><td colspan="4"></td></tr>
<tr><td>Predicted Flow Rate and Pressure at Point 1 – Step 10</td><td></td><td>Predicted Flow Rate and Pressure at Point 1 – Step 12</td><td></td></tr>
<tr><td>Predicted Flow Rate and Pressure at Point 2</td><td></td><td>Predicted Flow Rate and Pressure at Point 2</td><td></td></tr>
<tr><td>Predicted Flow Rate and Pressure at Point 3</td><td></td><td>Predicted Flow Rate and Pressure at Point 3</td><td></td></tr>
<tr><td>Predicted Flow Rate and Pressure at Point 4</td><td></td><td>Predicted Flow Rate and Pressure at Point 4</td><td></td></tr>
<tr><td>Predicted Flow Rate and Pressure at Point 5</td><td></td><td>Predicted Flow Rate and Pressure atPoint 5</td><td></td></tr>
<tr><td>Actual Flow Rate and Pressure at Point 1</td><td></td><td>Actual Flow Rate and Pressure at Point 1</td><td></td></tr>
<tr><td>Actual Flow Rate and Pressure at Point 2</td><td></td><td>Actual Flow Rate and Pressure at Point 2</td><td></td></tr>
<tr><td>Actual Flow Rate and Pressure at Point 3</td><td></td><td>Actual Flow Rate and Pressure at Point 3</td><td></td></tr>
<tr><td>Actual Flow Rate and Pressure at Point 4</td><td></td><td>Actual Flow Rate andPressure at Point 4</td><td></td></tr>
<tr><td>Actual Flow Rate and Pressure at Point 5</td><td></td><td>Actual Flow Rate and Pressure at Point 5</td><td></td></tr>
</table>

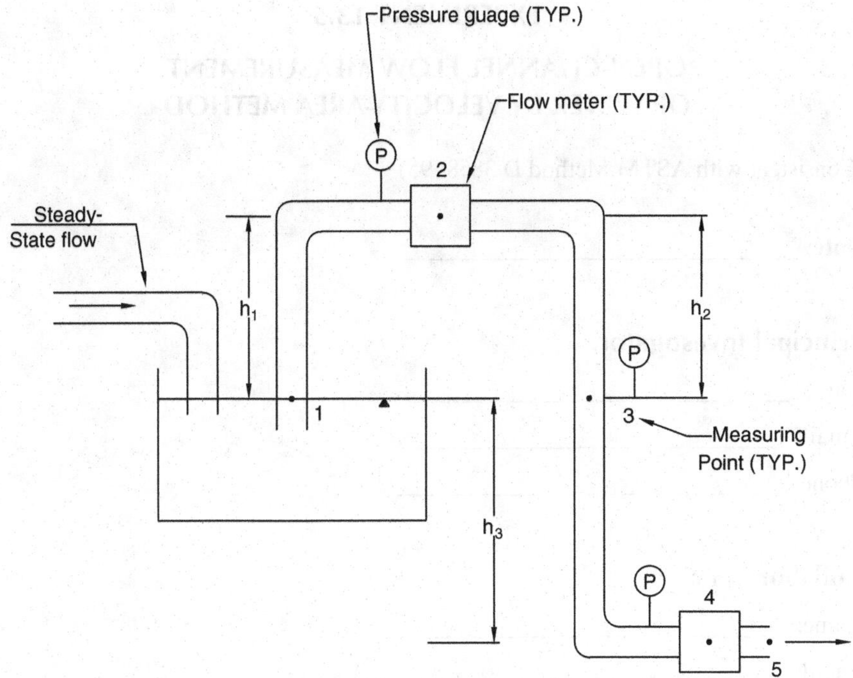

FIGURE 13.4.1 Siphon experiment device.

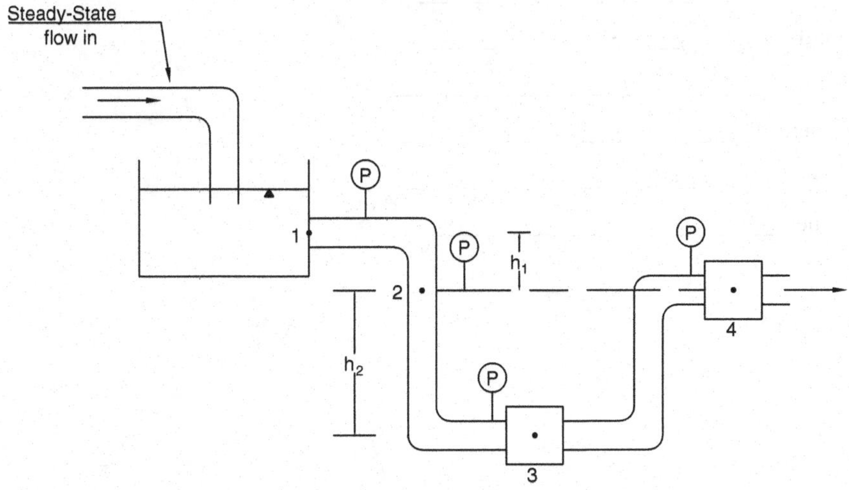

FIGURE 13.4.2 Inverted siphon device.

EXPERIMENT 13.5

OPEN-CHANNEL FLOW MEASUREMENT OF WATER BY VELOCITY-AREA METHOD

(Consistent with ASTM Method D 3858–95)

Date: ______________________________

Principal Investigator:

Name: ______________________________

Email: ______________________________

Phone: ______________________________

Collaborators:

Name: ______________________________

Email: ______________________________

Phone: ______________________________

Name: ______________________________

Email: ______________________________

Phone: ______________________________

Name: ______________________________

Email: ______________________________

Phone: ______________________________

Name: ______________________________

Email: ______________________________

Phone: ______________________________

Objective or Question to Be Addressed

This experiment demonstrates a method for measuring the volumetric flow rate of water in open channels using the flow velocity and the cross-sectional area to calculate the discharge. The methods described in this experiment are not suitable for velocities sufficient to move the flow measuring device or depth measuring tools out of a vertical position during measurement. In those cases, see ASTM Method D 3858–95 for how to adjust the measurements.

Theory behind Experiment

The bed of natural rivers and streams is seldom uniform, flat, or smooth. The banks are seldom straight and even, and the cross-sectional area of the stream is, therefore, not able to be accurately assessed easily. The concept behind this experiment is to accurately assess the depth and stream flow at several cross-sectional areas of the stream, as closely perpendicular to the stream flow as possible, and to integrate those cross-sectional areas and flow rates to calculate a reasonably close estimate of the actual flow rate at that location. The total flow or discharge measurement is calculated as the summation of the products of partial areas of the flow cross-section and their respective average velocities.

At the selected measuring location, the stream should be as straight as possible above and below the measuring section with the main thread of flow parallel to the banks. The stream should ideally be straight for at least three channel widths above and below the selected section. The streambed should be free of large rocks, piers, weeds, or other obstructions that will cause turbulence or create a vertical component in measured velocity. Water velocities and depths at the selected section must be consistent with the capabilities of the equipment available for making the measurement.

The accuracy of the flow rate computation is based on an integrated summation of many measurements. Therefore, the accuracy of the overall measurement is generally increased by increasing the number of partial cross-sectional measurements. The use of 25–30 partial cross-sections is generally recommended unless the channel is unusually irregular or complex. Fewer cross-sections may be used for smaller streams. The partial sections should be chosen so that each contains between about 3 % and 5 % of the total discharge for most channels.

Determination of the mean velocity in a given partial cross-section then becomes a sampling process throughout the vertical extent of that section, followed by analysis of the readings to determine a reasonable estimate of the average flow through that section. The recommended depths below the water surface at which flow velocity reading should be taken typically requires readings at 20 % (0.2) of the total depth and 80 % (0.8) of the total depth at each cross-section. For streams deeper than about 4 feet, a third reading at 60 % (0.6) of the total depth at that cross-section is useful. The more readings taken, the more accurate the result will be, but the variation with depth is usually such that the increase in accuracy for the increase in time to do all the readings is not justified.

The horizontal distance to any point in a cross-section must be measured from the same starting point on the same stream bank. Cableways, highway bridges, or foot bridges are commonly used to measure from because they tend to be perpendicular to the stream and provide access at appropriate cross-sectional locations. For small streams such spans are not common, and a line pulled tightly across the stream may be used to mark the cross-sectional line and flags placed on the line or rope to indicate the measurement locations.

Where the stream channel or cross-section is extremely wide, where no cableways or suitable bridges are available, or where it is impractical to string a tape or tagline, the distance from the initial point on the bank can be determined by optical or electrical distance meters, by stadia, or by triangulation to a boat or person located on the cross-section line.

The depth of the stream below the water surface at each cross-section, and the depth position of the current meter at that point, are usually measured by a rigid rod or by a weight suspended on a cable. Where a suspended weight is used, the selection of the proper weight is essential for the determination of the correct depth. A light weight will be carried downstream and incorrectly yield depth observations that are too large. A rule of thumb for the selection of proper sized weights is to use a weight slightly heavier, in pounds or kilograms, than the product of depth (ft) times the expected velocity (ft/sec) or 1.5 kg times the product of depth (m or ft) times the expected velocity (m/sec or ft/sec).

Potential Interferences and Interference Management Plan

Interference with this procedure is most likely to arise from the disturbance of the stream bottom during measurement of the flow velocities. Bottom disturbance can occur if the flow meter is dropped onto the bottom or if wading into the stream is necessary to accurately place the velocity meter. In addition, debris in the water may clog the velocity meter, skewing the readings.

The presence of an ice sheet on top of the water surface requires an adjustment to the way the water depth is computed. When a hole is cut in the ice through which to measure the flow, the total depth is measured from the water surface in the hole to the streambed. Then the distance from the water surface to the bottom of the ice layer is measured. The effective depth is computed by subtracting the thickness of the ice below the water surface from the total depth. In those cases where a thick slush layer exists below the ice cover, its thickness is determined by lowering the meter through it until it turns freely, then raising the meter until the rotor stops. The distance thus determined is then also subtracted from the overall depth of water. The partial section area computation is made by multiplying the effective depth times the width.

The presence of ice cover can have the effect of added channel roughness and resistance to flow. Therefore, the shape of the vertical-velocity profile is altered. When velocity is measured at the 0.2 and 0.8 depths, the observations are used as in an open-water measurement. However, if measurements are also made at the 0.6 depth, a coefficient of 0.92 is applied to the velocity observations to adjust for the added resistance of the ice sheet. An acceptable alternate procedure is to obtain a velocity observation at 0.5 depth and apply a coefficient of 0.88.

Normally, velocities at the 0.2 and 0.8 depths are sufficient for determining the mean vertical velocity at each cross-sectional area. Note, however, that the 0.2 depth velocity should be greater than the 0.8 depth velocity but not more than twice as great. If the 0.2 depth is less than the 0.8 depth, or if the 0.2 depth reading is more than twice the 0.8 depth reading, a third reading at the 0.6 depth should be made and included in the mean velocity calculation.

When the channel is less than about 2.5 feet (0.75 m) deep, the two-point measurement system (0.2 and 0.8 depth) does not work well. In that case a single reading at the 0.6 depth point is used as the average velocity for that cross-section.

Data to Be Collected with an Explanation of How They Will Be Collected (Ex.: "Water Temperature Data for Three Days Using a Continuous Data Logger")

The data to be recorded during this experiment are water depth measurements, made with a rigid rod or pole adequately marked for depth, or by an adequately weighted line, equally marked for depth. In addition, water flow velocities are measured using a flow meter attached to a rigid pole or other device to keep the flow meter vertical and located at the desired cross-sectional area of the stream.

The depth and velocity measurements are made at a series of pre-selected locations across the stream bed, perpendicular to the flow direction of the main channel. The preselected locations are marked on a marker line, and those locations are set by measurement from a set marker on one shore using a cloth or metal measuring tape.

All measured data are recorded on a data collection sheet at the time of measurement, and a sketch is made of the measuring location.

Using the attached data sheet, record at a minimum the following information:

- Names of all experimenters
- Date of experiment and date of sample collection
- Location of measurement (name of stream and where along the stream the measurements were made)
- Sketch of the measurement location
- Measurements made to locate the marker line stakes
- Measurements to each of the markers
- Depth and flow velocity of each vertical measurement made at each marker
- Calculations made to sum the flow at each cross-section
- Total flow calculated for the stream at the selected location

Tools, Equipment, and Supplies Required

- Current meter: usually a rotating element or electromagnetic type. The rotating-element type is the most common and is based on proportionality between the velocity of the water and the resulting angular velocity of the meter rotor. The instrument is placed in a predetermined location in the stream, vertically and horizontally, and the number or revolutions of the rotor during a measured interval of time is recorded to determine the flow velocity at that point.
- Counting system: The number of revolutions of the rotor is usually obtained from an internal counting device on the meter. The device operating manual should be consulted for the mechanism used by that specific meter. A stopwatch may be needed for manual counting systems.
- Channel width-measuring equipment: Steel tapes, metallic tapes, cloth tapes, or pre-marked taglines

- Depth-measuring equipment: rod or weighted cable
- Safety equipment: Safety lines, life vests, and waders

Safety

Care is needed working in and around streams to avoid slipping on wet surfaces or falling from bridges or causeways while measuring cross-sections or velocities. Wading into a stream can also present unseen dangers from holes in the stream bottom to snags, broken glass, sharp stone fragments, and other debris on the stream bottom.

Hard-soled waders, life jackets or vests, and a safety line are necessary when carrying out these measurements. For very small streams, fewer than 3 feet deep, and low velocity, safety lines may be deleted at the discretion of the experimenters.

At least two people are required to enter a stream or take a measurement, and only one should be in the stream at any one time. The second person needs to monitor the safety line and the general safety of the team.

Special care is needed when working on frozen streams to ensure that the ice is adequate to safely support the crew while conducting the measurements.

As with all experiments, eye protection and gloves are recommended during the conduct of this experiment.

Planned Experimental Procedures and Steps, in Order

1. Select a suitable site for making a discharge measurement.
2. Put on appropriate personal protective equipment and gather necessary tools and equipment in a convenient location.
3. Establish a cross-sectional line by placing a tight cord between two stakes on opposite sides of the stream, perpendicular to the mainstream flow. Using standard land surveying techniques, "tie down" the location of the two stakes from permanent control points on the shore, approximately 90 degrees apart from each other so that the exact stake locations can be reestablished later, if needed.
4. Starting from a fixed location on one shore, measure and record the location of the desired cross-section measurements and place a colored piece of tape (a marker) around the line at each location and draw a sketch of the arrangement, including the stake tie-down control points and distances, and the distance to each marker on the line from the starting point.
5. Measure and record the depth of the stream at each cross-sectional marker location.
6. Measure and record the flow velocity at each cross-sectional marker location in at least two depth locations (0.2 and 0.8 times the depth at that location) or at as many depth locations as the measuring plan calls for. Record the velocity and actual depth (in feet or meters) at each marker location.
7. After all measurements have been made and verified, remove the cross-section line.
8. Calculate the flow rate in accordance with the procedures in the data analysis section.

Data Collection Format and Data Collection Sheets Tailored to Experiment

See attached data collection sheet.

Data Analysis Plan

The flow rate is calculated from the following equation:

$$Q = \Sigma\ (av)$$

where: Q = total discharge (m^3/s, ft^3/s)
a = individual partial cross-sectional area (m^2, ft^2)
v = corresponding mean velocity of the flow at that cross-sectional area (m/s, ft/s)

In the procedure followed by this experiment, the average of the velocities measured at each cross-sectional area represents the mean velocity in that section. The cross-sectional areas are determined by the distance from the shore to half the distance between the first and second marker on the cross-section line, for the first partial section, times the depth at the first marker. The second partial area horizontal measurement is from half the distance from the first marker to the second marker plus half the distance from the second marker to the third, times the depth at the second marker. The third partial area goes from halfway between the second and third marker to halfway between the third and fourth marker times the depth at the third marker, and so forth to the final distance, which extends to the opposite bank.

The total flow cross-section is defined by the depths at the marker locations 1, 2, 3, 4, 5, . . . n and the mean velocity at those marker locations. The mean at each marker location is approximated by averaging the measured flow velocities at that marker.

The discharge for each partial cross-section is calculated from the following equations:

$$q_x = (v_x)\ (a_x)$$

where: q_x = discharge through partial section × (m^3/s, ft^3/s)
v_x = mean velocity at location × (m^2, ft^2)
a_x = distance from initial point to location × (m/s, ft/s)

The summation of the discharges for all of the partial sections is the total discharge of the stream.

Expected Outcomes

Every stream has a unique velocity profile, and there are no general estimates of the flow in any stream. Shallow streams may run faster or slower than deeper ones depending on the geometry of the stream. Therefore, the volumetric flow rates will vary equally as widely. Consideration should be given to whether the calculated values are reasonable based on the observations made at the time of measurement.

DATA COLLECTION SHEET OPEN-CHANNEL FLOW MEASUREMENT OF WATER BY VELOCITY-AREA METHOD								
Lead Experimenter								
Date of Experiment								
Name and Location of Stream and Location of Measurements on the Stream								
MEASUREMENTS								
Measurement Location Number	1	2	3	4	5	6	7	8
Distance from Shore								
Depth at Location								
Velocity at 0.2 × Depth								
Velocity at 0.5 × Depth								
Velocity at 0.8 × Depth								
CALCULATIONS								
Cross-Sectional Area	1	2	3	4	5	6	7	8
Calculated Area								
Mean Velocity at Area								
Calculated Flow Rate through Cross-Section								
Sum of Partial Flow Rates								

EXPERIMENT 13.6

OPEN-CHANNEL FLOW MEASUREMENT OF WATER WITH THIN-PLATE WEIRS

(Consistent with ASTM D 5242–21)

Date: ______________________________

Principal Investigator:

Name: ______________________________

Email: ______________________________

Phone: ______________________________

Collaborators:

Name: ______________________________

Email: ______________________________

Phone: ______________________________

Name: ______________________________

Email: ______________________________

Phone: ______________________________

Name: ______________________________

Email: ______________________________

Phone: ______________________________

Name: ______________________________

Email: ______________________________

Phone: ______________________________

Objective or Question to Be Addressed

This experiment describes a means for measuring the volumetric flow rate (but not the flow velocity) of water and wastewater in open channels using thin-plate rectangular or V-notch weirs.

Theory behind Experiment

Thin-plate weirs are overflow structures of specified geometric shape, typically square or V-notched. The geometry of these shapes means that the volumetric flow rate over these weirs is directly related to the measured depth (head) of water above the weir crest or vertex. By properly selecting the shape to use for each situation, a wide range of flow rates can be accurately measured with these simple devices. Flow rates ranging from about 0.008 ft^3/s (0.00023 m^3/s) to about 50 ft^3/s (1.4 m^3/s) are suggested as limits. The entire arrangement consists of an approach channel, the weir, and a head measuring device at the weir overflow.

The plate thickness in the direction of flow is limited to about 0.03 to 0.08 inches (about 1 to 2 mm) to minimize potential damage and to help avoid interference with the water jet flowing through the weir opening. The plate should be made of smooth, inflexible material, typically smooth metal or hard plastic. Upstream corners of the overflow section must be sharp and smooth, and the edges must be flat, smooth, and perpendicular to the weir face to ensure a uniform cross-section of flow through the weir. The weir plate must be vertical and perpendicular to the channel walls, and the overflow section must be symmetrical and located at the midpoint of the approach channel to yield accurate results.

With rectangular weirs, when the sidewalls and bottom of the approach channel are far enough from the edges of the notch for the contraction of the flow through the weir opening to be unaffected by those boundaries, the weir is termed "fully contracted". With lesser distances to the bottom or sidewalls, or both, the weir is "partially contracted".

When there are no side contractions and the rectangular weir crest extends across the channel, the weir is termed "full width" or "suppressed". In this case the approach channel must be rectangular, and the channel walls must extend at least 0.3 H (the depth of flow over the weir) downstream of the weir plate.

Potential Interferences and Interference Management Plan

Solids in the water flow, particularly wastewater flows, may accumulate upstream of the weir due to the reduction in flow velocity in the backwater that forms behind the weir. This can affect the approach conditions and the accuracy of the measurements over time, if not addressed regularly.

Data to Be Collected with an Explanation of How They Will Be Collected (Ex.: "Water Temperature Data for Three Days Using a Continuous Data Logger")

The data to be recorded during this experiment are water depth measurements, made with a rigid rod or pole adequately marked for depth, or by an adequately weighted line, equally marked for depth. In addition, the shape and dimensions of a weir are measured, including the angle of a V-notch weir, if used.

All measured data are recorded on a data collection sheet at the time of measurement, and a sketch should be made of the measuring location.

Using the attached data sheet, record at a minimum the following information:

- Names of all experimenters
- Date of experiment and date of sample collection
- Location of measurement (name of stream and where along the stream the measurements were made)
- Sketch of the measurement location
- Width of the channel at the location of the weir
- Dimensions of the weir and its location within the channel
- Depth of the water immediately upstream of the weir, the depth of water above the weir crest (if using a square weir) or above the vertex (if using a V-notch weir), and the depth of water below the weir
- Total flow calculated for the stream at the selected location

Tools, Equipment, and Supplies Required

- Weir measuring system consists of a thin-plate weir (rectangular or V-notch) and its immediate channel (the primary)
- Weir plate with a suitable opening. If the plate is thicker than 0.08 inches (2 mm) the downstream edges of the overflow section must be beveled at an angle of at least 45°
- Depth (head) measuring device (the secondary) for reading the depth of flow as a depth above the elevation of the crest or vertex of the notch

Note: the secondary device can be a simple scale. Alternatively, it can be an electronic instrument that continuously senses the depth, converts it to a flow rate, and displays or transmits a readout or record of the instantaneous flow rate or totalized flow, or both. The head measurement device should be located at a distance upstream of the weir equal to 4 H_{max} to 5 H_{max}, where H_{max} is the maximum head on the weir.

Note that this experiment provides best results when the crest lengths does not exceed about 4 ft (1.2 m) and head does not exceed about 2 ft (0.6 m).

- Suitable upstream approach channel that is of reasonably uniform cross-section and with sufficient banks to support the weir and head measuring device without leakage around the weir
- Suitable tailwater channel of essentially the same cross-section as the approach channel

For rectangular weirs

A rectangular overflow section can have either full or partial contractions or the side contractions may be suppressed.

The conditions for full contraction are as follows:

$$H/P \leq 0.5$$
$$H/L \leq 0.5$$

$$0.25\text{ ft } (0.08\text{ m}) \leq H \leq 2.0\text{ ft } (0.6\text{ m})$$
$$L \geq 1.0\text{ ft } (0.3\text{ m})$$
$$P \geq 1.0\text{ ft } (0.3\text{ m})$$
$$(B - L)/2 \geq 2H$$

where: H = measured head
P = crest height above the bottom of the channel
L = crest length
B = channel width

The partial contraction conditions are as follows:

$$H/P \leq 2$$
$$H \geq 0.1\text{ ft } (0.03\text{ m})$$
$$L \geq 0.5\text{ ft } (0.15\text{ m})$$
$$P \geq 0.3\text{ ft } (0.1\text{ m})$$

The weir should be set in the channel such that the tailwater level is at least 0.2 ft (0.06 m) below the crest of the weir.

For V-notch (triangular) weirs

The overflow section of a V-notch weir is an isosceles triangle oriented with the vertex downward. The most commonly used weirs are 90° ($\tan\theta/2 = 1$), 53.13° ($\tan\theta/2 = 0.5$) and 28.07° ($\tan\theta/2 = 0.25$).

The conditions for full contraction of V-notch weirs are as follows:

$$H/P \leq 0.4$$
$$H/B \leq 0.2$$
$$P \geq 1.5\text{ ft } (0.45\text{ m})$$
$$B \geq 3.0\text{ ft } (0.9\text{ m})$$
$$0.15\text{ ft } (0.05\text{ m}) \geq H \leq 1.25\text{ ft } (0.38\text{ m})$$

where: H = measured head
P = crest height above the bottom of the channel
L = crest length
B = channel width

The conditions for partial contraction of V-notch weirs are as follows:

- If the plate is thicker than 0.08 inches (2 mm), the downstream edge at the notch must be beveled at an angle of at least 60°.
- The approach flow should be as tranquil and uniformly distributed across the channel as possible. To accomplish this, a straight, rectangular approach channel length of ten channel widths, when the weir length is greater than half the channel width, is recommended where possible.

Safety

Risks associated with this experiment come from the nature of the weir and its placement in a stream of moving water. Thin plates may, by default, cause cuts if handled improperly. Cut-resistant gloves are recommended when working with these devices.

Wastewater may contain hazardous components that can be ingested, inhaled, or introduced subcutaneously through cuts and abrasions of the skin. Care is needed to avoid those contacts. In the event of ingestion, inhalation, or subcutaneous exposure, seek immediate medical assistance.

Installing a weir in an active streambed involves the risk of slipping and falling. That can lead to head injuries or broken bones. Care is needed when working in such conditions. In the event of a fall, particularly one involving a head injury, seek medical evaluation of the condition immediately because serious injuries do not always show up immediately.

As with all laboratory experiments, the use of gloves, eye protection, and a lab coat are recommended during the conduct of this experiment.

Planned Experimental Procedures and Steps, in Order

1. Put on appropriate personal protective equipment and gather all necessary tools and equipment in a convenient location.
2. Construct a weir of either a rectangular or V-notch shape and install it in a suitable stream in accordance with the details provided earlier.
3. Calibrate the head measuring systems if the installation is intended for long-term monitoring. See ASTM Test Method D 3858 for details on calibration of a weir and weir flow measuring system.
4. Carefully measure and record the values of H (head above the crest or the vertex of the notch), L (crest length for a rectangular weir), P (crest height above the bottom of the channel), B (channel width) using scales, rulers, a stilling well, or a similar device.
5. Calculate the flow rate from the equations provided in the data analysis section being sure to use the correct set of equations for the selected shape of the weir.

Data Collection Format and Data Collection Sheets Tailored to Experiment

See attached data collection sheet.

Data Analysis Plan

The flow rate, Q, over a *rectangular* weir that conforms to all requirements of ASTM Method D 5242–21 is determined from the Kindsvater-Carter equation:

$$Q = (2/3)\ (2\ g)^{1/2}\ C_e\ L_e\ (H_e)^{3/2}$$

where: g = acceleration due to gravity in compatible units
H_e and L_e = effective head and effective crest length respectively
C_e = discharge coefficient
The effective head, H_e, is related to the measured head, H, by:
$H_e = H + \delta H$

where δH is a constant equal to 0.003 ft (0.001 m) that is used to adjust for the effects of viscosity and surface tension. It is valid for water at ordinary temperatures (about 4–30 °C or about 39–85 °F).

The effective crest length, L_e, is related to the measured length, L, by:

$$L_e = L + \delta L$$

where δL is a function of the crest length to channel width ratio, L/B where L is the crest length and B is the channel width. Approximate values for δL for water at ordinary temperatures are given in Table 13.6.1. Interpolation between values will not significantly affect the flow rate determined.

The discharge coefficient, C_e, is a function of L/B and the head-to-crest height ratio, H/P. Its value is determined from Table 13.6.2. Interpolation between values will not significantly affect the flow rate determined.

TABLE 13.6.1
Values of δL

δL	L/B				
	0.2	0.4	0.6	0.8	1
in ft	0.0079	0.0091	0.011	0.013	0
in mm	2.4	2.8	3.3	4	0

TABLE 13.6.2
Values of C_e

H/P	L/B										
	0.0	0.1	0.2	0.3	0.4	0.5	0.6	0.7	0.8	0.9	1.0
0.0	0.585	0.588	0.589	0.590	0.591	0.592	0.594	0.596	0.598	0.599	0.601
0.2	0.585	0.588	0.589	0.590	0.592	0.594	0.598	0.602	0.607	0.612	0.616
0.4	0.584	0.587	0.588	0.591	0.594	0.597	0.602	0.608	0.617	0.624	0.631
0.6	0.584	0.587	0.588	0.591	0.595	0.599	0.605	0.613	0.626	0.637	0.646
0.8	0.583	0.586	0.587	0.592	0.597	0.601	0.609	0.619	0.635	0.649	0.661
1.0	0.583	0.586	0.587	0.592	0.598	0.603	0.613	0.625	0.645	0.662	0.676
1.2	0.582	0.585	0.586	0.593	0.600	0.606	0.617	0.631	0.654	0.674	0.691
1.4	0.582	0.585	0.586	0.593	0.601	0.608	0.620	0.636	0.663	0.687	0.705
1.6	0.581	0.584	0.585	0.593	0.602	0.610	0.624	0.642	0.673	0.700	0.720
1.8	0.581	0.584	0.585	0.594	0.604	0.612	0.628	0.648	0.682	0.712	0.735
2.0	0.581	0.583	0.584	0.594	0.605	0.615	0.632	0.654	0.691	0.725	0.750
2.2	0.580	0.582	0.583	0.595	0.607	0.617	0.635	0.660	0.701	0.737	0.765
2.4	0.580	0.581	0.582	0.595	0.608	0.619	0.639	0.663	0.710	0.750	0.780

The flow rate over a V-notch weir is determined from the following equations:

$$Q = (8/15)\,(2\,g)^{1/2}\,C_{et}\,\tan(\theta/2)\,(H_{et})^{5/2}$$

where: C_{et} and H_{et} are the discharge coefficient and effective head, respectively.
$H_{et} = H + \delta_{Ht}$
where δ_{Ht} is a function of notch angle as tabulated in Table 13.6.3.

The discharge coefficient, C_{et} is a function of the notch angle for fully contracted weirs only, and values are tabulated in Table 13.6.4.

TABLE 13.6.3
Values of δ_{Ht}

Notch Angle in Degrees	δ_{Ht} (ft)	δ_{Ht} (mm)
20	0.010	2.90
30	0.007	2.00
40	0.006	1.70
50	0.005	1.40
60	0.004	1.20
70	0.003	1.00
80	0.003	0.95
90	0.003	0.94
100	0.003	0.93

TABLE 13.6.4
Values of C_{et} for Fully Contracted V-Notch Weirs

Notch Angle, Degrees	C_{et}
20	0.592
30	0.588
40	0.581
50	0.579
60	0.578
70	0.577
80	0.578
90	0.579
100	0.581

For partially contracted weirs, only 90° notches may be used and the discharge coefficient is found from a complicated graph of analytical data. The values range from 0.575 to 0.60 with an average value of about 0.585. Use of that value for C_{et} will yield good results in most cases. For more precise work, see Figure 7 on page 5 of ASTM Method D 5242–21.

Expected Outcomes

The expected outcomes from this experiment will vary widely and will depend on the actual weir used, the value of the various measurements, and the precision of the measurements made.

DATA COLLECTION SHEET OPEN-CHANNEL FLOW MEASUREMENT OF WATER WITH THIN-PLATE WEIRS	
Lead Experimenter	
Date of Experiment	
Name of Stream and Where along the Stream the Measurements Were Made	
Dimensions and Shape of Weir	
Head above Crest or Vertex, H	
Crest Length for a Rectangular Weir, L	
Crest Height above Channel Bottom, P	
Channel Width, B	
Value of δL from Table 13.6.1	
Value of C_e from Table 13.6.2	
Value of δHt from Table 13.6.3	
Value of C_{et} from Table 13.6.4	

EXPERIMENT 13.7

DETERMINATION OF THE HORIZONTAL WATER FLOW RATE THROUGH A GEOSYNTHETIC SCREENING MATERIAL

(Consistent with ASTM Method D 8203–18)

Date: ______________________________

Principal Investigator:

Name: ______________________________

Email: ______________________________

Phone: ______________________________

Collaborators:

Name: ______________________________

Email: ______________________________

Phone: ______________________________

Name: ______________________________

Email: ______________________________

Phone: ______________________________

Name: ______________________________

Email: ______________________________

Phone: ______________________________

Name: ______________________________

Email: ______________________________

Phone: ______________________________

Objective or Question to Be Addressed

This experiment provides a test procedure for determining the horizontal water flow rate through a geosynthetic fabric under a constant-head pressure using potable water.

Theory behind Experiment

In this experiment, a sample of geosynthetic material is firmly secured across the opening of a reservoir created in the lab for this purpose. Potable water is then used to fill the reservoir until the flow running through the fabric into a collection reservoir (for recycling the water back to the inlet reservoir) is flowing at a constant head and steady state flow rate. A discharge volume, based on the steady state flow rate and time are measured to determine the flow rate through the fabric.

By dividing the area of the fabric exposed to the water flow into the calculated flow rate, a flow rate per unit of fabric area is also calculated.

This experiment is intended to simulate the flow through a geosynthetic fabric without sediment load. If field conditions encounter significant sediment in the runoff, the flow rate through the fabric is likely to decrease significantly. The experiment could then be run with water containing a similar suspended solids load as field conditions exhibit to determine the likely field developed flow rate and the time to failure of the fabric that could result in total clogging and overtopping of the fabric barrier.

Potential Interferences and Interference Management Plan

The fabric needs to be securely fastened to the reservoir outlet such that there is no leakage around the edges and the fabric is not folded or torn in any way. The fabric should not be stretched across the opening, as this could affect the ability of the fabric to function as designed and yield false results from the experiment.

Data to Be Collected with an Explanation of How They Will Be Collected (Ex.: "Water Temperature Data for Three Days Using a Continuous Data Logger")

Using the attached data sheet, record at a minimum the following information:

- Name of each experimenter
- Date of experiment
- Dimensions of inlet reservoir and receiving reservoir
- Dimensions of the calibrated collection container
- Source of test water
- Source and description of fabric being tested
- Collected volume and time to collect that volume at each depth and head setting
- Calculated flow rate in gal/min (L/min) at each head depth
- Calculated flow rate/unit of time/unit of fabric area in gal/min/ft^2 (L/min/m^2) at each head depth
- Calculated flow rate/unit of time/unit of length of the fabric in gal/min/ft (L/min/m) at each head depth

- Plot the flow per unit of flow per unit of time per unit of length of fabric on the y-axis against the depth of head behind the fabric at the time of measurement on the x-axis and draw a line of best fit between them.
- From the graph, estimate the maximum flow rate through the fabric at the maximum depth and head behind the fabric.

Tools, Equipment, and Supplies Required

- Water-tight inlet reservoir – constructed of waterproofed plywood, plexiglass, or similar material. The inlet reservoir is typically in the order of about 4 ft (1.2 m) wide by 8 ft (2.4 m) long and 2 ft (0.6 m) deep. The actual dimensions are not overly important but should be large enough for a good head to develop behind the fabric and a sufficient flow rate to develop to make calculation of the through flow rate to be easily done. Numbers in the cm^3/m^2, for example, are not likely to be useful when extrapolating to hundreds of m^2 of fabric in the field.
- A water-tight receiving reservoir, constructed of similar materials to those used for the inlet reservoir. The lower reservoir must be located below the inlet reservoir and is typically larger than the inlet reservoir to minimize the potential for back pressure from the water that has passed through the fabric. Again, the actual dimensions are not critical, but dimensions in the order of about 4 ft (1.2 m) wide by 20 ft (6.1 m) long by 4 ft (1.2 m) deep will generally allow the water to stabilize to a sufficiently smooth surface after flowing through the fabric that measurement of the water depth in that reservoir can be facilitated for volume calculations.
- A water recirculation system that will minimize the use of potable water by recirculating the water collected in the receiving reservoir back to the inlet reservoir in a manner that will not create significant currents in the water flow or discharge pressures against the fabric in the inlet reservoir. This will generally require a variable speed pump and piping system capable of maintaining a constant head behind the fabric being tested.
- A calibrated container of sufficient size to measure significant flows over a measurable time frame. Typically, a container of about 20 gal (75 L) is appropriate. The collection container needs to be the same width as the opening from the inlet reservoir so that the entire flow out of the inlet reservoir over a measured time fame can be captured.
- Stopwatch or other timing device
- A ruler of greater length than the depth of the water in the receiving reservoir, marked in suitably fine increments to accurately record the depth in the reservoir to the nearest tenth of an inch (0.25 cm)
- A constant supply of potable water or a sufficient supply to allow for filling the inlet reservoir and the receiving reservoir, along with the recirculation system
- A sample of material to be tested that is large enough to cover the opening of the inlet reservoir
- A piece of sturdy screen with about 0.25–0.5 inches (0.64–1.27 cm) openings large enough to cover the fabric sample

Safety

Construction of the test reservoirs will involve the cutting and fitting of wood or plastic material. The use of power tools or hand tools always provides opportunity for cuts, splinters, and abrasions. Care should be taken to minimize risks through the use of proper personal protective equipment.

Water on floors can be slippery, and care is needed to avoid falls.

As with all laboratory experiments, eye protection, gloves, and a lab coat are recommended for this experiment

Planned Experimental Procedures and Steps, in Order

1. Put on appropriate personal protective equipment and gather all necessary materials and supplies in a convenient location.
2. Construct the two reservoirs in accordance with the previous descriptions.
3. Install a sample of the geotextile fabric across the opening of the inlet reservoir in any manner that provides for no leakage around the fabric and prevents the fabric from pulling loose. A small strip of wood or plastic screwed to the inside walls of the inlet reservoir to clamp the fabric to the walls of the reservoir on four sides is suggested. Note that this will reduce the area of the opening and that area will need to be carefully measured to generate an accurate flow rate per unit of area later.
4. Place a piece of sturdy screen with about 0.25–0.5 inches (0.64–1.27 cm) openings against the face of the fabric on the downstream side to prevent distortion of the fabric during testing. The screen also needs to be securely fastened to the walls and bottom of the reservoir.
5. Place a dam across the open end of the inlet box during the initial filling of the inlet reservoir. The dam should be about 3/4 of the height of the fabric test sample and the width of the opening.
6. Allow clean tap water to flow into the inlet reservoir at a rate sufficient to raise the water level to approximately 3/4 the height of the fabric
7. As the water level in the inlet reservoir approaches the desired depth, slowly raise and remove the dam and increase the flow as necessary to develop a steady state flow condition with a constant head through the fabric.
8. Once steady-state, constant-head conditions are achieved, slide the calibrated collection container under the flow while starting the stopwatch.
9. Allow the collection container to collect between one third and two thirds of the depth of the container, then quickly, but without spillage, slide the collection container back out from under the flow and stop the stopwatch.
10. Allow the collected volume to settle.
11. Make sure that the collection container is level, then record the volume of water in the collection container based on the calibrations previously marked on it.
12. Adjust the flow rate to allow the water behind the fabric to recede to half the height of the fabric.

13, As the water level approaches the target depth, increase the flow enough to restabilize the flow rate at a steady-state, constant-head condition.
14. Repeat Steps 8–11.
15. Adjust the flow rate to allow the water behind the fabric to recede to one quarter of the height of the fabric.
16. Repeat Steps 8–11.

Data Collection Format and Data Collection Sheets Tailored to Experiment

See attached data collection sheet.

Data Analysis Plan

1. From the area and depth of the collected flow in the receiving reservoir, calculate the collected volume.
2. From the collected volume and time measurement to collect that volume, calculate the volumetric flow rate as the collected volume over the measured time and record that rate in terms of volume per unit of time for each depth setting.
3. Divide the volume per unit of time by the open area of the fabric through which the flow was occurring to yield the flow per unit of time per unit of fabric area, typically gal/min/ft^2 (L/min/m^2) for each depth setting.
4. Calculate the flow rate per unit of time per unit of length of the fabric in gal/min/ft (L/min/m) at each head depth setting.
5. Plot the flow rate in units of flow per unit of time per unit of length of fabric on the y-axis against the depth of head behind the fabric at the time of measurement on the x-axis and include a line of best fit for the data showing the equation of the line.
6. From the graph, estimate the maximum flow rate through the fabric at the maximum depth and head behind the fabric.

Expected Outcomes

The outcome for this experiment will depend on the fabric used, the material from which the fabric is constructed, and the weave tightness, plus the actual head on the fabric and the degree of turbidity, if any, in the water. In field conditions, the flow rate will decrease over time as silt and debris clog the surface of the fabric.

DATA COLLECTION SHEET DETERMINATION OF THE HORIZONTAL WATER FLOW RATE THROUGH A GEOSYNTHETIC SCREENING MATERIAL	
Lead Experimenter	
Date of Experiment	
Length of Inlet Reservoir	
Width of Inlet Reservoir	
Depth of Inlet Reservoir	
Calculated Volume of Inlet Reservoir	
Length of Receiving Reservoir	
Width of Receiving Reservoir	
Depth of Receiving Reservoir	
Calculated Volume of Receiving Reservoir	
Time to Fill Collection Container, First Pass	
Volume of Water Collected, First Pass	
Calculated Flow Rate through Fabric	
Time to Fill Collection Container, Second Pass	
Volume of Water Collected, Second Pass	
Calculated Flow Rate through Fabric	
Time to Fill Collection Container, Third Pass	
Volume of Water Collected, Third Pass	
Calculated Flow Rate through Fabric	
Estimated Maximum Flow Rate through Fabric	

Appendix

TABLE A.1
Collection, Preservation, and Holding Times for Selected Liquid Environmental Samples

Parameter	Epa Testing Methodology	Container/Volume To Collect	Preservatives, If Any	Transport/ Holding Conditions	Holding Times
Acid/Base Neutral Extractables	625	Two 1 L Teflon-lined Amber Glass	Cool to 4 °C Add NaSO to pH<2	Cool to 4 °C	7 days before extraction
Acid/Base Neutral Extractables	8270 C 8270 D	Two 1 L Teflon-lined Amber Glass	Cool to 4 °C	Cool to 4 °C	7 days before extraction
Acidity	SM 2310 B	250 mL Plastic	Cool to 4 °C	Cool to 4 °C	14 days
Alkalinity	SM 2320 B	250 mL Plastic	Cool to 4 °C	Cool to 4 °C	14 days
Ammonia	350.1 SM 4500 NH-BH	250 mL Plastic	HSO to pH< 2 Cool to 4 °C	Cool to 4 °C	28 days
Biological Oxygen Demand	SM 5210 B	1 L Plastic	Cool to 4 °C	Cool to 4 °C	48 hours
Boron	200.7 6010 C	250 mL Plastic	HNO_3 to pH<2 Cool to 4 oC	Cool to 4 °C	180 days
Bromide	300.0	250 mL Plastic	None	None	28 days
Chemical Oxygen Demand	410.4 SM 5220 D	250 mL Plastic	HSO to pH< 2 Cool to 4 °C	Cool to 4 °C	28 days
Chloride	9251 300.0 SM 4500 Cl-E	250 mL Plastic	Cool to 4 °C	Cool to 4 °C	28 days
Chlorine, Total Residual	SM 4500 Cl-G	250 mL Plastic	Cool to 4 °C	Cool to 4 °C	Analyze Immediately

(Continued)

TABLE A.1 (Continued)

Parameter	Epa Testing Methodology	Container/Volume To Collect	Preservatives, If Any	Transport/ Holding Conditions	Holding Times
Chlorinated Herbicides	8151 A	Two 1 L Teflon-lined Amber Glass	Cool to 4 °C	Cool to 4 °C	7 days before extraction
Chromium VI	7196 A SM 3500 Cr D	250 mL Glass or Plastic	Cool to 4 °C	Cool to 4 °C	24 hours
Cyanide	335.1 9010 B SM 4500 CN-CE	250 ml Plastic	NaOH to pH>12 Cool to 4 °C	Cool to 4 °C	14 days
Dissolved Organic Carbon	5310 B	40 mL VOA Vial with Teflon-lined Cap	Cool to 4 °C	Cool to 4 °C	28 days
Fluoride	300.0 SM 4500 F-B BC	500 mL Plastic	Cool to 4 °C	Cool to 4 °C	28 days
Formaldehyde	8315	1 L Amber Glass	Cool to 4 °C	Cool to 4 °C	3 days
Haloacetic Acids	552.2	250 mL Amber Glass Jar with Teflon Cap	Cool to 4 °C Add NH_4Cl	Cool to 4 °C	14 days
Hardness	SM 2340 B	100 mL Plastic	Cool to 4 °C Add HNO_3 to pH<2	Cool to 4 °C	6 months
Herbicides	8151 A	1 L Amber Glass Jar with Teflon Cap	Cool to 4 °C	Cool to 4 °C	7 days before extraction
Hexavalent Chromium	7196 A SM 3500 Cr-D	500 mL Plastic	Cool to 4 °C	Cool to 4 °C	24 hours
Mercury (TCLP)	1311 7470 A	4 oz (118 mL) Amber Glass Jar	Cool to 4 °C Add HNO_3 to pH<2	Cool to 4 °C	28 days before extraction
Nitrate	300.0 353.2 SM 4500 NO_3-F	250 ml Plastic	Cool to 4 °C	Cool to 4 °C	48 hours
Nitrate/Nitrite	300.0 353.2 SM 4500 NO_3-F	250 ml Plastic	Cool to 4 °C Add H_2SO_4 to pH<2	Cool to 4 °C	28 days

Nitrite	300.0 353.2 SM 4500 NO_3-F SM 4500 NO_2-B	250 ml Plastic	Cool to 4 °C	Cool to 4 °C	48 hours
Nitrogen, Ammonia	SM 4500 NH_3 B, C	500 mL Plastic	Cool to 4 °C Add H_2SO_4 to pH<2	Cool to 4 C°	28 days
Nitrogen, Total Kjeldahl	353.3/.1 SM 4500 N_{org}-C	250 ml Plastic	Cool to 4 °C Add H_2SO_4 to pH<2	Cool to 4 °C	28 days
Oil and Grease	1664 A	Two 1 L Amber Glass	Cool to 4 °C Add HCl to pH<2	Cool to 4 °C	28 days
Oil and Grease	1664 B	1 L Amber Glass with Teflon-lined Cap	Cool to 4 °C Add HCl or H_2SO_4 to pH<2	Cool to 4 °C	28 days
Orthophosphate	SM 4500 P-E	250 ml Plastic	Cool to 4 °C	Cool to 4 °C	48 hours
Polychlorinated Biphenols (PCBs)	608	Two 1 L Teflon-lined Amber Glass	Cool to 4 °C Add NaSO	Cool to 4 °C	7 days before extraction
Polychlorinated Biphenols (PCBs)	8082 8082 A	Two 1 L Teflon-lined Amber Glass	Cool to 4 °C	Cool to 4 °C	None before Extraction
Pesticides (Organochlorine)	608	Two 1 L Teflon-lined Amber Glass	Cool to 4 °C Add NaSO	Cool to 4 °C	7 days before extraction
Pesticides (Organochlorine)	8081 A 8081 B	Two 1 L Teflon-lined Amber Glass	Cool to 4 °C	Cool to 4 °C	7 days before extraction
Petroleum Hydrocarbon Identification	8015 B (Modified)	4 oz (118 mL) Teflon-lined Amber Glass Jar	Cool to 4 °C	Cool to 4 °C	14 days before extraction
Phenolics	420.4	1 L Glass	H_2SO_4 to pH<2 Cool to 4 °C	Cool to 4 °C	28 days
Phosphorous, Total	SM 4500 P-E	250 mL Plastic	Cool to 4 °C Add H_2SO_4 to pH<2	Cool to 4 °C	28 days
Polynuclear Aromatic Hydrocarbons (PAHs)	625 8270 D	Two 1 L Teflon-lined Amber Glass	Cool to 4 °C Add NaSO	Cool to 4 °C	7 days before extraction

(Continued)

TABLE A.1 (Continued)

Parameter	Epa Testing Methodology	Container/Volume To Collect	Preservatives, If Any	Transport/ Holding Conditions	Holding Times
Polynuclear Aromatic Hydrocarbons (PAHs)	8270 C 8270 C-SIM	Two 1 L Teflon-lined Amber Glass	Cool to 4 °C	Cool to 4 °C	7 days before extraction
Purgeables Using GCMS	524.2	40 mL VOA Vial with Teflon-lined Cap	Cool to 4 °C Add Ascorbic Acid and HCl to pH<2	Cool to 4 °C	14 days
Purgeables Using GCMS	624 8260 C	40 mL VOA Vial with Teflon-lined Cap	Cool to 4 °C Add HCl to pH<2	Cool to 4 °C	14 days
Salinity	SM 2520	150 mL Plastic	Cool to 4 °C	Cool to 4 °C	28 days
Semi-Volatiles (TCLP)	1311 8270 C 8081 A 8081 B 8151 A	8 oz (237 mL) Teflon-lined Amber Glass Jar	Cool to 4 °C	Cool to 4 °C	14 days before extraction
Silica	200.7 6010 C	150 mL Plastic	Cool to 4 °C	Cool to 4 °C	28 days
Sulfate	300.0 9038 SM 4500 SO_4 – E	250 mL Plastic	Cool to 4 °C	Cool to 4 °C	28 days
Sulfide	9030 B SM 4500 S_2-AD	Two 250 mL Plastic	ZnOAC, NaOH to pH>9 Cool to 4 °C	Cool to 4 °C	7 days
Sulfite	SM 4500 SO_3-B	100 mL Plastic	Cool to 4 °C	Cool to 4 °C	Analyze Immediately
Total Metals	200.7 200.8 6010 B 6010 C 6020 6020 A 7000 A	501 mL Plastic	HNO_3 to pH<2 Cool to 4 °C	Cool to 4 °C	180 days (28 days for Hg)
Total Metals (TCLP)	1311 6010 B 6010 C 6020 6020 A 7000 A	4 oz (118 mL) Amber Glass Jar	Cool to 4 °C	Cool to 4 °C	180 days before extraction
Total Organic Carbon	415.1 9060 SM 5310 C	Two 40 mL Amber Glass VOA Vials	Cool to 4 °C Add H_2SO_4 to pH<2	Cool to 4 °C	28 days
Total Petroleum Hydrocarbons	1664 A	4 oz (118 mL) Teflon-lined Amber Glass Jar	Cool to 4 °C	Cool to 4 °C	28 days

Total Petroleum Hydrocarbons	1664 B	1 L Amber Glass Jar with Teflon-lined Cap	Add HCL or H_2SO_4 to pH<2 Cool to 4 °C	Cool to 4 °C	28 days
Total Petroleum Hydrocarbons – GC/FID	8015 B (Modified)	4 oz (118 mL) Teflon-lined Amber Glass Jar	Cool to 4 °C	Cool to 4 °C	14 days before extraction
Total Phenol	420.1 9065 SM 510 ABC	Two 1 L Amber Glass	Cool to 4 °C Add H_2SO_4 to pH<2	Cool to 4 °C	28 days
Total Residual Chlorine	SM 4500 Cl-D	500 mL Plastic	Cool to 4 °C	Cool to 4 °C	24 hours
Total Solids	2540 C	250 mL Plastic	Cool to 4 °C	Cool to 4 °C	7 days
Total Dissolved Solids	SM 2540 C	500 mL Plastic	Cool to 4 °C	Cool to 4 °C	7 days
Total Suspended Solids	SM 2540 D	1 L Plastic	Cool to 4 °C	Cool to 4 °C	7 days
Total Volatile Solids	SM 2540 E	500 mL Plastic	Cool to 4 °C	Cool to 4 °C	7 days
Turbidity	180.1 2130 B	500 mL Plastic	Cool to 4 °C	Cool to 4 °C	48 hours
Volatile Organics	524.2	Two 40 mL Teflon-lined Amber Glass VOA Vials	Ascorbic Acid, HCl to pH<2 Cool to 4 °C	Cool to 4 °C	14 days
Volatile Organics	624	Two 40 mL Teflon-lined Amber Glass VOA Vials	NaSO Cool to 4 °C	Cool to 4 °C	7 days
Volatile Organics	8260 B	Two 40 mL Teflon-lined Amber Glass VOA Vials	HCl to pH<2 Cool to 4 °C	Cool to 4 °C	14 days
Volatiles (TCLP)	1311 8260 B	8 oz (237 mL) Teflon-lined Amber Glass Jar	Cool to 4 °C	Cool to 4 °C	14 days before extraction

Consolidated data based on published guidelines from various analytical laboratories. Verification of data with the testing lab prior to sampling is strongly recommended.

TABLE A.2
Collection, Preservation, and Holding Times for Selected Soil and Sediment Environmental Samples

Parameter	Epa Testing Methodology	Container/Volume To Collect	Preservative (If Any)	Transport/Holding Conditions	Holding Times
Acid/Base Neutral Extractables	8270 C	4 oz (118 mL) Glass Jar	Cool to 4 °C	Cool to 4 °C	14 days prior to extraction
Acid/Base Neutral Extractables Oil/NAPL Samples	8270 C	4 oz (118 mL) Glass Jar	Cool to 4 °C	Cool to 4 °C	None
Chrome VI	7196 A SM 3500 Cr – D	4 oz (118 mL) Glass Jar	Cool to 4 °C	Cool to 4 °C	21 days prior to digestion; 3 days after digestion
Chrome VI	SW 846 7196 A	8 oz (118 mL) Glass Jar	Cool to 4 °C	Cool to 4 °C	30 days
Cyanide	9010 B 9012 A 9014	4 oz (118 mL) Glass Jar	Cool to 4 °C	Cool to 4 °C	14 days
Diesel Range Organics (DRO)	8015 D (Mod)	8 oz (236 mL) Glass Jar with Teflon Cap	Cool to 4 °C	Cool to 4 °C	14 days
Flashpoint	1010 A 1030	8 oz (236 mL) Glass Jar with Teflon Cap	Cool to 4 °C	Cool to 4 °C	Analyze Immediately
Gasoline Range Organics (GRO)	8015 D (Mod)	8 oz (236 mL) Glass Jar with Teflon Cap	Cool to 4 °C Add 15 mL CH_3OH	Cool to 4 °C	14 days
Herbicides	8151 A	8 oz (236 mL) Glass Jar with Teflon Cap	Cool to 4 °C	Cool to 4 °C	14 days
Ignitability	1010	4 oz (118 mL) Glass Jar	Cool to 4 °C	Cool to 4 °C	180 days

Metals	6010 B 6020 A 7000	4 oz (118 mL) Glass Jar	Cool to 4 °C	Cool to 4 °C	180 days
Metals by TCLP	1311 6010 B 6010 C 6020 6020 A 7000 A	8 oz (236 mL) Glass Jar	Cool to 4 °C	Cool to 4 °C	180 days prior to extraction
Mercury	7471 A 7471 B	8 oz (236 mL) Amber Glass Jar	Cool to 4 °C	Cool to 4 °C	28 days
Mercury by TCLP	1311 7470 A 7474	8 oz (236 mL) Amber Glass Jar	Cool to 4 °C	Cool to 4 °C	28 days
Oil and Grease	1664 B	8 oz (236 mL) Glass Jar with Teflon Cap	Cool to 4 °C	Cool to 4 °C	28 days
Pesticides	8081 A 8081 B	8 oz (236 mL) Glass Jar with Teflon Cap	Cool to 4 °C	Cool to 4 °C	14 days prior to extraction
Polychlorinated Biphenols (PCBs) Homologs or Congeners	680 8270 C (M)	4 oz (118 mL) Glass Jar	Cool to 4 °C	Cool to 4 °C	14 days prior to extraction
Polychlorinated Biphenols (PCBs) Aroclors, Homologs or Congeners	8082 8082 A	8 oz (236 mL) Amber Glass Jar with Teflon Cap	Cool to 4 °C	Cool to 4 °C	14 days prior to extraction
Polynuclear Aromatic Hydrocarbons (PAHs)	8270 C 8270 C – SiM	4 oz (118 mL) Glass Jar	Cool to 4 °C	Cool to 4 °C	14 days prior to extraction
Polynuclear Aromatic Hydrocarbons (PAHs) Oil/NAPL Samples	8270 C 8270 C – SiM	4 oz (118 mL) Glass Jar	Cool to 4 °C	Cool to 4 °C	NA
Reactivity	Ch. 7 SW 846	8 oz (236 mL) Glass Jar	Cool to 4 °C	Cool to 4 °C	Analyze Immediately

(*Continued*)

TABLE A.2 (Continued)

Parameter	Epa Testing Methodology	Container/Volume To Collect	Preservative (If Any)	Transport/Holding Conditions	Holding Times
Semivolatiles by TCLP	1311 8081 A 8081 B 8270 C	8 oz (236 mL) Glass Jar	Cool to 4 °C	Cool to 4 °C	14 days prior to extraction
Sulfide	SM 4500 S – AD	4 oz (118 mL) Glass Jar	Cool to 4 °C	Cool to 4 °C No Headspace	28 days prior to extraction; 7 days after extraction
Total Organic Carbon (TOC)	9060	4 oz (118 mL) Glass Jar	Cool to 4 °C	Cool to 4 °C	28 days
Total Petroleum Hydrocarbons	8015 (Mod)	20 mL Glass Vial	Cool to 4 °C	Cool to 4 °C	NA
Volatile Organics	8260 B 5035 (High Level)	40 mL Amber Glass VOA Vial (5 gMinimum)	Cool to 4 °C, Add 5 mL MeOH	Cool to 4 °C	14 days
Volatile Organics	8261 B 5035 (Low Level)	Two 40 mL Amber Glass VOA Vial (5 gMinimum Each)	Cool to 4 °C, Add 5 mL water	Cool to 4 °C	48 hours to Freeze, 14 days to Analyze if Frozen

Consolidated data based on published guidelines from various analytical laboratories. Verification of data with the testing lab prior to sampling is strongly recommended.

TABLE A.3
Density and Dry Specific Gravity Estimates for Various Soil Types

Soil Type	Dry Unit Weight, g/cc	Porosity, Undisturbed, %	Porosity, Repacked, %	Dry Specific Gravity
Fine Sand	1.13–1.99	26.0–53.3	26.7–50.2	2.63
Medium Sand	1.27–1.93	28.5–48.9	27.2–52.3	2.65
Coarse Sand	1.42–1.94	30.9–46.4	37.0–42.0	2.67
Fine Gravel	1.60–1.99		25.1–38,5	
Medium Gravel	1.47–2.09		23.7–44.1	
Coarse Gravel	1.69–2.08		23.8–36.5	
Silt	1.01–1.79	33.9–61.1	41.0–56.0	<2
Clay	1.18–1.72	34.2–56.9	39.9–52.8	2.67–2.9

Source Adapted from: Summary of Hydrologic and Physical Properties of Rock and Soil Materials, as Analyzed by the Hydrologic Laboratory of the U.S. Geological Survey 1948–60 by D. A. MORRIS and A. I. JOHNSON https://pubs.usgs.gov/wsp/1839d/report.pdf Accessed November 17, 2020, and www.researchgate.net/figure/Specific-Gravity-for-Various-Soils_tbl4_281267333/Accessed June 1, 2022.

TABLE A.4
Typical Soil Porosity and Void Ratio Values for Selected Soils

Soil Type Or Description	Uniform Soil Classification System Designation (USCS)	Porosity [2]		Void Ratio [1]	
		Minimum Likely	Maximum Likely	Minimum Likely	Maximum Likely
Well-Graded Gravel, Sandy Gravel, Little or No Fines	GW	0.21	0.32	0.26	0.46
Poorly Graded Gravel, Sandy Gravel, Little or No Fines	GP	0.21	0.32	0.26	0.46
Silty Gravels, Silty Sandy Gravels	GM	0.15	0.22	0.18	0.28
Gravel	GW-GP	0.23	0.38	0.30	0.60
Clayey Gravels, Clayey Sandy Gravels	GC	0.17	0.27	0.21	0.37
Well-Graded Sands, Little or No Fines	SW	0.22	0.42	0.29	0.74
Coarse Sand	(SW)	0.26	0.43	0.35	0.75
Fine Sand	(SW)	0.29	0.26	0.40	0.85
Poorly Graded Sands, Gravely Sands, Little or No Fines	SP	0.23	0.43	0.30	0.75
Silty Sands	SM	0.25	0.49	0.33	0.98
Clayey Sands	SC	0.15	0.37	0.17	0.59
Inorganic Silts, Silty or Clayey Fine Sands with Slight Plasticity	ML	0.21	0.56	0.26	1.28
Uniform Organic Silt	(ML)	0.29	0.52	0.40	1.10
Inorganic Clays, Silty Clays, Sandy Clays with Low Plasticity	CL	0.42	0.68	0.41	0.69
Organic Silts and Organic Silty Clays with Low Plasticity	OL	0.42	0.68	0.74	2.26
Silty or Sandy Clay	(CL-OL)	0.20	0.64	0.25	1.80
Inorganic Silts with High Plasticity	MH	0.53	0.75	1.14	2.10
Inorganic Clays with High Plasticity	CH	0.39	0.59	0.63	1.45
Organic Clays with High Plasticity	OH	0.50	0.75	1.06	3.34

Source: Data adapted from:

1. Geotechdata.info, Soil Void Ratio, http://geotechdata.info/parameter/soil-void-ratio.html (as of November 16, 2013). Accessed November 17, 2020. 2. Geotechdata.info, Soil Porosity (Soil Porosity 2013), http://geotechdata.info/parameter/soil-porosity.html (as of November 16, 2013) Accessed November 17, 2020.

TABLE A.5
Density of Water and Temperature Coefficients (K) at Various Temperatures (°C and °F)

Temp. °C	Density (g/mL)	Temp. Coefficient (K)	Temp. °F	Temp. °C	Density (g/mL)	Temp. Coefficient (K)	Temp. °F
15	0.99910	1.00090	59.0	18.1	0.99858	1.00037	64.6
15.1	0.99909	1.00088	59.2	18.2	0.99856	1.00035	64.8
15.2	0.99907	1.00087	59.4	18.3	0.99854	1.00034	64.9
15.3	0.99906	1.00085	59.5	18.4	0.99852	1.00032	65.1
15.4	0.99904	1.00084	59.7	18.5	0.99850	1.00030	65.3
15.5	0.99902	1.00082	59.9	18.6	0.99848	1.00028	65.5
15.6	0.99901	1.00080	60.1	18.7	0.99847	1.00026	65.7
15.7	0.99899	1.00079	60.3	18.8	0.99845	1.00024	65.8
15.8	0.99898	1.00077	60.4	18.9	0.99843	1.00022	66.0
15.9	0.99896	1.00076	60.6	19	0.99841	1.00020	66.2
16	0.99895	1.00074	60.8	19.1	0.99839	1.00018	66.4
16.1	0.99893	1.00072	61.0	19.2	0.99837	1.00016	66.6
16.2	0.99891	1.00071	61.2	19.3	0.99835	1.00014	66,7
16.3	0.99890	1.00069	61.3	19.4	0.99833	1.00012	66.0
16.4	0.99888	1.00067	61.5	19.5	0.99831	1.00010	67.1
16.5	0.99886	1.00066	61.7	19.6	0.99829	1.00008	67.3
16.6	0.99885	1.00064	61.9	19.7	0.99827	1.00006	67.5
16.7	0.99883	1.00062	62.1	19.8	0.99825	1.00004	67.6
16.8	0.99881	1.00061	62.2	19.9	0.99823	1.00002	67.8
16.9	0.99879	1.00059	62.4	20	0.99821	1.00000	68.0
17	0.99878	1.00057	62.6	20.1	0.99819	0.99998	68.2
17.1	0.99876	1.00055	62.8	20.2	0.99816	0.99996	68.4
17.2	0.99874	1.00054	63.0	20.3	0.99814	0.99994	68.5
17.3	0.99872	1.00052	63.1	20.4	0.99812	0.99992	68,7
17.4	0.99871	1.00050	63.3	20.5	0.99810	0.99990	68.9
17.5	0.99869	1.00048	63.5	20.6	0.99808	0.99987	69.1
17.6	0.99867	1.00047	63.7	20.7	0.99806	0.99985	69.3
17.7	0.99865	1.00045	63.9	20.8	0.99804	0.99983	69.4
17.8	0.99863	1.00043	64.0	20.9	0.99802	0.99981	69.6
17.9	0.99862	1.00041	64.2	21	0.99799	0.99979	69.8
18	0.99860	1.00039	64.4	21.1	0.99797	0.99977	70.0
21.2	0.99795	0.99974	70.2	24.3	0.99723	0.99902	75.7
21.3	0.99792	0.99972	70.3	24.4	0.99720	0.99899	75.9
21.4	0.99792	0.99970	70.5	24.5	0.99717	0.99897	76.1
21.5	0.99789	0.99968	70.7	22.6	0.99764	0.99943	72.7
21.6	0.99786	0.99966	70.9	22.7	0.99761	0.99940	72.9
21.7	0.99784	0.99963	71.1	22.8	0.99759	0.99938	73.0
21.8	0.99782	0.99961	71.2	22.9	0.99756	0.99936	73.2
21.9	0.99780	0.99959	71.4	25	0.99705	0.99884	77.0
22	0.99777	0.99957	71.6	25.1	0.99702	0.99881	77.2
22.1	0.99775	0.99954	71.8	25.2	0.99700	0.99879	77.4

(Continued)

TABLE A.5 (Continued)

Temp. °C	Density (g/mL)	Temp. Coefficient (K)	Temp. °F	Temp. °C	Density (g/mL)	Temp. Coefficient (K)	Temp. °F
22.2	0.99773	0.99952	72.0	25.3	0.99697	0.99876	77.6
22.3	0.99770	0.99950	72.1	25.4	0.99694	0.99874	77.7
22.4	0.99768	0.99947	72.3	25.5	0.99692	0.99871	77.9
22.5	0.99766	0.99945	72.5	25.6	0.99689	0.99868	78.1
22.6	0.99764	0.99943	72.7	25.7	0.99687	0.99866	78.3
22.7	0.99761	0.99940	72.9	25.8	0.99684	0.99863	78.4
22.8	0.99759	0.99938	73.0	25.9	0.99681	0.99860	78.6
22.9	0.99756	0.99936	73.2	26	0.99679	0.99858	78.8
23	0.99754	0.99933	73.4	26.1	0.99676	0.99855	79.0
23.1	0.99752	0.99931	73.6	26.2	0.99673	0.99852	79.2
23.2	0.99749	0.99929	73.8	26.3	0.99671	0.99850	79.3
23.3	0.99747	0.99926	73.9	26.4	0.99668	0.99847	79.5
23.4	0.99745	0.99924	74.1	26.5	0.99665	0.99844	79.7
23.5	0.99742	0.99921	74.3	26.6	0.99663	0.99842	79.9
23.6	0.99740	0.99919	74.5	26.7	0.99660	0.99839	80.1
23.7	0.99737	0.99917	74.7	26.8	0.99657	0.99836	80.2
23.8	0.99735	0.99914	74.8	26.9	0.99654	0.99833	80.4
23.9	0.99732	0.99912	75.0	27	0.99652	0.99831	80.6
24	0.99730	0.99909	75.2	27.1	0.99649	0.99828	80.8
24.1	0.99727	0.99907	75.4	27.2	0.99646	0.99825	81.0
24.2	0.99725	0.99904	75.6	27.3	0.99643	0.99822	81.1
27.4	0.99641	0.99820	81.3	29.5	0.9958	0.99759	85.1
27.5	0.99638	0.99817	81.5	29.6	0.99577	0.99756	85.3
27.6	0.99635	0.99814	81.7	29.7	0.99574	0.99753	85.5
27.7	0.99632	0.99811	81.9	29.8	0.99571	0.99750	85.6
27.8	0.99629	0.99808	82.0	29.9	0.99568	0.99747	85.8
27.9	0.99627	0.99806	82.2	30	0.99565	0.99744	86.0
28	0.99624	0.99803	82.4	30.1	0.99562	0.99741	86.2
28.1	0.99621	0.99800	82.6	30.2	0.99559	0.99738	86.4
28.2	0.99618	0.99797	82.8	30.3	0.99556	0.99735	86.5
28.3	0.99615	0.99794	82.9	30.4	0.99553	0.99732	86.7
28.4	0.99612	0.99791	83.1	30.5	0.9955	0.99729	86.9
28.5	0.99609	0.99788	83.3	30.6	0.99547	0.99726	87.1
28.6	0.99607	0.99785	83.5	30.7	0.99544	0.99723	87.3
28.7	0.99604	0.99783	83.7	30.8	0.99541	0.99720	87.4
28.8	0.99601	0.99780	83.8	30.9	0.99538	0.99716	87.6
28.9	0.99598	0.99777	84.0	30.5	0.9955	0.99729	86.9
29	0.99595	0.99774	84.2	30.6	0.99547	0.99726	87.1
29.1	0.99592	0.99771	84.4	30.7	0.99544	0.99723	87.3
29.2	0.99589	0.99768	84.6	30.8	0.99541	0.99720	87.4
29.3	0.99586	0.99765	84.7	30.9	0.99538	0.99716	87.6
29.4	0.99583	0.99762	84.0	31	0.99533	0.99716	87.8

TABLE A.6
Specific Heat Capacity of Selected Materials

Material	J/(kg-°K)	J/(g-°C)	BTU/(lb-°F)	BTU/(lb-°C)
Air[5] (sea level, dry)		1.480		
Air[3] (sea level, dry, 0 °C)	1003.5			
Air[3] (typical room temp.)	1012			
Air, dry [6]	1015			
Air[4] (at 25 °C)		1.020		
Alcohol (ethyl)[2]	2400		0.58	
Alcohol (ethyl)[4]		2.460		
Alcohol (ethyl)[5]		2.440		
Aluminum[1]	921.096	0.921096	0.220	0.396
Aluminum[2] (at 20 °C)	900		0.215	
Aluminum[3]	897			
Aluminum[4] (at 25 °C)		0.900		
Aluminum[5]		0.897		
Aluminum [6]	900			
Asbestos [6]	800			
Asphalt[3]	920			
Asphalt[5]		0.920		
Brass (yellow)[1]	401.9328	0.401933	0.096	0.173
Brass[2] (at 20 °C)	380		0.092	
Brass[4] (at 25 °C)		0.380		
Brass[5]		0.375		
Brick[3]	840			
Brick[5]		0.840		
Carbon Steel[1]	502.416	0.502416	0.120	0.216
Cast Iron[1]	460.548	0.460548	0.110	0.198
Charcoal[5]		0.840		
Concrete[3]	880			
Concrete[4] (at 25 °C)		0.880		
Concrete[5]		0.880		
Concrete [6]	840			
Copper[1]	376.812	0.376812	0.090	0.162
Copper[3]	385			
Copper [6]	387			
Copper[4] (at 25 °C)		0.385		
Copper[5]		0.385		
Cork[5]		2.000		
Ethanol[3]	2440			
Ethylene Glycol[4] (at 25 °C)		2.200		
Glass[2] (at 20 °C)	840		0.20	

(*Continued*)

TABLE A.6 (Continued)

Material	J/(kg-°K)	J/(g-°C)	BTU/(lb-°F)	BTU/(lb-°C)
Glass[3] (Pyrex)	753			
Glass[5] (Pyrex)		0.753		
Glass[3,6] (silica)	840			
Glass[4] (at 25 °C)		0.840		
Granite[2] (at 20 °C)	790		0.19	
Granite[3]	790			
Granite[5]		0.790		
Ice[4] (0 °C)		2.010		
Ice[2] (−10 °C)	2050		0.49	
Ice[3] (−10 °C)	2050			
Iron[1]	460.548	0.460548	0.110	0.198
Iron[3]	412			
Iron[4] (at 25 °C)		0.444		
Iron[5]		0.449		
Iron [6]	452			
Lead[1]	125.604	0.125604	0.030	0.054
Lead[2,6] (at 20 °C)	128		0.0305	
Lead[3]	129			
Lead[4] (at 25 °C)		0.160		
Lead[5]		0.129		
Methane[3] (at 2 °C)	2191			
Methanol[3]	2140			
Nitrogen[3]	1040			
Oxygen[3]	918			
Sandy Clay[5]		1.381		
Steel [6]	452			
Vegetable Oil[4] (at 25 °C)		2.000		
Water[2] (at 20 °C)	4186		1.0000	
Water[5] (at 20 °C)		4.182		
Water[3] (at 25 °C)	4181.3			
Water[3] (at 100 °C)	4181.3			
Water[3]	4181.3			
Water [6]	4186.0			
Wood [6]	1700.0			
Zinc[1]	376.812	0.376812	0.090	0.162
Zinc[3]	387			
Zinc[4] (at 25 °C)		0.39		
Zinc[5]		0.388		

1. Engineersedge.com 2. Hyperphysics – Table 3. Wikipedia.org 4. ucdsb.on.ca 5. Engineerstoolbox.com 6. Texasgateway.org

TABLE A.7
Coefficients of Thermal Expansion of Selected Materials

Liquid	Volumetric Coefficient of Expansion (L/K, L/°C)	Linear Coefficient of Expansion (10^{-6} cm/cm/°C)
Acetic Acid	0.0011	
Acetone	0.00143	
Admiralty Brass		6.2
Alcohol, ethyl (ethanol)	0.00109	
Aluminum		7.3
Benzene	0.00125	
Bronze		6.6
Cast Iron		3.2
Calcium Chloride, 5.8 % Solution	0.00025	
Calcium Chloride, 40.9 % Solution	0.00046	
Carbon Disulfide	0.00119	
Carbon Tetrachloride	0.00122	
Chromium		1.8
Ductile Iron		3.3–3.4
Copper		5.4
Ether	0.00160	
Ethyl Acetate	0.00138	
Ethylene Glycol	0.00057	
R-12 Refrigerant	0.0026	
Hydrochloric Acid, 33.2 % Solution	0.00046	
Isobutyl Alcohol	0.00094	
Nickel		4.1
Oil (unused engine oil)	0.00070	
Olive Oil	0.00070	
Paraffin Oil	0.000764	
Petroleum	0.0010	
Phenol	0.0009	
Potassium Chloride, 24.3 % Solution	0.00035	
Red Brass		5.8
Silver		1.6
Sodium Chloride, 20.6 % Solution	0.00041	
Sodium Sulfate, 24 % Solution	0.00041	
Stainless Steel		2.9–5.3
Stainless Steel (common)		3.3
Tin		7.1
Turpentine	0.001000	
Water at Approx. 20 °C (68 °F)	0.000214	
Yellow Brass		6.3
Zinc		10.6

TABLE A.8
Average Moisture Content of Green Wood by Selected Species

	Moisture Content – %			Moisture Content – %	
Species	**Heartwood**	**Sapwood**	**Species**	**Heartwood**	**Sapwood**
Hardwoods			Softwoods		
Alder, Red	-	97	Baldcypress	121	171
Apple	81	74	Cedar, Eastern Red	33	-
Ash, Black	95	-	Cedar, Incense	44	213
Ash, White	46	44	Cedar, Western Red	58	249
Aspen	95	113	Cedar, Yellow	32	166
Beech, American	55	72	Douglas Fir, Coast Type	37	115
Birch, Paper	89	72	Fir, Balsam	88	173
Birch, Yellow	74	72	Fir, Grand	92	136
Cherry, Black	58	-	Fir, Noble	34	115
Chestnut, American	120	-	Fir, Pacific Silver	55	164
Cottonwood	162	146	Fir, White	98	160
Elm, American	95	92	Hemlock, Eastern	97	129
Hickory, Pignut	80	54	Hemlock, Western	85	170
Hickory, Red	69	52	Larch, Western	54	119
Magnolia	80	104	Pine, Loblolly	33	110
Maple, Silver	58	97	Pine, Lodgepole	41	120
Maple, Sugar	65	72	Pine. Ponderosa	40	148
Oak, California Black	76	75	Pine, Red	32	134
Oak, Northern Red	80	69	Pine, Sugar	98	219
Oak, Southern Red	83	75	Pine, Western White	62	148
Oak, White	64	78	Redwood, Old Growth	86	210
Sycamore, American	114	130	Spruce, Black	52	113
Tupelo, Black	87	115	Spruce, Engelman	51	173
Tupelo, Swamp	101	108	Spruce, Sitka	41	142
Walnut, Black	90	73	Tamarack	49	-

Source: Adapted from Glass and Zelinski 2015.

TABLE A.9
Average Moisture Content of Dried Lumber Correlated to Humidity

Average Humidity of Area	General Moisture Content Dry Wood Will Attain
19–25 %	5 %
26–32 %	6 %
33–39 %	7 %
40–46 %	8 %
47–62 %	9 %

Source: Adapted from Reliancetimber.com.

TABLE A.10
Compressive and Tensile Strength of Selected Wood at 12 % Moisture Content

Common Name	Compressive Strength Parallel to Grain (lbf/in^2)		Compressive Strength Perpendicular to Grain (lbf/in^2)		Tensile Strength Perpendicular to Grain (lbf/in^2)	
	lbf/in^2	kgf/m^2	lbf/in^2	kgf/m^2	lbf/in^2	kgf/m^2
Alder (Red)	5,820	4,091,865	440	309,351	420	295,289
Ash (White)	7,410	5,209,746	1,160	815,561	940	660,885
Aspen (Quaking)	4,250	2,988,046	370	260,136	260	182,798
Cedar (Atlantic White)	4,700	3,304,427	410	288,259	220	154,675
Cedar (Eastern Red)	6,020	4,232,479	920	646,824	n/a	n/a
Cedar (Aromatic)	5,200	3,655,962	590	414,811	270	189,829
Cedar (Yellow)	6,310	4,436,369	620	435,903	360	253,105
Douglas Fir (Western Coastal)	7,230	5,083,193	800	562,456	340	239,044
Douglas Fir (West Interior North)	6,900	4,851,180	770	541,364	390	274,197
Douglas Fir (West Interior South)	6,230	4,380,123	740	520,272	330	232,013
Basswood (American)	4,730	3,325,519	370	260,136	350	246,074
Beech (American)	7,300	5,132,408	1,010	710,100	1,010	710,100
Birch (Paper)	5,690	4,000,466	600	421,842	n/a	n/a
Birch (Yellow)	8,170	5,744,078	970	681,978	920	646,824
Cherry (Black)	7,110	4,998,825	690	485,118	560	393,719
Chestnut (American)	5,320	3,740,330	620	435,903	460	323,412
Elm (American)	5,520	3,880,944	690	485,118	660	464,026
Fir (Balsam)	5,280	3,712,207	404	284,040	180	126,553
Fir (White)	5,800	4,077,804	530	372,627	300	210,921
Hemlock (Eastern)	5,410	3,803,606	650	456,995	n/a	n/a
Hemlock (Western)	7,200	5,062,101	550	386,688	340	239,044
Hickory (Pignut)	9,190	6,461,209	1,980	1,392,078	n/a	n/a
Larch (Western)	7,620	5,357,390	930	653,855	430	302,320
Locust (Black)	10,180	7,157,248	1,830	1,286,617	640	449,965
Maple (Red)	6,540	4,598,075	1,000	703,070	n/a	n/a
Maple (Silver)	5,220	3,670,023	740	520,272	500	351,535
Maple (Sugar)	7,830	5,505,035	1,470	1,033,512	n/a	n/a
Oak (Red)	6,520	4,584,014	930	653,855	n/a	n/a
Oak (White)	7,440	5,230,838	1,070	752,284	800	562,456
Pine (Eastern White)	4,800	3,374,734	440	309,351	310	217,952
Pine (Jack)	5,660	3,979,374	580	407,780	420	295,289
Pine (Pitch)	5,940	4,176,233	820	576,517	n/a	n/a
Pine (Ponderosa)	5,320	3,740,330	580	407,780	420	295,289
Pine (Red)	6,070	4,267,632	600	421,842	460	323,412
Pine (Western White)	5,040	3,543,471	470	330,443	n/a	n/a
Poplar (Yellow)	5,540	3,895,006	500	351,535	540	379,658
Redwood (Old Growth)	6,150	4,323,878	700	492,149	240	168,737
Redwood (New Growth)	5,220	3,670,023	520	365,596	250	175,767
Spruce (Black)	5,960	4,190,295	550	386,688	n/a	n/a
Spruce (Red)	5,540	3,895,006	550	386,688	350	246,074
Sycamore (American)	5,380	3,782,514	700	492,149	720	506,210
Walnut (Black)	7,580	5,329,268	1,010	710,100	690	485,118
Willow (Black)	4,100	2,882,585	430	302,320	n/a	n/a

Source: Adapted from U.S. Department of Agriculture Wood Handbook, 1999.

TABLE A.11
Absolute Viscosity of Selected Fluids

Liquid	Absolute Viscosity at Normal Room Temperature (N s/m², Pa*s)	Absolute Viscosity at 4 °C (39.2 °F) (N s/m², Pa*s)
Air	1.983×10^{-5}	17.35×10^{-6}
Water	10^{-3}	1.5705×10^{-3}
Olive Oil	10^{-1}	N/A
Glycerol	1	N/A
Liquid Honey	10	N/A
Golden Syrup	100	N/A
Glass	10^{40}	N/A

Source: Adapted from: Engineering ToolBox. "Water – Dynamic (Absolute) and Kinematic Viscosity vs. Temperature and Pressure." [online], 2004. www.engineeringtoolbox.com/water-dynamic-kinematic-viscosity-d_596.html 5–27–22.

TABLE A.12
Absolute (Dynamic) Viscosity Unit Conversions

	cP	Pa*s	P	g/(cm*sec)	kg/(m*sec)	(N*s)/m²	lb/(ft*h)	lb/(ft*sec)	reyn
cP	1	0.001	0.01	0.01	0.001	0.001	2.41908	6.72197×10^{-4}	
Pa*s	1000	1	10	0.1	1	1	2419.08	0.672197	1.4504×10^{-4}
P	100	0.1	1	1	0.1	0.1	241.908	0.0672197	
g/(cm*s)	100	10	100	1	10	10	24190.88	6.72197	
kg/(m*s)	1000	1	10	0.1	1	1	2419.08	0.672197	
(N*s)/m²	1000	1	10	0.1	1	1	2419.08	0.672197	
lb/(ft*h)	0.41338	4.1338×10^{-4}	4.1338×10^{-3}	4.1338×10^{-5}	4.1338×10^{-4}	4.1338×10^{-4}	1	2.7778×10^{-4}	
lb/(ft*s)	1487.659	1.48816	14.8816	0.1487659	1.487659	1.487659	3600	1	
reyn		6894.76							1

cP = centipoise

Pa*s = pascal second

P = poise

N*s = newton second

Dynamic Viscosity (cP)/Density = Kinematic Viscosity (cSt)

TABLE A.13
Kinematic Viscosity Unit Conversions

	cSt	mm^2/s	ft^2/h	ft^2/s	in^2/s	m^2/h	m^2/s	St	cm^2/s
cSt	1	1	0.03875008	1.07639×10^{-5}	1.55×10^{-3}	0.0036	1×10^{-6}	0.01	1
mm^2/s	1	1	3.875008×10^{-2}	1.07639×10^{-5}	1.550003×10^{-3}	0.0036	1×10^{-6}	0.01	0.01
ft^2/h	25.806	25.8064	1	2.7778×10^{-4}	0.04	9.290304×10^{-2}	2.58064×10^{-5}	0.2580	0.25806
ft^2/s	92903.1	92903.1	3600	1	144	334.451	9.290304×10^{-2}	929.03	929.030
in^2/s	645.16	645.16	25	6.9444×10^{-3}	1	2.322576	6.4516×10^{-4}	6.4516	6.4516
m^2/h	277.78	277.78	10.7639	2.98998×10^{-3}	0.430556	1	2.7778×10^{-4}	2.7778	2.7778
m^2/s	1×10^{6}	1×10^{6}	38750.08	10.7639	1550003	3600	1	1×10^{4}	1×10^{4}
St	100	100	3.875008	1.07639×10^{-3}	0.1550003	0.36	1×10^{-4}	1	1
cm^2/s	100	100	3.875008	1.07639×10^{-3}	0.1550003	0.36	1×10^{-4}	1	1

cSt = centistoke
St = stoke

Bibliography

Ahmad, Hosni, Nur Pauzi, and Ahmad Sharif-fuddin. "Geotecyhnical Properties of Waste Soil from Closed Construction Dumping Area in Serdang, Selangor, Malaysia." *World Wide Web Journal of Geotechnical Engineering*, 2015. Accessed June 1, 2022. www.researchgate.net/figure/Specific-Gravity-for-Various-Soils_tbl4_281267333.

Amesweb.info. "Materials." n.d. www.amesweb.info/Materials/Linear-Thermal-Expansion-Coefficient-Metals.aspx.

Anscomb, F.J. "Graphs in Statistical Analysis." *American Statistician* 27 (1973): 17–21.

Beutler, M., K.H. Wiltshire, B. Meyer, C. Moldaenke, C. Luring, M. Meyerhofer, and U.P. Hansen. 2018. *Standard Methods for the Examination of Water and Wastewater*. Washington, DC: American Public Health Association.

Bevans, Rebecca. "A Guide to Experimental Design." July 3, 2020. www.scribbr.com/methodology/experimental-design/.

———. "An Introduction to Multiple Linear Regression." October 20, 2020. Accessed December 9, 2020. www.scribbr.com/statistics/multiple-linear-regression/.

Bordner, Robert, John A. Winter, and Pasquale Scarpino. "US EPA Environmental Monitoring and Support Laboratory." In *Microbial Methods for Monitoring the Environment*. Cincinnati, OH: US EPA, 1978.

"Chemistry 301." November 19, 2020. www.ch301.cm.utex.edu/data/section2.php?target=heat-capcities.php.

"Combinations and Permutations." n.d. Accessed December 4, 2020. www.mathsisfun.com/combinatorics/combinations-permutations.html.

"Engineering Perdue.edu." November 1997. Accessed January 4, 2021. https://engineering.purdue.edu/~frankenb/NU-prowd/sand.htm#:~:text=Sand%20filters%20usually%20are%20used,is%20usually%20colorless%20and%20odorless.

"Engineering ToolBox." 2005. Accessed May 27, 2022. www.engineeringtoolbox.com/water-dynamic-kinetic-viscosity-d_596.html.

———. "Metals – Temperature Expansion Coefficients." 2005. Accessed March 3, 2022. https://www.engineeringtoolbox.com/thermal-expansion-metals-d_859.html.

Engineeringtoolbox.com. "Thermal Conductivity." n.d. Accessed December 12, 2020. www.engineerstoolbox.com/thermal-conductivity-d_429.html.

Engineersedge.com. "Insulation Material Thermal Conductivity Chart." n.d. Accessed December 12, 2020. engineersedge.com/heat_transfer/insulation_material_thermal_conductivity_chart_13170.html.

"Experimentation." 1997–1998. www.yale.edu/Courses/1997-1998/101/expdes.htm.

Georgia State University. "Thermal Expansion Coefficients at 20 C." n.d. www.hyperphysics.phy-astr.gsu.edu/hbase/tables/thexp.html.

"Geotechdata.info." *Soil Ratio*, November 16, 2013. Accessed November 17, 2020. http://geotechdata.info/parameter/soil-ratio.html.

———. *Soil Porosity*, November 16, 2013. Accessed November 17, 2020. http://geotechdata.info/parameter/soil-porosity.html.

Glass, Samuel, and Samuel Zelinka. "Moisture Relation and Physical Properties of Wood." *Researchgate.net*. April 2010. Accessed March 13, 2022. www.researchgate.net/figure/1-Average-moisture-content-of-green-wood-by-species_tbl2_273459077.

"Glossary." n.d. Accessed January 9, 2021. http://techaliove.mtu.edu/modules/module001_alt/Glosasary.htm.

Green, D.W. "Encyclopedia of Materials: Science and Technology." In *Wood: Strength and Stiffness*. London: Elsevier Science, 2001. Accessed April 19, 2022. www.fpl.fs.fed.us/documents/pdf2001/ghreen01d.pdf.

"Hach Corporation." *Support.hach.com*, August 18, 2020. Accessed January 14, 2021. https://support.hach.com/app/answers/answer_view/a_id/1000336/~/what-is-the-difference-between-the-turbidity-units-ntu%2C-fnu%2C-ftu%2C-and-fau%3F-what.

Hopcroft, Francis J. *Conversion Factors for Environmental Engineers*. New York, NY: Momentum Press, 2017.

———. *Engineering Economics for Environmental Engineers*. New York, NY: Momentum Press. 2016.

———. *Presenting Technical Data to a Non-Technical Audience*. New York, NY: Momentum Press. 2019.

———. *Wastewater Treatment Concepts and Practices*. New York, NY: Momentum Press, 2015.

Hopcroft, Francis J., and Abigail J. Charest. *Experiment Design for Environmental Engineering*. 1st ed. Boca Raton, FL: CRC Press, 2022. doi:10.1201/9781003184249.

"The International Water Association." 2020. Accessed December 31, 2020. www.iwapublishing.com/news/coagulation-and-flocculation-water-and-wastewater-treatment.

Kenton, Will. "Multiple Linear Regression (MLR)." September 21, 2020. Accessed December 9, 2020. www.investopedia.com/terms/m/mlr.asp.

"Knowlton." *Workshop Companion*, 2019. Accessed April 19, 2022. workshopcompanion.com/knowlton/design/Nature_of_Wood/3_Wood_Strength/3_Wood_Stength.htm.

Metcalf & Eddy/AECOM. *Wastewater Engineering: Treatment and Resource Recovery*. Edited by George Tchobanoglous, H. David Stensel, Ryujiro Tsuchihashi, and Franklin Burton. 5th ed. New York, New York: McGraw Hill Education, 2014.

"Moisture Content of Wood." n.d. Accessed March 13, 2022. https://reliancetimber.com/moisture-content-of-wood/.

Morris, D.A., and A.I. Johnson. "Summary of Hydrologic and Physical Properties of Rock and Soil Materials, as Analyzed by the Hydrologic Laboratory of the U. S. Geologic Survey 1948–60." 1967. https://pubs.usgs.gov/wsp/1839d/report.pdf.

Ramakrishnaiah, C.R., and B. Prathima. "Hexavalent Chromium Removal by Vhemical Precipitation Method: A Comparative Study." *International Joirnal of Environmental Research and Development (Research India Publications)* 1, no. 1 (2011): 41–49. Accessed January 15, 2021. www.ripublication.com/ijerd.htm.

Ratnayaka, Don D., Malcolm J. Brandt, and K. Michael Johnson. *Water Supply*. 6th ed. London: Elsevier Science, 2009. Accessed January 4, 2021. www.sciencedirect.com/topics/engineering/sand-filter.

Ross, Robert. "Wood Handbook: Wood as an Engineering Material." General Technical Report FPL-GTR-282. US Department of Agriculture, Forest Service, Forest Service, Forest Products Laboratory US Department of Agriculture, 2021. Accessed May 15, 2022. www.fs.usda.gov/treesearch/pubs/62200.

"Safe Drinking Water Foundation." January 23, 2017. Accessed December 31, 2020. www.safewater.org/fact-sheets-1/2017/1/23/conventional-water-treatment.

"Soil Porosity." November 16, 2013. http://geotechdata.info/parameter/soil-porosity.html.

"Soil Void Ratio." November 16, 2013. http://geotechdata.info/parameter/soil-void-ratio.html.

"Specific Heat Capacity of Metals." n.d. Accessed November 19, 2020. www.engineersedge.com/materials/specific_heat_capacity_of_metals_13259.htm.

"Specific Heat Capacity Table." n.d. Accessed November 19, 2020. www2.ucdsb.on.ca/tiss/stretton/database/specific_heat_capacity_table.html.

"Specific Heat of Gases." n.d. Accessed November 19, 2020. hyperphysics.phy-astr.gsu.edu/hbase/kinetic/shegas.html#c4.

"Specific Heat of Some Common Substances." n.d. Accessed November 19, 2020. engineerstoolbox.com/specific-heat-capacity-d_391.html.

"Specific Heats and Molar Heat Capacities for Various Substances at 20 C." n.d. Accessed November 19, 2020. http://hyperphysics.phy-astr.gsu.edu/hbase/Tables/sphtt.html.

"Suez Water Technologies." 2021. Accessed January 4, 2021. www.suezwatertechnologies.com/handbook/chapter-06-filtration.

Sullivan, Lisa, and Wayne W. LaMorte. "Simple Linear Regression." n.d. Accessed December 9, 2020. https://sphweb.bumc.bu.edu/otlt/mph-modules/bs/bs704-ep713_multivariablemethods/index.html.

Sundararajan, K. "Design of Experiments – A Primer." n.d. Accessed November 2020. www.isixsigma.com/tools-templates/design-of-experiments-doe/design-experiments-%E2%90%93-primer/.

"Support Services." 2020. https://alphalab.com/index.php/support-services/holding-times#HH0.

"Table of Specific Heat Capacities." November 19, 2020. www.en.wikipedia.org/Table_of_specific_heat_capacities.

"Thermo Fisher.com." n.d. Accessed January 15, 2021. www.thermofisher.com/us/en/home/industrial/spectroscopy-elemental-isotope-analysis/spectroscopy-elemental-isotope-analysis-learning-center/trace-elemental-analysis-tea-information/atomic-absorption-aa-information.html#:~:text=Atomic%20absorption%2.

Tufte, E.R. *The Visual Display of Quantitative Information*. 2nd ed. Cheshire, CT: Graphics Press. 2001.

University of Chicago. "Thermal Coefficients." 2002. psec.uchicago.edu/thermal_coefficients/cte_metals_05517-90143.pdf.

"What Is Water Cement Ratio? – Guide & Calculation." 2009. Accessed April 6, 2022. www.civilology.com/water-cement-ratio/.

Yang, C., C.E. Brown, B. Hollebone, Z. Yang, P. Lambert, B. Fieldhouser, M. Landriault, and Z. Wang. "Chapter 4 – Chemical Fingerprints of Crude Oils and Petroleum Products." 2017. Accessed January 3, 2021. www.sciencedirect.com/science/article/pii/B9780128094136000047.

Zainuddin, Nur Ain, Tengku Azwan Raja Mamat, Hawaiah Imam Maarof, Siti Wahidah Puasa, and Siti Rohana Mohd Yatim. "Removal of Nickel, Zinc and Copper from Plating Process Industrial Raw Effluent Via Hydroxide Precipitation Versus Sulphide Precipitation." *IOP Conference Series: Materials Science and Engineering* 551 (2019): 012122.

Index

E

F

G

H

I

J

K

L

M

N

O

P

U

V

W